Advanced X-Ray Radiation Detection

Krzysztof (Kris) Iniewski

Editor

Advanced X-Ray Radiation Detection

Medical Imaging and Industrial Applications

 Springer

Editor
Krzysztof (Kris) Iniewski
Emerging Technologies CMOS Inc.
Port Moody, BC, Canada

ISBN 978-3-030-92991-6   ISBN 978-3-030-92989-3   (eBook)
https://doi.org/10.1007/978-3-030-92989-3

This Springer imprint is published by the registered company Springer Nature Switzerland AG
The registered company address is: Gewerbestrasse 11, 6330 Cham, Switzerland

Contents

About the Editor

Krzysztof (Kris) Iniewski is managing R&D development activities at Redlen Technologies Inc., a detector company based in British Columbia, Canada. During his 15 years at Redlen, he has managed development of highly integrated CZT detector products in medical imaging and security applications. Prior to Redlen, Kris held various management and academic positions at PMC-Sierra, University of Alberta, SFU, UBC, and University of Toronto.

Dr. Iniewski has published over 150 research papers in international journals and conferences. He holds 25+ international patents granted in the USA, Canada, France, Germany, and Japan. He wrote and edited 75+ books for Wiley, Cambridge University Press, McGraw-Hill, CRC Press, and Springer. He is a frequent invited speaker and has consulted for multiple organizations internationally.

Detectors for X-Ray Medical Imaging

Witold Skrzynski and Krzysztof (Kris) Iniewski

1 Introduction

Doctor's ability to diagnose and treat disease is defined by his or her ability to investigate and visualize the inner workings of living systems. Imaging technologies have become a central part of both everyday clinical practice and medical investigations. Their applications range from everyday clinical diagnosis to advanced studies of biological systems. Their relevance lays in many diverse areas in biology and medicine, such as diagnosis of cancer, assessment of the cardiovascular systems, or applications in neurosciences, to mention few examples.

Medical imaging represents an increasingly important component of modern medical practice, because early and accurate diagnosis can substantially influence patient treatment strategies and improve this treatment outcome and, by doing so, decrease both mortality in many diseases and improve life quality. It can also facilitate patient management issues and improve health-care delivery and the effectiveness of utilization of resources. The most current example is to use Computed Tomography (CT) to diagnose damage in lungs as a result of Covid-19 infection.

Depending on the physics and effects that are being used to obtain necessary imaging information, different modalities measure and visualize different characteristics of the investigated tissues or organs [1]. Attenuation of electromagnetic radiation lies behind X-ray imaging and computed tomography (CT); sound wave transmission and reflection are used in ultrasound (US) imaging; and magnetic

W. Skrzynski (✉)
Medical Physics Department, Maria Sklodowska-Curie National Research Institute of Oncology, Warsaw, Poland
e-mail: witold.skrzynski@pib-nio.pl

K. Iniewski
Redlen Technologies, Saanichton, BC, Canada

resonance imaging (MRI) uses magnetic field caused changes in hydrogen state of water molecules. The images created by most of these modalities display anatomy of the body, however some like Nuclear Medicine or functional MRI can represent body functions. The book focuses on detector for X-ray and gamma ray based imaging modalities as they both have common denominator: detection of number of photons and/or their energies.

Most medical X-ray imaging is based on a measurement of the X-ray beam attenuation. The attenuation is different for different tissues and may be additionally modified by the presence of contrast agents (e.g., intravenously administered iodine) if needed. There are several X-ray modalities widely used in medicine. Depending on their application, they can utilize different energies of radiation (e.g., lower energy in mammography than in radiography), or a different method of image creation (radiography vs computed tomography), but the basic scheme is always kept more or less the same; the patient is placed between a radiation source and radiation detector. The information from the detector is used to create an image.

In other medical imaging modalities, usually referred to as Nuclear Medicine, there is no X-ray tube, and the radiation source is directly injected inside the patient. The general term "nuclear medicine" encompasses several different imaging techniques, ranging from simple planar and whole-body studies to positron emission tomography (PET) and single-photon emission computed tomography (SPECT). All these diagnostic techniques create images by measuring electromagnetic radiation emitted by the tracer molecules which have been labeled with radioactive isotope and introduced into a patient's body. Additionally, nuclear medicine includes internal radiotherapy (IRT) procedures where radioactive high-dose injections are being used in cancer treatment.

2   X-ray and Gamma Ray Basics

2.1   *Ionization Radiation*

Ionizing radiation is radiation that is able to ionize atoms when passing through matter. Ionizing an atom occurs when radiation with enough energy removes bound electrons from their orbits leaving the atom charged. The most common types of ionizing radiation are alpha particles, protons, beta particles, neutrons, gamma rays, and X-rays.

Alpha particles are among the most massive types of radiation. They are nuclei of helium, and therefore, have two protons and two neutrons so they are positively charged. Alpha particles are emitted during the decay of unstable heavy atomic nuclei. As an example of application in medicine alpha particles are used to kill cancerous cells in a process called Theranostics. Proton is an elementary particle that is identical with the nucleus of a hydrogen atom, and they are positively charged. Protons can be accelerated to a high-energy with a particle accelerator.

Highly energetic protons are also used to kill cancerous cells in a procedure called External Proton Beam Therapy.

Beta particles are either electrons with negative charge or positrons which are positively charged and have a mass much lower than alpha particles and protons. The negative betas come from nuclei that have too many neutrons, while the positive betas are generated from nuclei having too many protons. Electron and positron annihilation process is used in the imaging modality called Positron Emission Tomography (PET).

Neutrons are neutral particles originating from the nucleus of an atom during nuclear reactions. They are rarely used in medical imaging. Gamma rays are electromagnetic radiation similar to radio waves or visible light, except that they have much higher energies (shorter wavelength). After a radioactive decay (alpha and beta), the new nucleus often has an excess of energy, that is usually released by the emission of gamma rays. Gamma rays are transmitted as photons and in many ways, they behave more like particles than as waves. Gamma rays are used in medical imaging modality called Single Photon Emission Tomography (SPECT).

Finally, X-rays are also electromagnetic radiations, similar to the gamma rays but they originate differently. X-rays originate in the orbital electron field surrounding the nucleus. This process takes place in its simple way when an electron is knocked out from its orbit, leaving the atom in an unstable state. Another electron of a higher energy falls down, filling the place of the missing electron and the energy difference between the two energies is released. Electrons may lose energy in the form of X- rays when they quickly decelerate upon striking a material. This is called Bremsstrahlung (or translating from German Breaking) Radiation. X-rays are frequently used in medical imaging in form of Radiography or Computed Tomography (CT). This chapter deals with applications of X-rays and gamma rays in medical imaging. But let us first review ways of interaction with matter.

2.2 X-Ray and Gamma Ray Interactions with Matter

Energetic photons such as X-rays and gamma rays interact with matter mainly by four basic processes: elastic scattering, photoelectric absorption, Compton scattering, and pair production. Elastic scattering, also called coherent scattering is a process in which there is a change of photons path without changing its energy. Photoelectric absorption is the process where the photon disappears after transferring all its energy to a photoelectron. Most photoelectric interactions occur in the K-shell of the atom because the density of the electron cloud is greater in this region. Compton scattering is a process in which some fraction of the photon energy is transferred to a free electron in the material. The path of the photon changes. Finally, pair production is a process in which the photon energy is converted into an electron–positron pair.

Only three of the four types play an important role in radiation measurements. These are photoelectric absorption, Compton scattering, and pair production. All these processes lead to a partial or complete transfer of the photon energy to electron energy.

Photoelectric absorption is, in most cases, the ideal process for detector operation. All of the energy of an incident photon is transferred to one of the orbital electrons of the atoms within the detector material usually to an electron of the K-shell. This photoelectron will have a kinetic energy equal to that of the incident photon energy minus the atomic binding energy of the ejected electron. The atom with a missing inner-shell electron emits a characteristic X-ray photon after an outer-shell electron is transferred to the incomplete inner shell. The emitted radiation is known as fluorescence. These fluorescent X-rays are unique fingerprint of the kind of atom that produces the fluorescence. This X-photon may interact with another atom by means of one or more processes until it loses all its energy or escapes from the material. Instead of emitting an X-ray photon the atom may emit an Auger electron with a characteristic energy. Auger electrons are of very low energy and can only travel a short distance within the matter. The Auger electron can escape from the material only when the absorption of the incoming photon takes place very close to the surface. In a semiconductor material as well as in other materials, the photoelectron will lose its kinetic energy via coulomb interactions with the semiconductor electrons, creating many electron-hole pairs.

Compton scattering is a collision between an incident photon and an orbital electron. The result is the creation of a recoil electron and a scattered photon. The incident photon energy is divided between the recoil electron and the scattered photon dependent on the scattering angle. In the case of multiple scattering, the Compton scattered photon interacts also with the radiation detector, either by additional Compton scatterings or by a photoelectric process. In this way all the energy of the incoming photon might end up in the detector. Physics of Compton effects is somewhat complex but useful detection apparatus in form of the Compton camera can be build.

Finally, for photon energies greater than twice the rest mass energy of the electron (1.022 MeV) pair production may occur and might actually be the dominant type of interaction at energies above several MeV. In this type of interaction, the energy of the incident photon is converted to an electron–positron pair. Afterwards the positron annihilates with an electron producing two 511 keV photons. In most imaging modalities several MeV photons are not used so this effect frequently can be neglected.

2.3   Radiation Detectors

Operation of any radiation detector depends on the manner in which the radiation to be detected interacts with the material of the detector. The result of the interaction then should be interpreted in order to extract information about the radiation. The interaction should cause a measurable change, typically change in electric charge, or current, that can be further processed by electronic circuits.

Table 1 List of some semiconductor materials used in radiation detection

Material	Germanium (Ge)	Silicon (Si)	Cadmium Zinc Telluride (CZT)	Thallium bromide (TlBr)	Mercury iodide (HgI$_2$)
Average Z	32	14	49	80	80
Density (g/cm^3)	5.32	2.33	5.78	7.56	6.4
Resistivity (Ohm-cm)	50	1E5	1E10	1E12	1E12

Radiation detection can be split into two principal techniques: indirect and direct detection. In indirect process detection is performed by scintillation materials and direct conversion is performed by semiconductors. Scintillators are materials that absorb photons of higher energy and release it as photons of lower energy, usually visible light. The material should capture as much of the energy of each photon is as possible in order to build an accurate photon energy spectrum and thus identify the material emitting the photons. The visible light signal is subsequently detected by light emitting diode or similar opto-electronics device.

In direct detection process semiconductor detector converts the X-ray or gamma ray photon energy into a number of electron-hole pairs creating electric charge. The signal can then be analyzed with the help of an external electronic circuit. Table 1 lists example of commonly used semiconductor materials used in imaging systems.

Concluding this section let us list basic requirement for radiation detectors. A detector can be used for measuring three different properties of the radiation: position of interaction, energy of the photon, and time of interaction. Position is obviously required to get an image of the source, by recording in detail the positions of the interactions of the incoming radiation. Energy is required to measure the spectrum of the source, i.e., a measurement of the energy deposited by each interacting photon. Finally, time of interaction might be helpful to decide on some details of physical processes happening in the detector.

3 Imaging Using X-Rays

3.1 Radiography

In the early days of X-ray imaging, glass plates with photographic emulsion were used as detectors, later replaced with films. The film is coated with a thin layer of photographic emulsion containing silver halide, which is not a very efficient X-ray absorber. The film performs much better in detecting visible light than X-rays. This is why the film has been quickly accompanied with an intensifying screen. The intensifying screen is a sheet of fluorescent material, converting the energy of each

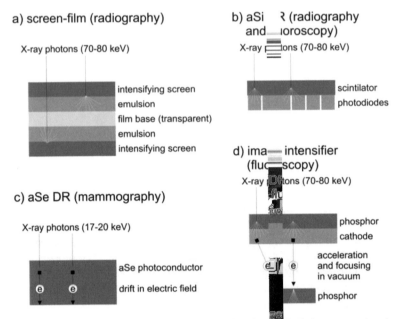

Fig. 1 Several detectors used in planar medical X-ray imaging (typical photon energies given for reference): (**a**) screen-film detector (radiography), (**b**) aSi detector (digital radiography), (**c**) aSe detector (digital mammography), (**d**) image intensifier (fluoroscopy)

absorbed X-ray photon to many visible light photons. When the screen is in direct contact with the film, the film gets exposed by visible light. A combined screen-film detector has a much higher sensitivity for X-ray detection than the film alone. The sensitivity depends on the material used on the screen. As an example, screens with rare earth elements more effectively absorb and convert X-ray radiation compared to calcium tungstate screens ($CaWO_4$). With screen-film detectors, it is possible to obtain a radiographic image with a reasonably low patient dose. In medicine, the film is almost always used with intensifying screens, placed permanently in a light-tight radiographic cassette. In most cases, two screens are placed on both sides of the film, which has two layers of emulsion on both sides (Fig. 1a). The light emitted from the intensifying screen can penetrate the film base reaching the second layer of emulsion and causing image blur. Compared with the light-only detector, the use of the screen-film decreases patient dose, but at the cost of a lower spatial resolution of the image. Similarly, a film with larger silver halide grains is more sensitive but has a lower spatial resolution. In mammography, a good spatial resolution is essential for good visualization of microcalcifications, and only one intensifying screen and one layer of emulsion are usually used. In dental radiography, small objects are imaged and a good spatial resolution is important, while radiation dose is not of much concern, and the film can be used alone [2].

Radiographic films require photochemical processing to convert latent images to visible images. It is a time-consuming process, it is not environmentally friendly, and has to be done in the darkroom. Various alternatives have been proposed over the years. Some of them gained initial interest but then were practically forgotten due to low performance, high cost, or impracticality. As an example, the xeroradiographic detector consists of a plate covered with an amorphous selenium photoconductor. Before use, the surface of the plate is uniformly positively charged. During exposure, electrons are excited in the photoconductor, causing local loss of positive charge. After exposure, a negatively charged toner is applied. The toner is attracted to the residual positive charge on the plate, then it can be moved to paper or plastic sheet, creating an image [3]. Interestingly, films that do not require photochemical processing and can be used in daylight conditions also exist. Radiochromic films become blackened directly by ionizing radiation, thanks to radiation-induced polymerization of diacetylene (dark as polymer, colorless as a monomer). Radiochromic films are not sensitive enough to be used for imaging, but they may be used as radiation detectors to monitor skin doses in interventional radiology [4].

Modern radiography has largely moved away from film detectors towards digital imaging. Computed radiography (CR) imaging plates and digital radiography (DR) flat-panel detectors do not need photochemical processing to get an image. They also have a much wider dynamic range then screen-film detector, e.g., $10^4:1$ instead of 10:1 [5], and offer possibilities for digital processing, enhancement, and storage of the images.

In the CR systems, imaging plates are used as a radiation detector. The plate is placed in a radiographic cassette and emits visible light when absorbing X-ray radiation, which makes it similar to the intensifying screen in screen-film systems. At the same time, a sort of latent image is created on the plate, which makes it similar to film. During exposure, some electrons are trapped. During readout in the CR scanner, the imaging plate is illuminated with red laser, causing trapped electrons to relax to their ground state. The relaxation is associated with emission of blue light (photoluminescence), which is measured with a photomultiplier and used as an information to create an image. The light signal is proportional to the number of trapped electrons, which is in turn proportional to the received dose of radiation. Proper choice of storage phosphor materials for CR plates is essential for good performance of the system. The efficiency of radiation-to-light conversion of the material is important, but other characteristics may be important as well. Some materials (e.g., $CsBr:Eu^{2+}$) perform better than others (e.g., $BaFX:Eu^{2+}$) because of their oriented, needle-like structure, which resembles a matrix of optical fibers. Such a structure reduces light scatter in the readout phase, thus resulting in less image blur even if the material layer is made thicker to achieve higher sensitivity [5].

In screen-film systems, the energy of the X-ray radiation is converted to the image with a time-consuming photochemical process. In CR systems photochemical processing is not needed, but the plate is still a passive detector—it operates without any active means but does not provide a direct readout. Scanning of CR plates takes time, an image is available a few minutes after exposure. In DR flat-panel detectors, the absorbed energy is converted to an electric charge within a detector, and the image is practically instantly displayed on the computer screen. In a vast majority of clinically used radiographic DR systems, the conversion is done with two steps, with intermediate conversion to light (Fig. 1b). First, X-ray radiation captured in a scintillation layer is converted to visible light. Secondly, the light is detected by a matrix of photodiodes. Electric charge is stored in a capacitor and then read out, amplified, digitized, and converted into an image. DR detectors with such indirect conversion are often referred to as aSi (amorphous silicon), although their amorphous silicon elements are not directly used in the detection of X-rays. This is done with the scintillation layer, which is made, e.g., of Gd_2O_2S or CsI. The materials differ in structure similarly as those used for CR imaging plates while Gd_2O_2S is granular, CsI has a needle-like structure. Direct-conversion DR systems also exist and are often used in mammography. The energy of the absorbed X-ray radiation is converted directly to electrical charge within the amorphous selenium (a-Se) photoconductor (Fig. 1c). Electrons excited to the conduction band move along electrical field lines, without spreading to the sides, and are collected by the capacitors, similarly as in indirect DR detectors [6].

Not only detectors have evolved since the beginning of medical X-ray imaging history. Modern X-ray tubes are very different from those used in the first experiments: they have a higher vacuum, hot filament, line focus, proper shielding, higher filtration, usually rotating target, sometimes flat electron emitter, and a liquid bearing [7]. Tubes were constructed specifically for fluoroscopic/angiographic applications (long exposures, pulsed beam), mammography (lower voltage, smaller focal spot), computed tomography (large g-force during rotation around a patient) [8]. Let the measure of improvement in the construction of X-ray radiation sources and detectors be a reduction in the average skin dose in pelvic radiographs: since 1896, it was reduced by a factor of about 400(!) [9].

3.2   Fluoroscopy

The screen-film detector integrates signals during whole exposure, which makes it suitable to capture an image of a stationary object. If the object moves, its image is blurred. Movements of the patient during imaging can be minimized with short exposure time and with immobilization. Patients can be asked to hold a breath, but it is not possible to stop internal movements such as heartbeat or peristalsis. In some situations, it is essential to visualize movements within a patient. Besides the movement of the patient's tissues, the flow of the contrast medium may be observed, movement of surgical instruments, or movement of implants, which are being placed in the patient. That is why a radiation detector is needed, which can provide a live view with a good time resolution.

In the early days of radiology, the live image was observed directly on the fluorescent screen. This resulted in large radiation exposure not only for the patient but also for the operator, who often kept the screen in hand. Since the image was not very bright, fluoroscopy could be only used in darkened rooms, which was inconvenient. In the 1950s a new image detector for fluoroscopy was introduced, namely X-ray image intensifier. The operation of the image intensifier consists of several steps. X-ray photons interact with a scintillation layer, converting each absorbed X-ray photon into many visible light photons (as with a simple fluorescent screen). The light strikes photocathode, knocking out electrons, which are then accelerated with electric potential and then fall again on another (smaller) fluorescent screen (Fig. 1d). As an effect, there are many more photons on the smaller screen, resulting in a much brighter (intensified) image with a smaller dose to a patient. The bright image can be observed in normal lighting conditions, recorded with a video camera, etc. Currently, image intensifiers are being replaced with flat-panel (DR) detectors with a fast readout [10].

3.3 Tomographic Imaging

In radiography, three-dimensional structures are visualized in a two-dimensional planar image. The third dimension is lost in the imaging process. Images of objects located at different depths overlap, making some anatomical structures invisible. From the early days of medical X-ray imaging researchers have tried to image selected planes within the patient [11]. Many experiments involved movement of the radiation source (tube) and/or detector (film) during the exposure. If the tube and detector are synchronously translated in opposite directions during exposure, only the objects located at one plane will be projected all the time at the same positions on the detector, resulting in a sharp image. For other objects, their projected image on the detector will move, resulting in blur. This idea was used in classical tomography to obtain a sharp image of anatomical structures located at the focal plane (Fig. 2). Typically, only the image of one plane was obtained in one exposure. Another film and other exposure had to be used to image structures located on another plane. It was not only time consuming, but also associated with a multiplication of patient exposure. This limitation can be overcome quite easily if a different detector is used. If the film is replaced with a flat-panel DR detector, tens of images at different angles can be acquired during the movement of the tube-detector system. If the images are simply summed up together, the same image as in classical tomography is obtained. If the images are shifted before summation, a different plane is sharply imaged. Thanks to an image detector with good time resolution and a very simple shift-and-add algorithm, images of several planes can be obtained from a single exposure. This solution, called digital tomosynthesis, is used clinically [12], although usually in a slightly different scheme (X-ray tube is moving over a segment of a circle, while the detector remains stationary).

Digital tomosynthesis is currently used mainly in breast imaging. Cross-sectional images of head and body are usually obtained with X-ray computed tomography (CT). During the examination, tube and radiation detectors rotate around a patient, and a cross-sectional image is reconstructed with mathematical algorithms. Despite similarities between digital tomosynthesis and computed tomography, the geometry

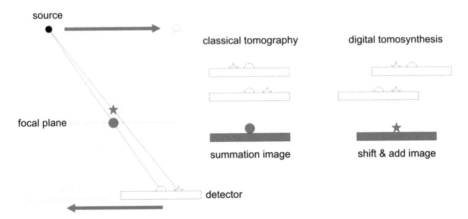

Fig. 2 Image acquisition in classical tomography (with film detector) and digital tomosynthesis (with flat-panel DR detector)

of the system is fundamentally different. In digital tomosynthesis, the image plane is perpendicular to the plane, in which the tube moves. In computed tomography image lays in the tube rotation plane.

In current multislice CT systems, there may be several rows of detectors (e.g., 64 rows with ca. 1000 detectors in each row). Detectors may perform 2–3 revolutions around a patient per second, registering the signal several hundred times during each rotation. Detectors for modern CT systems need not only to have good sensitivity and high dynamic range but also to be very fast and capable of working in large g-force conditions. For many years, xenon detectors have been used in CT, due to their good time characteristics (fast decay of signal). However, the low density of xenon—even pressurized—does not make it a very efficient detector for X-rays [13]. Currently, CT detectors are usually made of scintillators and photodiodes [8]. The scintillators are chosen to have a high light output, but also a low afterglow. The performance of the detectors, and thus also image quality, can be improved not only by choice of a better scintillator or more efficient photodiode but also by moving the analog-to-digital electronics as close as possible to the detectors, to minimize electronic noise [14].

3.4 Spectral Imaging

Attenuation coefficients for X-rays depend on the energy of photons. This dependence is different for different materials and tissues. This fact is used in bone densitometry (DXA, dual-energy X-ray absorptiometry), in which bone mineral density is calculated by comparing attenuation of radiation for two different energies

[15, 16]. The typical radiographic image provides information about the attenuation of the radiation by the tissues, but it does not bring direct information about the density of the tissue. Identical attenuation can be observed, e.g., for the thicker bone of lower density, as well as for thinner bone of higher density. In bone densitometry, quantitative information on bone density is calculated from absorption data measured for two different energies of radiation. Similar approach can be used to obtain other information. In contrast-enhanced spectral mammography, two exposures are made for two different beam energies (different tube voltage and different filtration). Both exposures are made after administration of an intravenous contrast agent, but detector data obtained for two energies may be processed to create virtual non-contrast image [17]. Possibility of using two energies to determine electron density and effective atomic number of the examined tissues has already been proposed for the first commercially available computed tomography system [18]. Currently, spectral CT allows for virtual monochromatic imaging, creation of virtual non-contrast images, better quantification of iodine content, and differentiation of renal stones based on their atomic composition [19, 20]. In some spectral computed tomography systems, the examination has to be performed twice to obtain data for two energies. Since the datasets are not obtained at the same time, patient movement can be an issue. Other spectral CT systems are capable of truly simultaneous imaging for two different energies, thanks to duplication of the tube-detectors system or fast alternating tube voltage switching [21].

All the spectral imaging methods mentioned so far are based on the use of two radiation beams with two different energies. Another approach is also possible, based on a single beam, and provided that detector separately measures signals in two (or more) energy ranges. In one design of a spectral CT scanner, a layered detector is used, with two layers sensitive to two different energy ranges [21]. Another possibility is the use of photon-counting detectors (PCD), e.g., cadmium telluride semiconductors [22, 23]. The signal from most radiation detectors used in medical X-ray imaging is simply proportional to the total absorbed energy of radiation. In PCD, each detected X-ray photon generates a separately measured electrical pulse. The height of each pulse can be compared with the threshold, or several different thresholds, to assign it to one of the energy bins. For each pixel, PCD provides information on the number of pulses separately for each energy bin. This allows for truly simultaneous acquisition of separate images for several energy ranges during one exposure. If several contrast agents are administered to a patient before an examination, with different radiation absorption characteristics (e.g., iodine and gadolinium), the virtual reconstruction of images with individual contrast agents is possible. Besides energy-resolving capabilities, photon-counting detectors have no electronic noise, which allows better imaging [14, 24].

4 Imaging Using Gamma Rays

4.1 Nuclear Medicine

In diagnostic X-ray imaging, the energy of X-ray radiation is optimized to achieve good visibility of anatomical structures with patient dose as low as possible. Depending on the application, tube voltage may be in the order of 25–45 keV in mammography, several dozen keV in radiography and fluoroscopy, and up to 120–140 keV in computed tomography (CT). This electric potential is used to accelerate electrons within the X-ray tube, and it sets a limit for the maximum energy of emitted photons (e.g., 140 keV in CT). Most of the photons reaching detectors have lower energy. However, imaging with photons of higher energy is also used in medicine. They are frequently emitted as gamma rays using radioactive sources, not from X-ray tube and the corresponding imaging modalities are referred to as Nuclear Medicine.

Nuclear medicine imaging study may be divided into two classes:

- single-photon imaging, which measures photons emitted directly from a radioactive nucleus, usually referred as SPECT (Single Photon Emission Tomography)
- coincidence imaging of pairs of 511-keV photons emitted in exactly opposite directions, usually referred as PET (Positron Emission Tomography)

PET images are acquired by cameras containing several small detectors arranged in multiple rings around the patient, while single-photon imaging studies use Anger cameras, with multiple large-area detectors. New version of SPECT camera is being build that utilizes Compton scatter and is therefore called Compton cameras [25].

4.2 Single Photon Emission Tomography (SPECT)

In nuclear medicine imaging, a radioisotope is administered to the patient. After its uptake within the tissues, γ-radiation produced in the radioisotope's decay is detected from the outside of the patient. Tc-99m, which is the most widely used radioisotope for imaging, emits monochromatic γ quanta with an energy of 140 keV. Other isotopes used in SPECT are listed in the Table 2 below, emit gamma rays in a 70–440 keV range.

4.3 Positron Emission Tomography (PET)

In positron emission tomography (PET) imaging, a radioisotope is administered to the patient, which emits β^+ particles (positrons). Emitted positron annihilates with an electron, resulting in simultaneous emission of two 511 keV photons in opposed

Table 2 Radioisotopes used in SPECT imaging

Isotope	Energy (keV)	Half-life	Clinical applications
Technetium (Tc-99m)	140	6 h	Main SPECT isotope, heart, brain, liver, lungs, cancer imaging
Gallium (Ga-67)	93, 185, 300	3.3 days	Abdominal infections, cancer imaging
Indium (In-123)	171, 245	2.8 days	Infections, cancer imaging
Iodine (I-131)	364	8 days	Thyroid
Thallium (Tl-201)	70, 167	3 days	Myocardial perfusion
Lutetium (Lu-177)	113, 208	6.6 days	Prostate cancer imaging
Actinium (Ac-225)	440	10 days	Emerging tracer, future applications

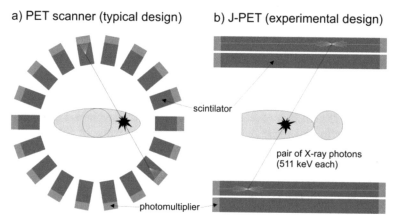

Fig. 3 Signal detection in positron emission tomography (PET) systems: (**a**) typical PET scanner design, (**b**) J-PET (experimental design)

directions. Radiation detectors are distributed around the patient (Fig. 3a). If two of them coincidentally detect two photons, it is assumed that the photons originate from the same annihilation event. It is also assumed that the annihilation event has occurred on the line connecting the two detectors (LOR, line of response). This information is a base for image creation.

Scintillators, which are used in PET as detectors, need to have a good time response with a very fast decay of light pulse to allow coincidence detection. They also need to have a good detection efficiency for 511 keV photons, which is usually associated with high density and high atomic number. Materials used in radiography (e.g., CsI, the density of 4.5 g/cm^3) or computed tomography (e.g., CWO, the light decay time of 14.5 μs) are not well suited for that task, compared to, e.g., LSO or LYSO scintillators (density 7–7.5 g/cm^3, decay time ca. 40 ns) [26]. With a fast scintillator and a fast light detector (e.g., silicon photomultiplier), it is even possible to estimate time between detection of the two coincidence photons, and to approximately determine the position of annihilation along the LOR. Inclusion of the additional data in image reconstruction improves image quality in the so-called time-of-flight (TOF) PET scanners [27].

Potentially, detectors with a much worse absorption efficiency could be also used in PET scanners. Compared to crystals commonly used in PET scanners, plastic scintillators have a low density (approximately 1 g/cm³). Additionally, 511 keV photons usually do not leave all of their energy in the plastic scintillator during an interaction. However, the light pulse generated at each interaction is extremely short (1–2 ns), which allows us to obtain information on the place of interaction within the detector. Low-density plastic scintillator, used in a J-PET prototype in a non-typical geometrical setup (Fig. 3b), allows obtaining results comparable as for typical PET scanners [28, 29].

5 Other Imaging Modalities

5.1 High-Energy Photon Therapy

In oncology, radiotherapy with high-energy photon beams is used to treat patients with cancers. The maximum energy of photons in the therapeutic radiation beam is much higher in diagnostic radiology, with a typical value of 6 MeV. Although the beam is not optimized for imaging, it can be used for that task. The images obtained with the therapeutic beam (so-called portal images) can be used to verify the realization of the therapy, to check the alignment of the realized radiation field with a planned field, and to verify the proper positioning of the patient. Originally, portal imaging was performed with films, then various electronic portal imaging devices (EPIDs) were introduced. In one design, a fluorescent screen and a vidicon camera were used. The camera could not be placed directly on the beamline, because it would be hit with high-energy X-rays. Therefore, the light from the fluorescent screen was deflected with a mirror (Fig. 4a). In other designs, a matrix of liquid-filled ion chambers was used as the detector. An ion chamber filled with air, or even with a pressurized gas would have a low detection efficiency for high-energy photons. Currently EPIDs are usually constructed similarly to radiographic flat-panel DR detectors, with a scintillator layer and an array of photodiodes. The main difference between the DR detector and EPID is the presence of an additional metal layer (e.g., 1 mm of copper) in front of the scintillation layer (Fig. 4b). The metal layer is responsible for a generation of recoil electrons, which interact in the scintillator [30–32]. Additionally, detector elements (pixels) are usually significantly bigger in EPIDs than in radiography detectors (e.g., 400–800 μm vs 100 μm). Although bigger pixels lower detector's spatial resolution, such design allows each pixel to gather a higher signal in a shorter time.

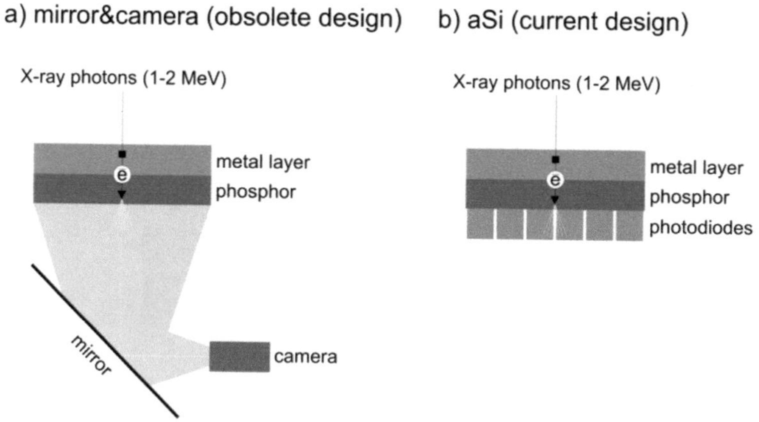

a) mirror&camera (obsolete design)   b) aSi (current design)

X-ray photons (1-2 MeV)   X-ray photons (1-2 MeV)

metal layer
phosphor

metal layer
phosphor
photodiodes

mirror

camera

Fig. 4 Electronic portal imaging detectors (EPIDs) used in radiotherapy with high-energy X-ray photons (typical photon energies given for reference): (**a**) mirror and camera system (obsolete design), (**b**) aSi system (current design)

5.2 Scattered Radiation

In traditional medical X-ray imaging, scattered photons add noise to an attenuation-based image. This is why in radiography anti-scatter grids (ASG) are generally used, eliminating a large portion of the scattered photons. At the same time, an anti-scatter grid reduces the signal at the detector. That usually needs to be compensated with a higher patient dose compared to a no-grid setup. Other scatter-reduction methods include air-gap, meaning the increased distance between patient and detector, and narrow beam geometry. If a narrow beam is used (pencil or slit), the trajectory of scattered photons will fall outside the beam and detector area. Slit-beam geometry is natural for CT scanners but may be also used in other imaging modalities, such as mammography [33].

In nuclear medicine imaging, monochromatic photons are used (radioisotope-emitted gamma rays). Detectors in gamma cameras can discriminate energy of each absorbed γ-ray quantum. A signal from scattered quanta, which has lower energy, can be eliminated with energy windowing, and only the monochromatic peak is used for image creation. That approach cannot be used in radiography. Even in the case of photon-counting detectors, the scattered photons cannot be separated over the wide spectrum of photons emitted from the X-ray tube. It could be possible if a monochromatic beam was used, e.g., created with a conventional X-ray source combined with a mosaic crystal [34].

Photons, that are scattered in directions other than towards detector, obviously do not have an impact on the quality of transmission image. While they are usually only of radiological protection concern, they could also be used for imaging. X-ray backscatter systems are used at many airports for passenger screening. Passenger, in a standing position, is scanned with a pencil radiation beam (Fig. 5a). Backscatter

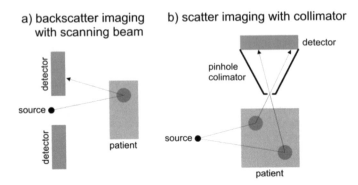

Fig. 5 Examples of scatter-based schemes for X-ray imaging: (**a**) backscatter imaging with scanning beam, (**b**) scatter imaging with collimator

radiation is detected, allowing to image items hidden under clothes [35] with a low dose [36]. A similar scheme has also been considered for the imaging of humans [37]. A scanning beam has also been used, to limit volume in each scattering occurs. If a large-area beam was used, the whole irradiated volume would be the source of scattered radiation, and it would be hard to distinguish between signals emitted by different organs. A similar situation can be observed in nuclear medicine imaging, where the whole patient is the source of radiation emitted by the radiopharmaceutical product present in his tissues. In nuclear medicine, this is solved by the use of slit collimators or pinhole collimators. This approach could be adapted to scatter-based imaging in lung radiotherapy (Fig. 5b). Lung tumor has a much higher density than surrounding lung tissue and is a major source of radiation scatter. A slit X-ray kilovoltage beam may be used for imaging [38], as well as the megavoltage therapeutic beam itself [39]. Experiments show that scatter-based images may provide better tumor contrast as compared to transmission images obtained with EPID [39].

5.3 *Forced Fluorescence*

In the description of image formation in X-ray imaging, a photon that interacts through the photoelectric effect is referred to as removed from the beam. This is true because such a photon transfers all of its energy to the electron. On the other side, a new X-ray photon may be emitted from the atom as a result of electron ejection (forced fluorescence). Human tissues are composed mainly of elements with a low atomic number, so the resulting monochromatic characteristic radiation has an energy of a few keV and does not reach the detector. The situation can change, if the patient is administered a contrast agent containing a high atomic number element, such as gold. The energy of characteristic radiation for gold is close to 70 keV. Theoretically, it is possible to create an image showing the capture of a contrast agent based on the registration of its forced fluorescence. In X-ray fluorescence computed tomography (XFCT), many detectors would be placed around the patient, and energy windowing and collimators with parallel holes would be used [40, 41] Such fluorescence can be also caused by different beams, e.g., protons [42].

5.4   Phase-Contrast Imaging

In the above description, the X-ray beam was treated as a flux of particles (photons). At the same time, it can be described as a wave, with phase and trajectory changing while traveling through matter. The refraction of X-rays is much lower than the refraction of visible light, but it is observable. Possibility to obtain medical images dependent on refraction and phase-change has been shown in experiments (phase-contrast imaging). One possibility is to pass the radiation coming from the patient through two grates differing in spatial frequency [43]. As a result, we obtain a stronger or weaker signal not in the shadow of the entire structure, but in places where the wave refracted on the patient's tissues. The image may appear similar to an attenuation-based X-ray image with an edge detection filter applied. In recent years, it has been shown that attenuation-based images and phase-contrast images of the mammary gland can be obtained simultaneously, during a several-second exposure with a dose similar to regular mammography [44]. It has also been shown that phase-based mammographic images may be obtained at higher beam energies than traditionally used in mammography, with simultaneous improvement in lesion detectability and a patient dose saving compared to traditional attenuation-based imaging [45].

5.5   Multi-Mode Imaging Modalities

X-ray and gamma ray imaging modalities can be combined together or with other imaging modalities like MRI or ultrasound. Combination of multiple modalities creates even more powerful imaging solution albeit at the higher costs of imaging equipment and related operating cost of administrating the procedure. Example includes combining SPECT with CT to create SPECT functional image superimposed on the CT anatomical image, or PET-MRI where functionality of PET is superimposed on MRI image. Other combinations, SPECT-MRI, PET-ultrasound, etc. have been also developed. Due to associated cost (the most sophisticated scanner can cost over $4M a piece) this type of equipment is rarely used in practice except in leading research hospitals in developed countries. More details about them can be found in selected books [1, 25, 46]. Here we are just providing one example of recent SPECT-CT scanner that performs dual-modality function as it is excellent conclusion to the X-ray and gamma ray detection chapter.

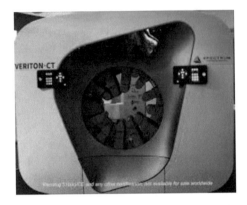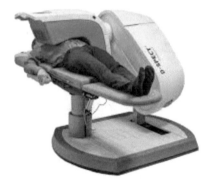

Fig. 6 SPECT-CT (VERITON-CT) (left) vs. SPECT-only (D-SPECT) (right) SPECT cameras from Spectrum Dynamics (https://spectrum-dynamics.com/)

One way of overcoming extremely high costs of dual-modality equipment is to provide anatomical image using low cost CT systems, e.g., with small number of slices, like 16 or 64 instead of state-of-the-art 320 high end CT machines while providing functions by Nuclear Medicine modality. An example of such development is VERITON SPECT-CT scanner that was announced at the EANM 2017 tradeshow in Vienna, Austria and fully launched in SNMI trade show in Philadelphia, USA in June 2018. It is world's first CZT detector based multi-organ scanner that offers unparalleled sensitivity, image quality, and diagnostic accuracy. Detectors specifically configured for each organ for optimal results.

The purpose of the VERITON-CT scanner is to provide full body diagnostics driven primary by oncology needs. As a result, the technical requirements for VERITON are somewhat different than for D-SPECT cardiology camera although both use CZT detector technology. One obvious difference is requirement in VERITON to have CT image, in order to localize precisely activity spots. Another is the detection area, VERITON-CT requires larger field of view to fit the whole body as illustrated in Fig. 6. The third reason is a range of isotopes used in imaging, while Tc99m is a workhorse contrast used in SPECT full body imaging might benefit from other isotopes (possibly used simultaneously with Tc99m). The required energy range of detection is 70 keV (for TI-201 isotope) to 364 keV (for I-131 isotope).

6   Summary

X-ray imaging is a well-known imaging modality that has been used for over 100 years since Roentgen discovered X-rays based on his observations of fluorescence. His initial results were published in 1895, and reports of diagnoses of identified fractures shortly followed. A year later, equipment manufacturers started selling X-ray equipment.

Today, X-ray and its three-dimensional (3D) extension, computed tomography (CT), are used commonly in medical diagnosis, and are present in all hospitals around the globe [25, 46]. As X-rays are high-energy photons their generation creates incoherent beams that experience insignificant scatter when passing through various media. As a result, X-ray imaging is typically based on through transmission and analysis of the resulting X-ray absorption data.

Various types of passive and active radiation detectors are used in medical imaging. In some of the active detectors energy of X-ray radiation is converted directly to electric charge, and then to a digital image. In many systems, an intermediate conversion of the signal carriers takes place, e.g., to visible light. Different detector materials and designs can be chosen depending on the energy of radiation. The perfect detector could be characterized with high detection efficiency, high spatial, time, and energy resolution. Usually, there is an interplay between the parameters, and depending on the application one may want to, e.g., increase detector's sensitivity at the cost of its spatial resolution. While X-ray imaging in medicine is well-established, the availability of new detectors opens new possibilities. On the other hand, experimental imaging modalities create new possibilities for use of the existing radiation sources and detectors.

References

1. Iniewski Krzysztof (2009), Medical imaging: principles, detectors, and electronics, Wiley. ISBN: 978-0-470-39164-8. doi: https://doi.org/10.1088/0031-9155/61/20/7246
2. Sprawls P (2018) Film-screen radiography receptor development: A historic perspective. Med Phys Int 6:56–81
3. Udoye CI, Jafarzadeh H (2010) Xeroradiography: stagnated after a promising beginning? A historical review. Eur J Dent 04:095–099. doi: https://doi.org/10.1055/s-0039-1697816
4. delle Canne S, Carosi A, Bufacchi A, et al (2006) Use of GAFCHROMIC XR type R films for skin-dose measurements in interventional radiology. Phys Medica 22:105–110
5. Cowen AR, Davies AG, Kengyelics SM (2007) Advances in computed radiography systems and their physical imaging characteristics. Clin Radiol 62:1132–1141. doi: https://doi.org/10.1016/j.crad.2007.07.009
6. Samei E, Flynn MJ (2003) An experimental comparison of detector performance for direct and indirect digital radiography systems. Med Phys 30:608–622. doi: https://doi.org/10.1118/1.1561285
7. Behling R (2018) X-ray tubes development- IOMP history of medical physics. Med Phys Int 6:8–55
8. Shefer E, Altman A, Behling R, et al (2013) State of the Art of CT Detectors and Sources: A Literature Review. Curr Radiol Rep 1:76–91. doi: https://doi.org/10.1007/s40134-012-0006-4
9. Kemerink GJ, Kütterer G, Kicken PJ, et al (2019) The skin dose of pelvic radiographs since 1896. Insights Imaging 10:39. doi: https://doi.org/10.1186/s13244-019-0710-1
10. Balter S (2019) Fluoroscopic Technology from 1895 to 2019 Drivers: Physics and Physiology. Med Phys Int 7:111–140
11. Webb S (1992) Historical experiments predating commercially available computed tomography. Br J Radiol 65:835–837. doi: https://doi.org/10.1259/0007-1285-65-777-835
12. Machida H, Yuhara T, Tamura M, et al (2016) Whole-body clinical applications of digital tomosynthesis. Radiographics 36:735–750. doi: https://doi.org/10.1148/rg.2016150184

13. Fuchs T, Kachelrie M, Kalender WA (2000) Direct comparison of a xenon and a solid-state CT detector system: measurements under working conditions. IEEE Trans Med Imaging 19:941–948. doi: https://doi.org/10.1109/42.887841

14. Lell MM, Wildberger JE, Alkadhi H, et al (2015) Evolution in computed tomography: the battle for speed and dose. Invest Radiol 50:629–644. doi: https://doi.org/10.1097/RLI.0000000000000172

15. Cameron JR, Sorenson J (1963) Measurement of bone mineral in vivo: An improved method. Science (80) 142:230–232. doi: https://doi.org/10.1126/science.142.3589.230

16. Mazess RB, Peppler WW, Gibbons M (1984) Total body composition by dual-photon (153Gd) absorptiometry. Am J Clin Nutr 40:834–839. doi: https://doi.org/10.1093/ajcn/40.4.834

17. Patel BK, Lobbes MBI, Lewin J (2018) Contrast Enhanced Spectral Mammography: A Review. Semin Ultrasound CT MR 39:70–79. doi: https://doi.org/10.1053/j.sult.2017.08.005

18. Rutherford R, Pullan BR, Isherwood I (1976) Measurement of effective atomic number and electron density using an EMI scanner. Neuroradiology 11:15–21. doi: https://doi.org/10.1007/BF00327253

19. Grajo JR, Patino M, Prochowski A, Sahani D V (2016) Dual energy CT in practice. Appl Radiol 45:61–2

20. Karçaaltıncaba M, Aktaş A (2011) Dual-energy CT revisited with multidetector CT: review of principles and clinical applications. Diagn Interv Radiol 17:181–194. doi: https://doi.org/10.4261/1305-3825.DIR.3860-10.0

21. Johnson TRC (2012) Dual-energy CT: general principles. AJR Am J Roentgenol 199:3–8. doi: https://doi.org/10.2214/AJR.12.9116

22. Leng S, Bruesewitz M, Tao S, et al (2019) Photon-counting detector CT: system design and clinical applications of an emerging technology. RadioGraphics 39:729–743. doi: https://doi.org/10.1148/rg.2019180115

23. Taguchi K, Iwanczyk JS (2013) Vision 20/20: Single photon counting x-ray detectors in medical imaging. Med Phys 40:100901. doi: https://doi.org/10.1118/1.4820371

24. Barber WC, Nygard E, Iwanczyk JS, et al (2009) Characterization of a novel photon counting detector for clinical CT: count rate, energy resolution, and noise performance. In: Proc. SPIE

25. Farncombe Troy, Iniewski Kris, (2013), Medical imaging technology and applications, CRC Press, ISBN 9781466582620

26. Nassalski A, Kapusta M, Batsch T, et al (2007) Comparative study of scintillators for PET/CT detectors. IEEE Trans Nucl Sci 54:3–10. doi: https://doi.org/10.1109/TNS.2006.890013

27. Vandenberghe S, Mikhaylova E, D'Hoe E, et al (2016) Recent developments in time-of-flight PET. EJNMMI Phys 3:. doi: https://doi.org/10.1186/s40658-016-0138-3

28. Kowalski P, Wiślicki W, Shopa RY, et al (2018) Estimating the NEMA characteristics of the J-PET tomograph using the GATE package. Phys Med Biol 63:. doi: https://doi.org/10.1088/1361-6560/aad29b

29. Moskal P, Rundel O, Alfs D, et al (2016) Time resolution of the plastic scintillator strips with matrix photomultiplier readout for J-PET tomograph. Phys Med Biol 61:2025–2047. doi: https://doi.org/10.1088/0031-9155/61/5/2025

30. Baily NA, Horn RA, Kampp TD (1980) Fluoroscopic visualization of megavoltage therapeutic x ray beams. Int J Radiat Oncol Biol Phys 6:935–939. doi: https://doi.org/10.1016/0360-3016(80)90341-7

31. Munro P (1995) Portal imaging technology: Past, present, and future. Semin Radiat Oncol 5:115–133. doi: https://doi.org/10.1016/S1053-4296(95)80005-0

32. Woźniak B, Ganowicz M, Bekman A, Maniakowski Z (2005) A comparison of the dosimetric properties of the Electronic Portal Imaging Devices (EPIDs) LC250 and aS500. Reports Pract Oncol Radiother 10:249–254. doi: https://doi.org/10.1016/S1507-1367(05)71097-X

33. Lai CJ, Shaw CC, Geiser W, et al (2008) Comparison of slot scanning digital mammography system with full-field digital mammography system. Med Phys 35:2339–2346. doi: https://doi.org/10.1118/1.2919768

34. Green FH, Veale MC, Wilson MD, (2016) Scatter free imaging for the improvement of breast cancer detection in mammography. Phys Med Biol 61:7246–7262.

35. Kaufman L, Carlson J (2011) An evaluation of airport x-ray backscatter units based on image characteristics. J Transp Secur 4:73–94. doi: https://doi.org/10.1007/s12198-010-0059-7

36. Stepusin EJ, Maynard MR, O'Reilly SE, et al (2017) Organ Doses to Airline Passengers Screened by X-Ray Backscatter Imaging Systems. Radiat Res 187:229–240. doi: https://doi.org/10.1667/RR4516.1

37. Towe BC, Jacobs AM (1981) X-ray backscatter imaging. IEEE Trans Biomed Eng BME 28:646–654. doi: https://doi.org/10.1109/TBME.1981.324755

38. Yan H, Tian Z, Shao Y, et al (2016) A new scheme for real-time high-contrast imaging in lung cancer radiotherapy: a proof-of-concept study. Phys Med Biol 61:2372–2388. doi: https://doi.org/10.1088/0031-9155/61/6/2372

39. Redler G, Jones KC, Templeton A, et al (2018) Compton scatter imaging: A promising modality for image guidance in lung stereotactic body radiation therapy. Med Phys 45:1233–1240. doi: https://doi.org/10.1002/mp.12755

40. Bazalova M, Kuang Y, Pratx G, Xing L (2012) Investigation of X-ray fluorescence computed tomography (XFCT) and K-edge imaging. IEEE Trans Med Imaging 31:1620–1627. doi: https://doi.org/10.1109/TMI.2012.2201165

41. Jones BL, Cho SH (2011) The feasibility of polychromatic cone-beam x-ray fluorescence computed tomography (XFCT) imaging of gold nanoparticle-loaded objects: a Monte Carlo study. Phys Med Biol 56:3719–3730. doi: https://doi.org/10.1088/0031-9155/56/12/017

42. Bazalova-Carter M, Ahmad M, Matsuura T, et al (2015) Proton-induced x-ray fluorescence CT imaging. Med Phys 42:900–907. doi: https://doi.org/10.1118/1.4906169

43. Olivo A, Speller R (2007) A coded-aperture technique allowing x-ray phase contrast imaging with conventional sources. Appl Phys Lett 91:111–114. doi: https://doi.org/10.1063/1.2772193

44. Scherer K, Willer K, Gromann L, et al (2015) Toward clinically compatible phase-contrast mammography. PLoS One 10:6–12. doi: https://doi.org/10.1371/journal.pone.0130776

45. Omoumi FH, Ghani MU, Wong MD, et al (2020) The potential of utilizing mid-energy X-rays for in-line phase sensitive breast cancer imaging. Biomed Spectrosc Imaging 1–14. doi: https://doi.org/10.3233/BSI-200204

46. Iwanczyk Jan, Iniewski Kris (2015), Radiation detectors for medical imaging, CRC Press, ISBN 9781498704359

Low-Dimensional Semiconductor Materials for X-Ray Detection

Zhiwen Jin, Zhizai Li, Wei Lan, and Qian Wang

1 Introduction

X-ray belongs to high-energy radiation with strong penetrating ability, and it can display the internal structure of the scanned objects without destruction. Since it was first discovered by Wilhelm Röntgen in 1895, it has been diffusely adopted in modern defense security, medical detection, industrial non-destructive testing, food inspection, etc. [1–4]. The basic principle of non-destructive testing is that X-ray ionizes the subjects by interaction with the atomic electrons to accurately detect the internal information of target object with different radiant properties (photon intensity/dose rate, direction, wavelength and phase, etc.) [5]. Due to the high-energy radiation, X-ray shows a wavelength between 0.01 and 10 nm, which is difficult to be sensed by ordinary photographic substrates. Effective materials should be explored to absorb it and convert it into detectable information.

To date, the available strategies for X-ray detectable technology are roughly divided into two kinds: indirect conversion of X-ray photons into low-energy visual light by scintillators, and direct conversion into electrical signals by semiconductors [6, 7]. The indirect scintillator-based system is the dominated X-ray detecting method with lots of superiorities, such as high light yield, high-energy resolution, flexible option and quick conversion rate [8, 9]. Inorganic scintillators ($CaWO_4$, Tl^+, $Bi_4Ge_3O_{12}$, etc.) with heavy elements (Bi, Ce, Eu, Tl, Pb, W, etc.) have been most widely explored because of their relatively high atomic number for excellent X-ray stopping power, and superior scintillation performance of X-ray conversion [10, 11]. Although more than a century of efforts have been made in crystal engineering, constituents engineering and shaped design (bulk

Z. Jin (✉) · Z. Li · W. Lan · Q. Wang
School of Physical Science and Technology, Lanzhou University, Lanzhou, China
e-mail: jinzw@lzu.edu.cn; lizhz15@lzu.edu.cn; lanw@lzu.edu.cn; qianwang@lzu.edu.cn

© The Author(s), under exclusive license to Springer Nature Switzerland AG 2023
K. (Kris) Iniewski (eds.), *Advanced X-Ray Radiation Detection*,
https://doi.org/10.1007/978-3-030-92989-3_2

single crystal, nanoparticle, fiber, film, etc.), several shortcomings, complicated preparation processes, high cost, and lateral spread of light-scattering crosstalk need to be conquered before application [12–15]. Moreover, the severe transmission loss originating from weak optical coupling between scintillator and thin-film transistors/complementary metal-oxide semiconductor (TFT/CMOS) in indirect imaging device further limits the imaging resolution [16].

The other X-ray detecting strategy is called direct X-ray, which uses semiconductors to directly transform the radiation into electrical signals without additional process. This method has the advantages of wide linear response range, fast pulse rise time, high-energy resolution, and high-spatial resolution, and is compatible with indirect X-ray detection [9, 16]. Notably, materials should meet the primary standards of outstanding stopping X-ray ability, superior mobility-lifetime products ($\mu\tau$), and high resistivity to achieve effective X-ray detection. The traditional semiconductors, silicon (Si), amorphous cadmium zinc telluride (CZT), mercury (II) iodide (HgI_2), and selenium (a-Se) are successful developments in pursuit of high X-ray performance [17–23]. However, some intrinsic weaknesses, such as small resistivity (~10^4 Ω cm), low X-ray absorption cross-section (e.g., a-Se for photon energies >50 keV), lower $\mu\tau$ (e.g., a-Se, $\mu\tau$ of 10^{-7} cm^2 V^{-1}), toxic elements and serious current loss (e.g., HgI_2), and high processing temperature (e.g., Si, CZT), limit their further application. Therefore, exploring alternative X-ray scintillating and semiconductor materials is of great significance for a wide range of commercial applications.

Recently, low-dimensional (LD) semiconductors have emerged as competitive candidates for X-ray detection, because of the strong quantum confinement effects and efficient photophysical properties (high exciton binding energy, large resistivity, and strengthened stability) [24]. For scintillator-based indirect X-ray, the LD semiconductors exhibit better scintillation than the three-dimensional (3D) at room temperature (RT), mainly because the higher exciton binding energy is beneficial to inhibit thermal quenching [25]. Moreover, the quantum wells of LD semiconductors confine the electrons and holes generated by ionizing radiation, and increase the overlap of the electron and hole wave function, thereby increasing the intensity of exciton oscillation and shortening the lifetime of exciton radiation [26]. Notably, the nomenclature of 3D and LD (two-dimensional (2D), one-dimensional (1D), and zero-dimensional (0D)) herein represents the structural dimensionality of semiconductor at the molecular level. Moreover, the LD semiconductors (especially halide perovskite semiconductors) with relatively large resistivity and high activation energies are beneficial to lower drifted baseline and improve stability against decomposing by hydrophobic organic spacer layer [27, 28]. In short, inorganic ($La_2Hf_2O_7$ nanoparticles (NPs), $NaLuF_4$:Tb, $Cs_2Hg_6S_7$, and Ga_2O_3, etc.), organic (P3HT:PCBM, polymer poly (9,9-dioctyfluorene)), and halide perovskite ($CsPbBr_3$, $Cs_3Bi_2I_9$, and Cs_2TeI_6, etc.) are widely explored, and some of them exhibit wonderful performances. Thus, it is necessary to summarize the advanced works focusing on LD semiconductors for X-ray detection.

In this book chapter, the working mechanism and essential parameters of the direct and indirect X-ray detector will be first introduced. Then, advanced works based on LD semiconductors are systematically presented in both indirect and direct X-ray detection, as well as their application in X-ray imaging. Finally, the remaining challenges and our perspectives in the field of LD semiconductor-based X-ray detection are also summarized, mainly focusing on material design, stability, device structure, and device performance.

2 Background of X-Ray Detection

To search ideal materials for effective X-ray detection, the detectable mechanism and important parameters in X-ray detectors are introduced in both scintillators and direct detection, as shown in Fig. 1a, b. Further, based on the recent hotspots of LD semiconductor materials, the key advantages and some superior properties of X-ray detection are highlighted.

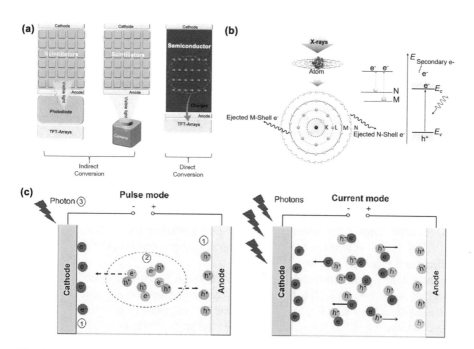

Fig. 1 (**a**) Diagram of X-ray detector in scintillator and direct modes. (**b**) Schematic illustration of the transformational mechanism from X-ray photon to current. Reproduced with permission [94]. Copyright 2019, Wiley-VCH Publications. (**c**) Two working mode of pulse and current modes in direct X-ray detection [31]. Reproduced with permission. Copyright 2021, American Chemical Society Publications

2.1 Scintillator

The scintillators are explored to cooperate with the photographic-film based X-ray detector in the early period to conquer its low efficiency, which can largely absorb and convert the high-energy radiations into visible light for further detection by the arrayed device (e.g., charge-couple device (CCD) or photodiode). Notably, physical processes such as photoelectric effect, Compton scattering, Tomson scattering, and electron pair effect occur during the interaction between X-ray and scintillators [29]. The photoelectric effect and Compton scattering effect dominate when the radiative energy is lower than 1 MeV, while the electron pair effect increases generated carriers when energy is higher than 1.02 MeV [6].

The Mechanism of Scintillator Detection

Totally speaking, the scintillating process could be roughly divided into three main stages, including converting incident radiation, transport of charge carriers, and scintillation [30]. During the first stage, the scintillators tend to absorb the incident X-ray (100 eV < hv <100 keV), converting it into a tremendous amount of hot electrons and free holes. Moreover, these generated charge carriers further interact with electrons, phonons, and plasmas to induce a large number of secondary electrons and holes [31].

During the second stage, the generated carriers tend to migrate to the luminescent centers, which suffer from several processes such as energy loss, electron and hole capture. The energy loss mainly happens when the migrating charge carriers are captured by the defects, including native defects (e.g., vacancy, interstitial, grain boundaries) and impurities (dopants, interfacial, and surface trap states) [32, 33]. Therefore, passivation strategy, optimization of synthesis route, and improvement of crystal quality are significant in improving the luminescent intensity [34–36]. During the final scintillation stage, the electron–hole (e–h) pair emits ultraviolet/visible (UV/Vis) emission through the emission center (usually impurity/doped ions in the lattice) and radiation combination.

Importantly, the scintillator of X-ray detection can be classified into internal and external scintillators [7, 37]. For the former, the luminescence centers consist of inherent defects of structural units/hosts, and the electrons directly transport from the conduction band to the valence band. However, the latter uses appropriate ions (external dopants) to create new luminescence centers and doping energy levels. Therefore, the extrinsic scintillators usually show a longer emission wavelength than the corresponding intrinsic one due to their narrower electron transition gap.

The Merit Parameters of Scintillator Detection

First of all, the essential parameters of light yield in scintillator are introduced, because the larger the light yield, the higher the sensitivity of X-ray detector, and the clearer the image resolution. The parameter is defined as the number of emissive photons in scintillator when it absorbs each 1 MeV ionizing radiation. The corresponding number of ultraviolet/visible (UV/Vis) photons transformed by scintillator could be obtained by the following formula: [32]

$$N_{Ph} = \frac{E}{\beta E_g} \times SQ \qquad (1)$$

where E is the energy of incident X-ray, E_g represents the bandgap of materials, S and Q stand for the quantum efficiencies during transport stage (electrons and holes to luminescent center) and scintillation (radiative recombination) stages. Notably, β is a constant between 2 and 3 (usually 2.5) and is defined as the average energy for one thermalized e–h pair [38]. However, the actual value is smaller than the calculated value, because some of the loss happens on scintillator and photoelectric acceptor. Besides, the morphology of scintillators, including their geometry, shape, thickness, porosity, and size, largely limits their transfer efficiency.

Besides, X-ray stopping ability is used to evaluate the capability of materials for completely absorbing radiation. The X-ray attenuation formula is adopted to predict X-ray stopping ability [39–41]:

$$I = I_0 \exp(-\mu_m d) = I_0 \exp(-\alpha \rho d) \qquad (2)$$

$$\alpha \propto \frac{Z^4}{AE^3} \qquad (3)$$

where μ_m is the linear absorption coefficient, ρ, d, and α are the materials density, thickness, and the X-ray absorption cross-section, respectively. Z is the atomic number of materials, A is the atomic mass, and E is the energy of X-ray photon.

Whether they are indirect scintillation or direct semiconductor, the materials with $\rho \geq 5$ g/cm^3 and Z value > 40 are needed to entirely absorb X-ray radiation [16]. Therefore, recent researches are focused on high-Z elements (Bi, Ce, Eu, Pb, and Tl) materials, especially in inorganic Lanthanides based scintillating system. Additionally, the response-decay time, radiation/environment stability, linearity of light response, and energy resolution are also important in scintillation detection.

2.2 Direct Detection

The direct X-ray detection utilizes semiconductor films to be complementary to scintillation detection. The e–h pair appears when X-ray interacts with the semiconductor films, which could be collected by device electrodes after the application of an external electric field and directly readout with an external circuit for analysis.

The Mechanism of Direct Detection

There are two working modes named current mode and pulse mode (voltage mode) depending on its operating principle, as shown in Fig. 1c, d [1, 31]. The current mode is used in receiving high-energy photons for high-photoflux application (e.g., dosimeter and medical imaging), and generating excess free charges by photoelectric and Compton scattering effects. The mode ensures sufficient signal-to-noise ratio (SNR), smooth image, and rapid frame rate to achieve fast image collection in imaging.

Differently, the pulse mode is used in the relatively low-photoflux or photon-insufficient application (e.g., photon counters and spectroscopy). In this mode, each photon arrives at the acceptor by turn and generates e–h pairs, which are collected by electrodes under small bias. Due to the appearance of the small signal, the charge-sensitive preamplifier is typically applied to integrate the current signals into the voltage signal. Then, the shaping amplifier and multichannel analyzers are employed to further better match the signal with the communication channel.

The Merit Parameters of Direct Detection

For X-ray detection and its application, the most important key parameter is the sensitivity, which is defined as the capability of detector to collect charge per unit area under the exposure of X-ray. For X-ray application in medical treatment, the total radiation accumulation in a single chest computed tomography (CT) is equal to the two-year natural irradiation dose accumulation. Moreover, the patients are at higher risk for radiation-induced malignancies when the dose is beyond 100 mSv, and almost 2% of cancers are related to the high-dose radiation exposure during CT scan [42]. To well predict the health condition, several parameters are used to evaluate the efficiency of X-ray detection, including sensitivity, noise, lowest detectable dose (LoD) rate, and resistivity.

The formula of sensitivity is as follows: [43]

$$R_s = \frac{I_p - I_d}{D \times A} \tag{4}$$

where R_s is the sensitivity, I_p is generated photocurrent, I_d is dark current, D is irradiation dose rate, and A is the effective area of X-ray detector. In addition, noise is defined as the lowest level of a detectable signal, which is consistent with inherent uncertainty or fluctuation and thus has a significant impact on the sensitivity. The noise consists with the intrinsic uncertainties or fluctuations, including shot noise

(i_{shot}), thermal noise ($i_{thermal}$), flicker noise ($i_{1/f}$), and generation-recombination noise (i_{g-r}) [44, 45].

Another significant parameter of LoD largely limits the X-ray detectable application (e.g., X-ray diagnostics, the LoD needs 5.5 μC Gy$_{air}$ s^{-1}) [46, 47]. This detection limitation was defined by the International Union of Pure Applied Chemistry (IUPAC) as the real signal value that was three times higher than noise [27, 48]. Notably, all these parameters could be modified by resistivity, because large resistivity leads to small dark current and low noise signal during the process of X-ray detection. The resistivity value of commercial application should be above 10^9 Ω cm to keep dark current in small value [49].

Also, some parameters are used to evaluate the imaging ability of materials. The modulation transfer function (MTF) is the most significant one, which defines the transmission capacity of the input signal to modulate the spatial frequency (unit: line pairs per millimeter (lp/mm)). The formula shows as follows [17, 50]:

$$MTF(f) = T_w(f) \times T_{tr}(f) \times T_a(f) \tag{5}$$

$$T_w(f) = P_{pe}T_{pe} + P_k T_k \tag{6}$$

where T_{pe} and T_k are primary photoelectron and k-fluorescence reabsorption, respectively. P_{pe} and P_k are the probability of primary photoelectron releasing carriers and the probability of k-fluorescence reabsorption, respectively. $T_w(f)$ is a weighted MTF derived from the main photoelectron and k-fluorescence reabsorption. $T_{tr}(f)$ is the effect of carrier trapping on MTF, and $T_a(f)$ is the MTF induced by pixel electrode. Quality of image (spatial resolution) is evaluated when MTF is 0.2, which is usually obtained through the slanted-edge method [51, 52].

3 The Development of LD Semiconductor in X-Ray Detection

Recently, the traditional scintillator materials, including NaI:Tl, CsI:Tl, CdWO$_4$, and Bi$_4$Ge$_3$O$_{12}$, exhibit intrinsic shortcomings of high cost, poor physical properties, low conversion efficiency, and strict fabricated condition. In contrast, like NaYF$_4$ nanoparticles, bismuth oxide (Bi$_2$O$_3$) nanoparticles, and LD halide perovskite, reducing the dimension of LD semiconductors shows unexpected physical properties.

Therefore, in this chapter, some of the advanced works focusing on LD materials are proposed for indirect and direct X-ray detection from the classifications of inorganic, organic, and novel LD halide perovskite.

3.1 Low-Dimensional Scintillator X-Ray Materials

The primary advantage of LD semiconductors is the quantum confinement effect. The binding energy in 2D system is four times larger than the corresponding 3D component, which is beneficial to suppress the dark current and improve the detectable efficiency [53]. Besides, the overlapping region of electron and hole wave functions is enlarged in LD system, resulting in a strengthened excitonic intensity and shortened excitonic radiative lifetime [25, 26]. Moreover, LD scintillators (e.g., 2D perovskite scintillator) show enhanced environment and thermal stability, as well as large Stokes shift to reduce self-absorption [54, 55]. Notably, the ratio of self-absorption is greater than 50% for conventional perovskite nanocrystals (NCs). Typically, increasing Stokes shift is beneficial to suppress self-absorption, and it can be realized by doping, constructing thick shells, and reducing dimension [13].

Inorganic LD Semiconductor

The inorganic scintillators are the most mature and earliest materials in high-energy radiation detection for their high atomic number (Z_{high}) elements, wide bandgap, superior thermal stability, and easy synthesis. Besides, the raw materials are rich enough to be easily available, thus, most of them have been adopted in commercial applications, including YAG:Ce, Y_2O_3:Eu, Y_2O_2S:Eu, and $LaPO_4$:Ce,Tb. Among them, rare-earth (RE) ions doping, especially lanthanides ions, is the most popular strategy to modify its photophysical properties. CaI_2 is a low-density and low cross-section scintillator, LuI_3 is fragile for synthesis as single crystal (SC), and Gd_2SiO_5 itself does not possess scintillated property. However, after the RE ions doping, CaI_2:Eu^{2+} and LuI_3:Ce^{3+} show strengthened light yield of more than 100,000 photons/MeV, while Gd_2SiO_5:Ce scintillation performance is also improved because of the parity-allowed electric dipole 5d 5d$\rightarrow$4f transition of the Ce^{3+} ion [56, 57].

Based on RE ions doping, the inorganic LD scintillators have achieved the device's fabrication of color tunability. Recently, Gupta et al. introduced the various lanthanides ions (Eu^{3+}, Tb^{3+}, and Dy^{3+}) and regulated their relative ratio to further control the emission light from greenish-white to yellowish-white. They found that for singly doping Eu^{3+}, Tb^{3+}, and Dy^{3+} lanthanides ions, the $La_2Hf_2O_7$ NPs scintillators presented red, green, and yellowish-blue light emission, while the $La_2Hf_2O_7$ NPs scintillators displayed a color tunability of warm white light in triply doping. Notably, the radioluminescence (RL) emission spectra of doped-$La_2Hf_2O_7$ NPs scintillators showed in Fig. 2a–c that all of them possessed better RL properties because of the introduction of the heavy Hf atom [58].

Moreover, most of the inorganic scintillators are integrated as a flat-panel detector, which is expensive and unsuitable to 3D imaging especially in irregular and curved objects. To overcome these problems, recently, Ou et al. invented a flat-panel-free imaging method, X-ray luminescence extension imaging (Xr-LEI), by

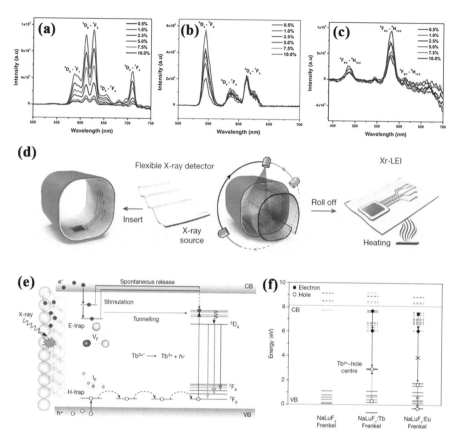

Fig. 2 (**a**) RL emission spectra for $La_2Hf_2O_7:Eu^{3+}$ NPs, (**b**) $La_2Hf_2O_7:Tb^{3+}$ and (**c**) $La_2Hf_2O_7:Dy^{3+}$ inorganic scintillators. Reproduced with permission. Copyright 2019, Elsevier Inc Publications [58]. (**d**) Schematic illustrates of flexible X-ray detector for 3D electronic imaging. (**e**) The radioluminescence mechanism of long-lived persistent Tb^{3+} doped $NaLuF_4$ NCs. (**f**) Energy diagram of lanthanide 4f levels (red line) versus to host bands (black line) [59]. Reproduced with permission. Copyright 2021, Nature Publishing Group

using lanthanide-doped (terbium (Tb^{3+})) $NaLuF_4$ lanthanide-dope nanoscintillator to achieve radiography on largely curved and irregular 3D objects. Also, the multicolor radioluminescence modulation in the ultraviolet-visible to the near-infrared range can be obtained by different activator doped (e.g., Nd^{3+}, Tm^{3+}, Dy^{3+}, Tb^{3+}, Er^{3+}, Ho^{3+}, Sm^{3+}, and Pr^{3+}) into the $NaLuF_4$ host materials. They found that $NaLuF_4$:Tb (15 mol%) nanoscintillator showed more attractive X-ray performance, which was mainly ascribed to its large atomic number, low photon energy, and small surface quenching. Moreover, the gain mechanism of lanthanide ions doping was detailedly researched. They pointed that high X-ray energy could dislocate fluoride anions (F^-) into the interstitial sites, and induced fluoride vacancies and interstitials, accompanying with many energetic electrons trapped in Frenkel defect-associated trap states, as shown in Fig. 2d. Finally, they developed a flexible X-ray detector for Xr-LEI, which was fabricated by integrating NaLuF$_4$:Tb (15 mol%) nanoscintillator with poly(dimethylsiloxane) (PDMS) with a large area (16 cm × 16 cm × 0.1 cm). Xr-LEI-based images displayed high spatial resolution, especially for the stretchable X-ray detector, higher than 20 lp/mm, as shown in Fig. 2e [59].

Organic LD Composite Semiconductor

The organic LD composite materials show advantages including easy manufacturing, flexibility, high grade of transparency (most materials), and low cost [9]. One of the limitations in scintillators is the uncontrollable optical crosstalk, which comes from the light emission transportation isotropically. The structural design (e.g., needle-type structure, 1D structure) could partly reduce its influence. However, there is rare research on it for their small density and Z value.

Novel Halides Perovskite LD Semiconductor

Recently, halide perovskites with superior properties, including tunable bandgap, high carrier mobility, moderate defect tolerance, and small exciton binding energy, are widely explored in solar cells, light-emitting diodes, and photodetectors [60–62]. Recently, because of the introduction of high B-site atomic number element (monovalence, Cu^+ or Ag^+; bivalence, Pb^{2+} or Sn^{2+}; trivalence Bi^{3+} or In^{3+}) and the emergence of strongly bound excitons, the LD halide perovskites are expended into the high-energy radiative detection [63–65].

In 2016, Birowosuto et al. detailedly explored the X-ray scintillation properties in the different 3D MAPbI$_3$, MAPbBr$_3$, and 2D (EDBE)PbCl$_4$ halide perovskite SCs. They found that all the scintillators exhibited high light yield exceeding 120,000 photons/MeV, but 3D halide perovskites showed a strong thermal quenching effect, and low light yield (<1000 photons/MeV) at room temperature. Instead, the 2D (EDBE)PbCl$_4$ perovskite SC possessed larger exciton binding energy of ~360 meV than 3D components (2–70 meV), which enhanced its ability against thermal quenching [66].

CsPbBr$_3$ perovskite nanocrystals scintillators are the most widely used halide perovskite, which was first used for X-ray detection in 2018 by Heo and his cooperators for its high photoluminescence quantum yields (PLQY) over 95% [67]. Then, the fabricated method was continually optimized for better scintillation property. Zhang et al. used a modified coprecipitation method at RT to obtain the high light yield (~21,000 photons/MeV) and concentration (150 mg/ml) CsPbBr$_3$ colloid quantum dot scintillators, which exhibited strong RL and enduring stability under X-ray illumination. Moreover, high-concentration CsPbBr$_3$ colloid quantum dot scintillators can induce the self-assembly process of perovskite nanosheets for compact films with the necessary thickness. Based on the CsPbBr$_3$ film with sufficient thickness, the corresponding imaging system provides a high spatial resolution of 0.21 mm [14].

However, the severe reabsorption issues in LD scintillator materials limit X-ray detect performance (especially spatial resolution). Li et al. reported an all-perovskite tandem device for X-ray detection, which consisted of $CsPbBr_3$ perovskite NC scintillators and perovskite photodetectors (PDs). $CsPbBr_3$ NC of specific thickness was used to ensure quick energy transition, and the reabsorption was largely reduced. Besides, the thick $CsPbBr_3$ films were engineered by dodecyldimethylammonium bromide (DDAB) to obtain high PLQY (>50%) and RL efficiency. Finally, based on the engineered $CsPbBr_3$ NCs and outstanding photophysical performance perovskite photodetectors, an advanced X-ray sensitivity of 54,684 $\mu C\ Gy_{air}^{-1}\ cm^{-2}$ was obtained [13]. Recently, a wonderful spatial resolution of 15.0 lp/mm was obtained by Ma et al. by a small amount of europium (Eu) doped into $CsPbBr_3$ QD to improve crystallization and uniform distribution. Because of the reduction of light scattering, the corresponding X-ray exhibited fantastic performance [68].

Besides, the stability of halide perovskite NCs/quantum dots (QDs) is poor especially for their superior sensitivity to the ambient environment (irradiation, heat, and moisture), which tends to induce aggregation and phase transition. Moreover, scintillator materials are usually high-temperature sintering, which typically induces stable problems for its agglomeration or large bulk crystal [69]. Therefore, Cao et al. devoted an "emitter-in-matrix" concept to improve the stability and corresponding X-ray performance of $CsPbBr_3$ NCs by embedding it inside the Cs_4PbBr_6 crystal matrix as $CsPbBr_3@Cs_4PbBr_6$ system. This improvement was ascribed to that the Cs_4PbBr_6 crystal matrix improving the X-ray stopping ability and optical transparency for light output. Finally, the large X-ray imaging screen (240 mm × 320 mm) based on $CsPbBr_3@Cs_4PbBr_6$ system displayed large linearity of dose rate from 93.75 $\mu C\ Gy_{air}\ s^{-1}$ to 1340.37 $\mu C\ Gy_{air}\ s^{-1}$, a 3 ns quick decay time, and clear X-ray images, as shown in Fig. 3a [12].

Moreover, the color of halide perovskite was adjustable by modifying their components with different halide ions and showed a large attraction in visualization tools during X-ray radiography [70, 71]. Qiu et al. used the tunable bandgap of $CsPbCl_xBr_{3-x}$ nanocrystal scintillators to construct a flexible device of multicolor X-ray scintillation, as shown in Fig. 3b. Then, the desirable luminescence color could be well controlled by adjusting the bandgap energy of $CsPbCl_xBr_{3-x}$ nanocrystal scintillators to further influence on the trapping and recombination of e–h pairs. After solution-processability, the $CsPbCl_xBr_{3-x}$ nanocrystal scintillators compositive thin-film scintillator is fabricated, they obtained a device with superior X-ray performance, including low detection limitation (13 $nGy_{air}\ s^{-1}$), fast response time (decay time of 44.6 ns), and superior X-ray resolution (MTF = 0.72, spatial resolution = 2 lp/mm). The X-ray images of Apple iPhone electronic circuits have been shown in Fig. 3c [7]. Then, Xie et al. recently integrated lanthanide-doped upconversion nanoparticles (UCNPs) and halide perovskite of $MAPbI_3$ into construing a perovskite/lanthanide nanotransducer. As shown in Fig. 3d, the $NaYF_4$:Yb/Tm UCNPs were coated with a layer of mesoporous silica ($mSiO_2$) shell and $MAPbI_3$ nanocrystal in-situ synthesized inside of the $mSiO_2$. The

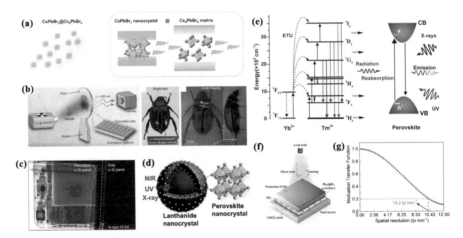

Fig. 3 (**a**) Schematic representation the mechanism of Cs_4PbBr_6 matrix for improving light output. Reproduced with permission [12]. Copyright 2019, American Chemical Society Publications. (**b**) Schematic demonstration of the imaging process for biological samples, and (**c**) in an Apple iPhone [7]. Copyright 2018, Nature Publishing Group. (**d**) Schematic demonstration of a lanthanide–perovskite nanocrystals for X-ray, UV, and NIR photons detection. (**e**) Schematic demonstration of energy-transfer upconversion process for UCNPs@mSiO$_2$@MAPbBr$_3$ lanthanide–perovskite nanocrystals [15]. Reproduced with permission. Copyright 2021, Wiley-VCH Publications. (**f**) Multilayered design in Rb$_2$AgBr$_3$ flat-panel X-ray imaging, and (**g**) corresponding spatial resolution in different MTF values [72]. Reproduced with permission. Copyright 2020, Wiley-VCH Publications

functional mechanism is that NaYF$_4$:Yb/Tm UCNPs which suffers energy-transfer upconversion processes and emits UV–vis light, then radiates reabsorption process from lanthanide nanocrystals to perovskite nanocrystals that happens under NIR excitation shown in Fig. 3e. Importantly, they embedded the perovskite/lanthanide nanotransducer into the PDMS to fabricate an imaging screen with a large area of 8 cm × 8 cm, and the device showed an excellent linear response at different accelerated voltage from 20 to 50 kV and a superior resolution of 20 lp/mm [15].

However, the toxic Pb element has limited halide perovskite as a green candidate material for X-ray detection, therefore, choosing alternative low-toxic, even nontoxic elements (Sn, Bi, Ge, and Sb, etc.) to replace the Pb element and constructing LD halide perovskites becomes an important research topic. Cao et al., recently used Sn^{2+} to replace the toxic Pb^{2+} and fabricated 2D tin halide perovskite ((C$_8$H$_{17}$NH$_3$)$_2$SnBr$_4$) scintillator with an emission wavelength of 596 nm, which showed a high near-unity PLQY of 98% under ultraviolet light irradiation. Moreover, they used polymethyl methacrylate (PMMA) to package the scintillator and formed the perovskite-polymer composite with a large area (100 × 100 mm^2) for detection. Moreover, the detector based on perovskite-polymer composite presents a sensitivity of 104.23 μC Gy$_{air}^{-1}$ cm^{-2} and a high image resolution [30]. Zhang et al. applied a cooling crystallization strategy to synthesize a nontoxic metal Rb$_2$AgBr$_3$ halide scintillator single crystal with 1D rod-like, which displayed novel defect-bound excitons radioluminescence, rapid scintillation, and highlight yield (25,600 photons/MeV) wide photon energies (1 keV–100 MeV). The large stokes shift (214 nm) of Rb_2AgBr_3 scintillator was beneficial to reduce self-absorption and improve its physical properties. Besides, after combining with close-spaced sublimation (CSS) method by heating Rb_2AgBr_3, they finally obtained highly preferred orientation and uniform radioluminescence Rb_2AgBr_3 film with a 250 µm thickness and 25 cm^2 size. A state-of-art spatial resolution of 10.2 lp/mm (MTF = 0.2) showed after integrating it with a commercial CMOS panel and excellent dynamic X-ray imaging ability, which was shown in Fig. 3f [72]. In another work, they expanded the research toward Cs–Cu–I system for a 1D $CsCu_2I_3$. This novel material possessed 1D crystal structure, which allowed low lateral light-scattering crosstalk problems for better spatial resolution. Moreover, the congruent-melting properties, Rb_2AgBr_3, making itself suit to CSS method for thick film preparation for efficiently absorbing X-ray. Finally, the oriented structured $CsCu_2I_3$ thick film with a large area (25 cm^2) is obtained by CSS and seed-screening methods, and the corresponding X-ray imager displayed a high spatial resolution of 7.5 lp/mm [73].

Wei et al. displayed a heterometallic $Cs_4MnBi_2Cl_{12}$ all-inorganic SC scintillators with a $[BiCl_6]^{3-}$-$[MnCl_6]^{3-}$-$[BiCl_6]^{3-}$ triple-layered structure. Because the energy transfer happened between the $[BiCl_6]^{3-}$ octahedron donor and $[MnCl_6]^{3-}$ acceptor, $Cs_4MnBi_2Cl_{12}$ scintillators exhibited strong orange emission with a high PLQY of 25.7%, which was almost 51 times larger than the $CsMnCl_3 \cdot 2H_2O$ (PLQY = 0.5%) without Bi alloying. For the reason that heavy atom of Bi was introduced and superior RL intensity than commercial NAI:Tl scintillators, $Cs_4MnBi_2Cl_{12}$ scintillators showed more favorable for X-ray detection, as shown in Fig. 4a. Moreover, a large-area module (10 cm × 10 cm) X-ray imager based on $Cs_4MnBi_2Cl_{12}$ single crystal powders could recognize the detailed information of objects [74].

Also, 0D materials for X-ray detection have been explored recently. Their dispersive electronic structure is beneficial to locate excitons and increase excitons binding energy (*Eb*) for high-intensity PLQYs [75]. Morad et al. developed a series of 0D $Bmpip_2MBr_4$ (Bmpip = 1-butyl-1-methyl-piperidinium) hybrid halide perovskite scintillators, where M is Ge^{2+}, Sn^{2+}, and Pb^{2+} with electronic spectra of ns^2. All the scintillators showed singlet exciton and triplet emission (self-trapped exciton) and exhibited broadband emitters in the visible spectral region after the excitation (above 3.9 eV) shown in Fig. 4b. Moreover, they fabricated $Bmpip_2PbBr_4$ and $BmpipSnBr_4$ as pellets for comparing X-ray performance with commercial NaI:Tl scintillator, and found that all of them exhibited similar intensities under 50 kV Ag tube excitation shown in Fig. 4c [63]. Then, Xu et al. developed an organic metal halide 0D phosphonium manganese (II) bromide hybrid $(C_{38}H_{34}P_2)MnBr_4$ at room temperature by a simple solvent diffusion method. The reproducible $(C_{38}H_{34}P_2)MnBr_4$ single crystal with a size beyond 1 in was obtained, which showed excellent stability, intense green emission, and a PLQY of ~95%. After combining the $(C_{38}H_{34}P_2)MnBr_4$ fine powders with PDMS, the composited film displayed better flexibility and even emission of green light under UV irradiation.

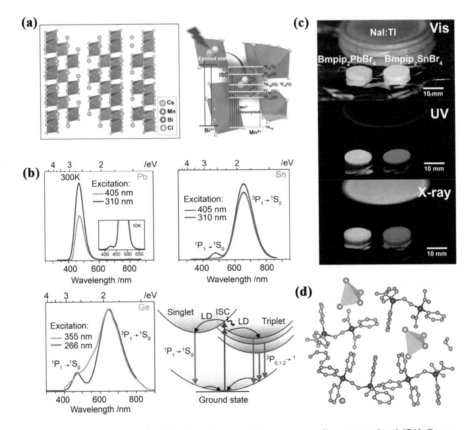

Fig. 4 (**a**) Crystal structure of $Cs_4MnBi_2Cl_{12}$, and the corresponding energy level [74]. Reproduced with permission. Copyright 2020, Elsevier Inc Publications. (**b**) The showing both singlet and triplet emission bands of $Bmpip_2MBr_4$ (M = Pb, Sn, Ge) in PL spectra with different wavelength. (**c**) The luminescence properties of NaI:Tl commercial scintillator, $Bmpip_2PbBr_4$ pellet, and $Bmpip_2SnBr_4$ pellet under ambient light, UV irradiation, and X-ray irradiation [63]. Reproduced with permission. Copyright 2019, American Chemical Society Publications. (**d**) Crystal structure of $(C_{38}H_{34}P_2)MnBr_4$ SCs [76]. Reproduced with permission. Copyright 2019, Nature Publishing Group

The corresponding X-ray imager system showed great linearity for the dose rate (36.7 μC Gy_{air} s^{-1} to 89.4 μC Gy_{air} s^{-1}), moderated light yield (66,256 photons/MeV) and superior LoD (461.1 nC $Gy_{air\,s}^{-1}$) which was shown in Fig. 4d [76].

Besides, $Cs_3Cu_2I_5$ was a 0D semiconductor in Cs–Cu–I system with advantages of lead-free, high luminescence stability, and high PL (79,279 photons/MeV) [8]. Zhou et al. have recently used $Cs_3Cu_2I_5$ NCs as X-ray detected materials, and found strong blue light emissions of high PLQY of 59%, and a large Stokes shift and long photoluminescence life have emerged. They ascribed this improvement to the functional of the self-trapped exciton. The electronic configuration of Cu^+ $3d^{10}$

has changed to Cu^{2+} $3d^9$ after absorbing photons, resulting in Jahn–Teller distortion and recombination of the excited-state structure. Besides, the constructed CT system based on $Cs_3Cu_2I_5$ NCs showed clear images of a snail sample, indicating $Cs_3Cu_2I_5$ NCs were successfully applied in X-ray CT measurement [77]. Subsequently, the PLQY of $Cs_3Cu_2I_5$ NCs was improved by Lian et al., who introduced indium iodide (InI_3) to modify the injection process in which it suppressed the large colloid $Cs_3Cu_2I_5$ NCs appeared at high temperature and influenced its stability. This additive made $Cs_3Cu_2I_5$ NCs show a stable blue emission at 445 nm, and also made a high PLQY reached 73.7%. The corresponding X-ray detector was coupled with a commercial silicon photomultiplier and displayed better linear in a range of 13.4–94.1 μC Gy_{air} s^{-1} [8].

3.2 LD Direct X-Ray Materials

Although directing the X-ray detection was simple, high sensitivity, and high spatial resolution, the advanced performance including the LoD and sensitivity was typically inferior to the scintillator-based detector, which could be ascribed into the high dark current and uncontrolled baseline drift induced by ion migration. Fortunately, reducing dimensionality was beneficial to large activity energy, dark resistivity, and to improve stability for further X-ray performance, thus this chapter focus on the advantage of LD X-ray materials and their application in this field.

Inorganic LD Semiconductor

Although inorganic semiconductors have shown high density and large Z by completely absorbing X-ray, one remained but urgent problem was that its energy band broadening because of the atomic potentials which were spread with the increasing Z, which induced the wavefunction overlap in neighboring atoms, causing band broadening and low bandgap [78]. In 2011, Androulakis et al. developed a strategy named dimensional reduction to conquer the E_g-Z limitation. As shown in Fig. 5a, they introduced M_2Q into a binary network of AQ (M = alkali metal ion; A = Hg, Cd; Q = S, Se, Te) for Q^{2-} ions doping into the cubic lattice of target materials and forming anionic $[A_xQ_y]^{n-}$ frameworks in a lower dimensionality. Significantly, the LD inorganic materials have showed an admirable performance. For examples, $Cs_2Hg_6S_7$ showed a resistivity of ~2 × 10^6 Ω cm (growth direction) and $\mu\tau$ of 0.81 × 10^{-3} cm^2 V^{-1}, and $Cs_2Hg_3Se_4$ showed a resistivity of ~1.1 × 10^9 Ω cm (growth direction) and $\mu\tau$ of 0.78 × 10^{-3} cm^2 V^{-1} [79].

Recently, β-Ga_2O_3 has possessed a large density (6.44 g/cm^3), high breakdown electric field (8 MV/cm), excellent radiative resistance, and superior thermal stability. Therefore, it is suitable for high-energy detection. Ga_2O_3 was firstly reported as a flexible X-ray detector by Liang et al. in 2018, who used a radio

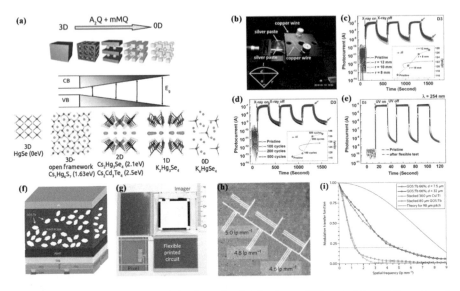

Fig. 5 (**a**) Schematic illustration of the dimensional reduction and LD materials [79]. Reproduced with permission. Copyright 2011, Wiley-VCH Publications. (**b**) The bending test of flexible a-Ga$_2$O$_3$, and (**c–e**) corresponding photoelectric performance [80]. Reproduced with permission. Copyright 2019, American Chemical Society Publications. (**f**) The device structure of hybrid X-ray sensor with an a-Si:H backplane and a hybrid frontplane. (**g**) Photograph of the hybrid organic X-ray imager and micrograph of an individual pixel, and (**h**) corresponding spatial resolution. (**i**) The MTF value of hybrid organic X-ray image sensor with different layer thickness [86]. Reproduced with permission. Copyright 2015, Nature Publishing Group

frequency magnetron sputtering system to deposit amorphous a-Ga$_2$O$_3$ on the polyethylene naphthalate (PEN) substrate at room temperature. After precisely controlling the oxygen flux, they observed that X-ray photocurrent has increased under a small oxygen gas environment. Combining with density functional theory (DFT) calculation, they showed the improvement of the current as a long time for neutralizing more ionized oxygen vacancies states, as shown in Fig. 5b–e [80]. Besides, the pure β-Ga$_2$O$_3$ showed a high effective electron concentration of 10^{17} cm^{-3}, which tends to induce a large dark current. Therefore, Chen et al. reported a β-Ga$_2$O$_3$:Mg SC with high resistivity of ~6.4 × 10^{11} Ω cm, and a low dark current of 0.28 nA. They ascribed the enlarged resistivity to Mg atoms substituted the octahedral Ga atoms as acceptors, which compensated the free electron in the crystal. Moreover, the lowest detectable X-ray dose rate was improved from ~69.5 μC Gy$_{air}$ s^{-1} to 695 μC Gy$_{air}$ s^{-1} in a quick response time, because neutralization of the ionized oxygen vacancies needs a long decay time while Mg^{2+} doping largely suppressing the oxygen vacancy concentration [81].

Organic LD Composite Semiconductor

Compared with the traditional solid-state detector, the performance of organic composite semiconductor depends on the high-energy absorption and subsequent ionization process. However, the low-Z limits its further application of X-ray detection. Recently, a new strategy of integrating semiconducting polymer with heavy atoms (strong high-energy photon absorption) has been widely explored for conquering the low-Z problem [82–84]. Mills et al. introduced heavy atoms of metallic tantalum and electrically insulating Bi_2O_3 to enhance the sensitivity from 141 nC Gy_{air}^{-1} cm^{-3} to 468 nC Gy_{air}^{-1} cm^{-3} for Bi_2O_3 NP (57 wt.%) load device [85]. Then, Büchele developed a new strategy by introducing suitable nanoparticles (terbium-doped gadolinium oxysulfide, GOS:Tb) into the organic composite P3HT:PCBM to adjust its absorption spectrum from near-infrared to X-ray region. The bulk heterojunction (BHJ) between phenyl-C61-butyric acid methyl ester (PCBM) and hole conductor can absorbed the re-emitted photons from the scintillator particles, resulting in small optical crosstalk and high resolution of 4.75 lp/mm at MTF = 0.2. The device and imaging performance were showed in Fig. 5f–i [86]. Then, Thirimanne et al. have invented an organic BHJ and Bi_2O_3 NP composites to improve its X-ray performance, which largely reduced the energy loss by dissociating the generated bound e–h pairs. Furthermore, they fabricated it on the soft substrate and received a high sensitivity of 300 μC Gy_{air}^{-1} cm^{-2} [87].

However, the common understand and influence of nanoparticle doping in X-ray detectors lack. Therefore, recently, Ciavatti et al. based on polymer poly(9,9-dioctyfluorene) blended with Bi_2O_3 NPs to research its gain mechanism. They found that it was mainly the high-Z NPs doing that influenced the X-ray absorption of components, while larger NPs loading was bad to morphology and also its carrier behavior including carriers injection and transport [82].

Novel Halides Perovskite LD Semiconductor

Commonly, higher sensitivity has the smaller risk for human body when exposed to X-ray irradiation. However, the limited thickness of halide perovskites is unsuitable to completely absorb X-ray (typical 2–3 nm for halides perovskites), while usual thicknesses are in a range from hundreds of micrometers to several millimeters which depends on the fabricated method (film, wafer, and single crystal, etc.) [16]. Instead, applying large bias on the opposite electrodes is a common strategy to effectively extract charges to reduce the exposure time during medical treatment. Nevertheless, the large bias induces a polarization effect to happen, resulting in a large dark current and dark current drift phenomenon. Therefore, great effects are done in reducing halide perovskite dimensionality to control its dark current in a low value when pursuing high sensitivity.

Xu et al. were the first to propose a 2D Ruddlesden–Popper (RP)-type $(BA)_2CsAgBiBr_7$ (BA = butylammonium) perovskite single crystal as X-ray detection and exhibited unprecedented photophysical properties [88]. Subsequently,

Li et al. novelly synthesized a large-dimensional-size 2D single crystal of 4-fluorophenethylammonium lead iodide ((F-PEA)$_2$PbI$_4$) during a simple solution process. The dipole interaction in the organic molecule of (F-PEA)$_2$PbI$_4$ improved the ordering of benzene rings and strengthened the supramolecular electrostatic interaction between electron-deficient F atoms and neighbor benzene rings. Moreover, the deficient F atoms was a supramolecular anchor, which effectively blocks the ions migration path, and tailoring the Fermi level toward the middle of the bandgap to a record resistivity of 1.36×10^{12} Ω cm. The benefit of the deficient F atoms introduction made the X-ray device display a sensitivity of 3402 μC Gy$_{air}$$^{-1}$ cm^{-2} under hard X-rays (120 keV$_p$) and a LoD of 23 nC Gy$_{air}$ s^{-1} [28]. Similarly, Xiao et al. recently systematically researched the organic molecules of BA and branched isobutylamine (i-BA)) for the purpose of understanding their differences in adjustable anisotropic structure, optoelectronic properties, and X-ray detection performance. They found that a shorter interlayer distance and a barrier height of i-BA were beneficial to charge collection. Finally, they obtained superior X-ray sensitivity of 13.26 mC Gy$_{air}$$^{-1}$ cm^{-2} under a low electric field of 2.53 V mm^{-1} [89].

In addition to the RP-type 2D halide perovskite, lately, Dion–Jacobson (DJ) type components have been widely explored for their much shorter distance during the adjacent inorganic layers, stable structure, and efficient charge transport. Shen et al. synthesized a centimeter-sized DJ-typed BDAPbI$_4$ (BDA = NH$_3$C$_4$H$_8$NH$_3$) perovskite SC with a low trap state of 3.1×10^9 cm^{-3}, the suitable bandgap of 2.37 eV, and excellent X-ray performance (sensitivity of 242 μC Gy$_{air}$$^{-1}$ cm^{-2} and LoD of 430 nC Gy$_{air}$ s^{-1}), as shown in Fig. 6a–c [90]. Wang et al. reported a novel 2D lead-free iodide double perovskite SC named (DFPIP)$_4$AgBiI$_8$ (DFPIP = 4,4-difluoropiperidinium), which B-site metal ions were located by the alternating of Ag$^+$ and Bi^{3+} ions. Besides, inorganic layers were consisted of corner-sharing AgI$_6$ and BiI$_6$ octahedra, where bilayers DFPIP cation was embed between the adjacent inorganic layers. This unique structure made it exhibit excellently ferroelectricity properties with a high Curie temperature (T_c) of 422 K and large spontaneous polarization (P_S) of 10.5 μC cm^{-2}. Moreover, the corresponding X-ray detector showed better sensitivity of 188 μC Gy$_{air}$$^{-1}$ cm^{-2} and a LoD of 3.13 μC Gy$_{air}$ s^{-1} because of the improved mobility-lifetime products ($\mu\tau$) beyond the order of 10^{-5} [91].

1D halide perovskites have also been widely explored for their superior physical properties, such as enhanced stability and small ion migration path. For example, Tao et al. fabricated a new 1D lead-free bismuth-iodide hybrid of (H$_2$MDAP)BiI$_5$(1, H$_2$MDAP = N-methyl-1,3-diaminopropanium) with a large size of $9 \times 7 \times 4$ mm^3 with the top-seeded solution and growth method [92]. Zhang et al. reported a superior physical properties 1D CsPbI$_3$ SCs with large resistivity of 7.4×10^9 Ω cm, high charge mobility (electron = 44.3 ± 3.3 cm^2 V^{-1} s^{-1}, hole = 30.4 ± 1.9 cm^2 V^{-1} s^{-1}), and moderated X-ray performance [93].

Additionally, the 0D perovskite including QDs and NCs could be compatible with the inexpensive inkjet printing method, and could be further applied in large-scale X-ray arrays fabrication. This method could print perovskite homogenously

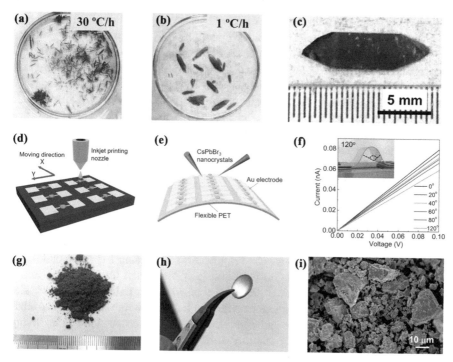

Fig. 6 (a) Photograph of (BDA)PbI$_4$ SCs grows in a ramp of 30 °C/h, (b) 1 °C/h, and (c) final centimeter-sized (BDA)PbI$_4$ SCs [90]. Reproduced with permission. Copyright 2020, Wiley-VCH Publications. (d) Schematic demonstration of inkjet printing method for photoelectric devices. (e) Flexible X-ray detector arrays based on CsPbBr$_3$ NCs, and (f) corresponding IV test by bending it different angles under X-ray illumination [94]. Reproduced with permission. Copyright 2019, Wiley-VCH Publications. (g) Photograph of MA$_3$Bi$_2$I$_9$ powders, (h) polycrystalline pellets, and (i) SEM image [97]. Reproduced with permission. Copyright 2020, Wiley-VCH Publications

on various substrates, including rigid and flexible, and was potential for soft X-ray detection [94]. Therefore, perovskite QDs and NCs are ideal materials in flexible devices, large-scale X-ray detection. Liu et al. developed a solution to fabricate CsPbBr$_3$ QDs, and exhibited a moderated sensitivity in both rigid and flexible-based X-ray detections which was shown in Fig. 6d–f. Under the synchrotron soft-X-ray beamline, the LoD of 0.0172 mGy$_{air}$ s^{-1}, and high sensitivity of 1450 μC Gy$_{air}$$^{-1}$ cm^{-2} were obtained under a small bias voltage of 0.1 V [94].

Nowadays, a new LD semiconductors with a formula of A$_3$B$_2$X$_9$ have been widely researched due to their superior physical properties which contained small dark current, controllable ion migration for reducing baseline drift, and large activity energy [95, 96]. In the early stages, A$_3$B$_2$X$_9$ perovskites focused on organic components. And the first 2D layered perovskite-like single crystal named (NH$_4$)$_3$Bi$_2$I$_9$ SC was reported by Zhuang et al. in 2019, which exhibited amazing anisotropic X-ray performance at different crystal directions. This anisotropic behavior was beneficial to suppress the ion migration, and boosted the X-ray detector performance with the LoD of 55 nGy_{air} s^{-1} (perpendicular direction) and 8.2×10^3 μC Gy_{air}^{-1} cm^{-2} (parallel direction) [27]. Importantly, 0D was the lowest dimensionality, where semiconductors have shown the strongest quantum confinement effect, softest lattice, and also highest electron–phonon coupling [75]. On the contrary, $A_3B_2X_9$ can also expand into 0D materials for its unique structure and physical properties. As reported, perovskite wafers showing high materials utilization, enough thickness and scale [16, 34]. In that case, Tie et al. have developed a 0D $MA_3Bi_2I_9$ polycrystalline pellets as a thickness enough wafer by scalable cold isostatic-pressing. And the perovskite wafer exhibited a high resistivity of 2.28×10^{11} Ω cm, the low dark current concentration of ~10^7 cm^{-3} as well as a high mobility of ~2 cm^2 V^{-1} s^{-1}. The corresponding X-ray displayed the lowest detectable dose rate of 9.3 nGy_{air} s^{-1}, and superior sensitivity of 563 μC Gy_{air}^{-1} cm^{-2}, as shown in Fig. 6g–i [97].

Nonetheless, the organic components (NH_4^+ and MA^+) have shown limitable thermal and moisture stability [49] which made the all-inorganic $A_3B_2X_9$ perovskites, including $Cs_3Bi_2I_9$ and Cs_2TeI_6, become promising LD semiconductors for further high-stability radiation detection [98]. Currently, Cs_2TeI_6 turns into a deficient perovskite with isolated $[TeI_6]^{2-}$ octahedra, and exhibits 0D perovskite structure. The all-inorganic components and high atomic number of Cs_2TeI_6 lead-free halide perovskite have the endurance radiation and thermal stability for X-ray detection [99].The outstanding stability and solution-processable ability make it suitable for flexible X-ray detection. Guo et al. fabricated thick Cs_2TeI_6 by using the electrospraying method and expended it to integrate with flexible polyimide (PI) for the purpose of fabricating a flexible X-ray detector. They compared the films morphology evolution in different flexible and rigid substrates and found that the contact angle was smaller in the flexible substrate (21–22°) than that rigid one (~40.5°). They pointed that a small contact angle could result in easy nucleation due to small formation energy. Besides, the smaller growth stress and lattice mismatch between the perovskite layer and substrate might assist grain growth larger. Eventually, based on the electrosprayed Cs_2TeI_6 with PI substrate, the fabricated flexible X-ray detectors displayed superior bending stability without cracking. At the same time, they have also showed excellent X-ray performance with the LoD small than 0.17 μC Gy_{air} s^{-1} and moderated a spatial resolution [100]. Furthermore, $Cs_3Bi_2I_9$ constitutes a pair of $[BiI_6]^{3-}$ octahedra sharing with common faces for formation $[Bi_2I_9]^{3-}$ dioctahedral cluster, and Cs^+ fills between $[Bi_2I_9]^{3-}$ and the voids between the dioctahedral. Recently, Zhang et al. applied a nucleation-controlled solution method to control the size and quality of $Cs_3Bi_2I_9$ SC because the fabricated $Cs_3Bi_2I_9$ SC possessed low trap density (1.4×10^{10} cm^{-3}), high $\mu\tau$ product (7.97×10^{-4} cm^2 V^{-1}), ultra-low dark current noise (102 pA), and high stability. Finally, excellent X-ray performance was obtained with high X-ray sensitivity 1652.3 μC Gy_{air}^{-1} and LoD of 130 nC Gy_{air} s^{-1} [101].

However, the quantitative relation between the structural dimensionality of $A_3B_2X_9$ and their components still remains a mystery. Besides, the internal mechanism is unclear. The tolerance factor is an unsuitable application in this system.

Lately, Xia et al. devoted themselves to study the suitable structural descriptor for $A_3B_2X_9$ perovskites. They found two kinds of A-site cations that determined their performance. One was fitted in the half cube of bismuth halide octahedra, the other remained to determine the spacing between the neighboring two layers. Based on this discipline, they created a new 2D perovskite of $Rb_3Bi_2I_9$ single crystal with high atomic number ($Z = 61.6$), excellence carrier mobility (2.27 cm^2 V^{-1} s^{-1}) as well as low trap states (8.43×10^{10} cm^{-3}). All these parameters worked together to obtain the record lowest detectable limitation of 8.32 nGy$_{air}$ s^{-1} [102].

4 Conclusion and Perspective

The LD semiconductor materials have been successfully developed into inorganic, organic, and developed the halide perovskites for high-performance X-ray detection and imaging. In this chapter, we highlight the significance of LD semiconductor materials in X-ray detection according to their unique structures and photophysical properties. First, we systematically introduce the mechanisms of scintillators and direct X-ray detection and discuss their key parameters. Then, we emphasize the breakthrough as well as their advantages and limitations of both scintillators and direct X-ray detection in different kinds of LD semiconductors which includes inorganic, organic, and halide perovskites. For the last chapter, we, respectively, propose the challenges and promising directions of future development for scintillators and direct X-ray detection.

4.1 LD Scintillators X-Ray Detection

- **Dealing with severe self-absorption problem.** The self-absorption problem leads to a small Stokes shift, which induces the overlapping of absorption, emission spectra, and the self-absorption of photons [31]. To solve these problems, we proposed several strategies including innovating material fabrication (enlarging the Stokes shift), employing doping rare metal, and designing device which could effectively reduce the self-absorption problem. Yu et al. developed Mn^{2+} dopants into 2D halide perovskite, which showed a small photon energy (~2 eV) than the 2D perovskite host (~3.1 eV), and reduced the photon loss induced by self-absorption [55]. Li et al. used a device design method by developing different size LD halide perovskite ($CsPbBr_3$ NCs) with a certain ratio, which performed as an efficient energy to minimize the reabsorption problem [13]. Besides, such as emitter-in-matrix device design [12], and novel LD materials are also developed [8, 72].
- **Developing high-Z and superior radiative-endurance organic LD materials.** Organic scintillators possess a lot of advantages such as lightweight, flexible, and environmentally friendly characteristics. Besides, it is suitable to apply flexible

wearable electronic instruments. Importantly, the low-density and poor Z value problem should be well addressed first. Embed with high-Z materials shows promising for further application and their radiative resistance and material stability improvement. Furthermore, mechanism of organic materials embedded with inorganic ones should also be deeply understood.

- **Increasing exciton binding energy.** The large exciton binding energy is beneficial to suppress the thermal quenching that functions as a high light yield scintillator fabrication. Although LD materials, especially halide perovskite, possess large exciton energy, there is a preparative problem in high-quality and large-scale 2D perovskite SC fabrication due to their complicated components [65, 103]. Traditional method for 2D SC fabrication is inverse temperature crystallization (ITC), which shows advantages of low reaction time and easy control [104]. However, some limitations such as the introduction of impurities and difficultly control needed facet during crystal growth [105–107]. Therefore, novel strategies for reproducible and perfect SC need further exploration.

- **Exploring the mechanism of lanthanides ions doping in LD materials.** The lanthanides elements such as Eu, Bi, Ce, etc. doping are widely explored in LD semiconductor materials including inorganic and halide perovskite because of their unique luminescent property [11 , 33]. Moreover, the lanthanides ions doping in scintillators can also induce large Stokes shift to avoid self-absorption problems while the related mechanisms are limited and their function is blurry [31]. Therefore, it is promising to carry out related research on its internal mechanism for further boosting of the light properties of scintillators.

4.2 Direct X-ray detection

- **Developing LD materials with enough thickness, flexible, and large area.** Since the LD semiconductors in X-ray detection have developed for several years, there are abundant breakthroughs have been done. For instance, the highest spatial resolution for a direct X-ray detector in LD halide perovskite was obtained by Ma et al. by 15.0 lp/mm [68]. And the centimeter-sized molecular TMCMCdCl$_3$ (TMCM$^+$, trimethylchloromethyl ammonium) was obtained by Ma et al. with a superior LoD of 1.06 μC Gy$_{air}$ s^{-1} under 10 V bias [108]. Also, 2D halide perovskite SC [28, 109], centimeter-size 2D lead-free perovskite SC [91], metal-free perovskite SC [110, 111], and open-framework semiconductor [112] are developed nevertheless the flexible devices with large thickness and area still have limited research. Some of the advanced design in 3D materials could be applied in LD materials, such as perovskite-filled membranes (PFMs) method [113], printed perovskite inks method [114], metal-semiconductor–insulator ladder-like configuration [115].

- **Fabricating low ionic migration, high-quality, and reproducible LD materials.** The uncontrollable ionic migration in a direct X-ray detector could induce high dark current and noise, which largely restricted the sensitivity and spatial resolution. Furthermore, the quality and reproducibility of LD materials influenced the performance of the X-ray detector. For example, the resistivity of recently reported halide perovskite shows a wide range from 10^7 Ω cm to 10^{12} Ω cm, which also shows a difficulty for commercial application [16]. Some advanced strategies have been successfully applied in 3D materials such as passivated layer introduction [116, 117], post-treatment with UV-O$_3$ treatment [35], annealing process [118], and structural design [34, 119], etc.

- **Developing inorganic LD halide materials.** The inorganic halide perovskites with enduring, large Z of Cs$^+$ ions introduction are beneficial to improve their X-ray stopping ability and stability [120]. Moreover, most of the inorganic SCs possess larger resistivity than the organic-inorganic components, which could effectively suppress the dark current [16]. Importantly, the inorganic halide perovskites QDs/NCs, especially CsPbBr$_3$, could be used as ink for solution-processable perovskite films and shows large interest in flexible panels detector in a large area [7, 13, 94].

References

1. H. Wei, J. Huang, *Nat. Commun.* 2019, *10*, 1066.
2. G. A. Armantrout, *Nuclear Instruments and Methods in Physics Research Section A: Accelerators, Spectrometers, Detectors and Associated Equipment* 1982, *193*, 41-47.
3. D. R. Shearer, M. Bopaiah, *Health Physics* 2000, *79*, 20-21.
4. G. Zentai, M. Schieber, L. Partain, R. Pavlyuchkova, C. Proano, *J. Cryst. Growth* 2005, *275*, 1327-1331.
5. G. Kakavelakis, M. Gedda, A. Panagiotopoulos, E. Kymakis, T. D. Anthopoulos, K. Petridis, *Adv. Sci.* 2020, *7*, 2002098.
6. P. Huang, S. Kazim, M. Wang, S. Ahmad, *ACS Energy Lett.* 2019, *4*, 2960-2974.
7. Q. Chen, J. Wu, X. Ou, B. Huang, J. Almutlaq, A. A. Zhumekenov, X. Guan, S. Han, L. Liang, Z. Yi, J. Li, X. Xie, Y. Wang, Y. Li, D. Fan, D. B. L. Teh, A. H. All, O. F. Mohammed, O. M. Bakr, T. Wu, M. Bettinelli, H. Yang, W. Huang, X. Liu, *Nature* 2018, *561*, 88-93.
8. L. Lian, M. Zheng, W. Zhang, L. Yin, X. Du, P. Zhang, X. Zhang, J. Gao, D. Zhang, L. Gao, G. Niu, H. Song, R. Chen, X. Lan, J. Tang, J. Zhang, *Adv. Sci.* 2020, *7*, 2000195.
9. F. Zhou, Z. Li, W. Lan, Q. Wang, L. Ding, Z. Jin, *Small Methods* 2020, *4*, 2000506.
10. O. D. I. Moseley, T. Doherty, R. Parmee, M. Anaya, S. D. Stranks, *J. Mater. Chem. C* 2021, *9*, 11588–11604.
11. L. Lu, M. Sun, Q. Lu, T. Wu, B. Huang, *Nano Energy* 2021, *79*, 105437.
12. F. Cao, D. Yu, W. Ma, X. Xu, B. Cai, Y. M. Yang, S. Liu, L. He, Y. Ke, S. Lan, K.-L. Choy, H. Zeng, *ACS Nano* 2019, *14*, 5183–5193.
13. X. Li, C. Meng, B. Huang, D. Yang, X. Xu, H. Zeng, *Adv. Opt. Mater.* 2020, *8*, 2000273.
14. Y. Zhang, R. Sun, X. Ou, K. Fu, Q. Chen, Y. Ding, L.-J. Xu, L. Liu, Y. Han, A. V. Malko, X. Liu, H. Yang, O. M. Bakr, H. Liu, O. F. Mohammed, *ACS Nano* 2019, *13*, 2520-2525.
15. L. Xie, Z. Hong, J. Zan, Q. Wu, Z. Yang, X. Chen, X. Ou, X. Song, Y. He, J. Li, Q. Chen, H. Yang, *Adv. Mater.* 2021, *33*, 2101852.
16. Z. Li, F. Zhou, H. Yao, Z. Ci, Z. Yang, Z. Jin, *Mater. Today* 2021, *48*, 155–175.
17. W. Que, J. A. Rowlands, *Medical Physics* 1995, *22*, 365-374.
18. W. Que, J. A. Rowlands, *Phys. Rev. B* 1995, *51*, 10500-10507.
19. S. O. Kasap, *Journal of Physics D: Applied Physics* 2000, *33*, 2853.

20. M. Schieber, H. Hermon, A. Zuck, A. Vilensky, L. Melekhov, R. Shatunovsky, E. Meerson, H. Saado, *Nuclear Instruments and Methods in Physics Research Section A: Accelerators, Spectrometers, Detectors and Associated Equipment* 2001, *458*, 41-46.
21. C. Szeles, *Physica Status Solidi B: Basic Solid State Physics* 2004, *241*, 783–790.
22. Y. M. Ivanov, V. M. Kanevsky, V. F. Dvoryankin, V. V. Artemov, A. N. Polyakov, A. A. Kudryashov, E. M. Pashaev, Z. J. Horvath, *Physica Status Solidi C: Current Topics in Solid State Physics* 2003, *0*, 840-844.
23. V. F. Dvoryankin, G. G. Dvoryankina, A. A. Kudryashov, A. G. Petrov, V. D. Golyshev, S. V. Bykova, *Technical Physics* 2010, *55*, 306-308.
24. X. Liu, T. Xu, Y. Li, Z. Zang, X. Peng, H. Wei, W. Zha, F. Wang, *Sol. Energy Mater. Sol. Cells* 2018, *187*, 249-254.
25. K. Shibuya, M. Koshimizu, K. Asai, H. Shibata, *Appl. Phys. Lett.* 2004, *84*, 4370-4372.
26. K. Shibuya, M. Koshimizu, H. Murakami, Y. Muroya, Y. Katsumura, K. Asai, *Jpn. J. Appl. Phys.* 2004, *43*, 1333-1336.
27. R. Zhuang, X. Wang, W. Ma, Y. Wu, X. Chen, L. Tang, H. Zhu, J. Liu, L. Wu, W. Zhou, X. Liu, Y. Yang, *Nat. Photon.* 2019, *13*, 602-608.
28. H. Li, J. Song, W. Pan, D. Xu, W.-A. Zhu, H. Wei, B. Yang, *Adv. Mater.* 2020, *32*, 2003790.
29. Y. Su, W. Ma, Y. Yang, *Journal of Semiconductors* 2020, *41*, 051204.
30. J. Cao, Z. Guo, S. Zhu, Y. Fu, H. Zhang, Q. Wang, Z. Gu, *ACS Appl. Mater. Interfaces.* 2020, *12*, 19797–19804.
31. Y. Zhou, J. Chen, O. M. Bakr, O. F. Mohammed, *ACS Energy Lett.* 2021, *6*, 739-768.
32. M. Nikl, *Measurement Science and Technology* 2006, *17*, 37-54.
33. F. Maddalena, L. Tjahjana, A. Xie, Arramel, S. Zeng, H. Wang, P. Coquet, W. Drozdowski, C. Dujardin, C. Dang, M. Birowosuto, *Crystals* 2019, *9*, 88.
34. W. Pan, H. Wu, J. Luo, Z. Deng, C. Ge, C. Chen, X. Jiang, W.-J. Yin, G. Niu, L. Zhu, L. Yin, Y. Zhou, Q. Xie, X. Ke, M. Sui, J. Tang, *Nat. Photon.* 2017, *11*, 726-732.
35. H. Wei, Y. Fang, P. Mulligan, W. Chuirazzi, H.-H. Fang, C. Wang, B. R. Ecker, Y. Gao, M. A. Loi, L. Cao, J. Huang, *Nat. Photon.* 2016, *10*, 333-339.
36. C. Ji, S. Wang, Y. Wang, H. Chen, L. Li, Z. Sun, Y. Sui, S. Wang, J. Luo, *Adv. Funct. Mater.* 2019, *30*, 1905529.
37. L. Gao, Q. Yan, *Solar RRL* 2019, *4*.
38. M. Nikl, A. Yoshikawa, *Adv. Opt. Mater.* 2015, *3*, 463-481.
39. NIST, 2018, https://www.nist.gov/pml/x-ray-mass-attenuation-coefficients.
40. K. D. G. I. Jayawardena, H. M. Thirimanne, S. F. Tedde, J. E. Huerdler, A. J. Parnell, R. M. I. Bandara, C. A. Mills, S. R. P. Silva, *ACS Nano* 2019, *13*, 6973-6981.
41. R. D. Evans, A. Noyau, *582*, The Atomic Nucleus Vol. 582 (McGraw-Hill, 1955).
42. E. C. Lin, *Mayo Clin Proc* 2010, *85*, 1142-1146.
43. H. Zhang, Y. Yang, X. Wang, T.-L. Ren, Z. Gao, R. Liang, X. Zheng, X. Geng, Y. Zhao, D. Xie, J. Hong, H. Tian, *IEEE Trans. Electron Devices* 2019, *66*, 2224-2229.
44. L. Li, H. Chen, Z. Fang, X. Meng, C. Zuo, M. Lv, Y. Tian, Y. Fang, Z. Xiao, C. Shan, Z. Xiao, Z. Jin, G. Shen, L. Shen, L. Ding, *Adv. Mater.* 2020, *32*, 1907257.
45. R. D. Jansen-van Vuuren, A. Armin, A. K. Pandey, P. L. Burn, P. Meredith, *Adv. Mater.* 2016, *28*, 4766-4802.
46. L. Basiricò, A. Ciavatti, B. Fraboni, *Adv. Mater. Technol.* 2020, *6*, 2000475.
47. F. Deschler, D. Neher, L. Schmidt-Mende, *APL Materials* 2019, *7*, 080401.
48. M. Thompson, S. L. R. Ellison, R. Wood, *Pure Appl. Chem.* 2002, *74*, 835-855.
49. Y. Liu, Y. Zhang, Z. Yang, J. Cui, H. Wu, X. Ren, K. Zhao, J. Feng, J. Tang, Z. Xu, S. Liu, *Adv. Opt. Mater.* 2020, *8*, 2000814.
50. D. M. Panneerselvam, M. Z. Kabir, *Journal of Materials Science: Materials in Electronics* 2017, *28*, 7083-7090.
51. E. Sameib, M. J. Flynn, *Medical Physics* 1998, *25*, 102-113.
52. S. Yamamoto, K. Kamada, A. Yoshikawa, *Radiation Measurements* 2018, *119*, 184-191.
53. M. Akatsuka, N. Kawano, T. Kato, D. Nakauchi, G. Okada, N. Kawaguchi, T. Yanagida, *Nuclear Instruments and Methods in Physics Research Section A* 2020, *954*, 161372.

54. A. Xie, F. Maddalena, M. E. Witkowski, M. Makowski, B. Mahler, W. Drozdowski, S. V. Springham, P. Coquet, C. Dujardin, M. D. Birowosuto, C. Dang, *Chem. Mater.* 2020, *32*, 8530-8539.
55. D. Yu, P. Wang, F. Cao, Y. Gu, J. Liu, Z. Han, B. Huang, Y. Zou, X. Xu, H. Zeng, *Nat. Commun.* 2020, *11*, 3395.
56. M. D. Birowosuto, P. Dorenbos, C. W. E. van Eijk, K. W. Krämer, H. U. Güdel, *J. Appl. Phys.* 2006, *99*, 123520.
57. H. Mohammadi, M. R. Abdi, M. H. Habibi, *J. Lumin.* 2020, *219*, 116849.
58. S. K. Gupta, J. P. Zuniga, M. Abdou, M. P. Thomas, M. De Alwis Goonatilleke, B. S. Guiton, Y. Mao, *Chem. Eng. J.* 2020, *379*, 122314.
59. X. Ou, X. Qin, B. Huang, J. Zan, Q. Wu, Z. Hong, L. Xie, H. Bian, Z. Yi, X. Chen, Y. Wu, X. Song, J. Li, Q. Chen, H. Yang, X. Liu, *Nature* 2021, *590*, 410-415.
60. H. Huang, F. Zhao, L. Liu, F. Zhang, X.-g. Wu, L. Shi, B. Zou, Q. Pei, H. Zhong, *ACS Appl. Mater. Interfaces.* 2015, *7*, 28128-28133.
61. D. Chen, S. Yuan, X. Chen, J. Li, Q. Mao, X. Li, J. Zhong, *J. Mater. Chem. C.* 2018, *6*, 6832–6839.
62. M. Leng, Y. Yang, Z. Chen, W. Gao, J. Zhang, G. Niu, D. Li, H. Song, J. Zhang, S. Jin, J. Tang, *Nano Lett.* 2018, *18*, 6076-6083.
63. V. Morad, Y. Shynkarenko, S. Yakunin, A. Brumberg, R. D. Schaller, M. V. Kovalenko, *J. Am. Chem. Soc.* 2019, *141*, 9764-9768.
64. K. Shibuya, M. Koshimizu, Y. Takeoka, K. Asai, *Nuclear Instruments and Methods in Physics Research B* 2002, *14*, 207–212.
65. N. Kawano, M. Koshimizu, G. Okada, Y. Fujimoto, N. Kawaguchi, T. Yanagida, K. Asai, *Sci. Rep.* 2017, *7*, 14754.
66. M. D. Birowosuto, D. Cortecchia, W. Drozdowski, K. Brylew, W. Lachmanski, A. Bruno, C. Soci, *Sci. Rep.* 2016, *6*, 37254.
67. J. H. Heo, D. H. Shin, J. K. Park, D. H. Kim, S. J. Lee, S. H. Im, *Adv. Mater.* 2018, *30*, 1801743.
68. W. Ma, T. Jiang, Z. Yang, H. Zhang, Y. Su, Z. Chen, X. Chen, Y. Ma, W. Zhu, X. Yu, H. Zhu, J. Qiu, X. Liu, X. Xu, Y. M. Yang, *Adv. Sci.* 2021, *8*, 2003728.
69. L. Wang, K. Fu, R. Sun, H. Lian, X. Hu, Y. Zhang, *Nano-Micro Lett.* 2019, *11*, 52.
70. C. Wang, H. Lin, Z. Zhang, Z. Qiu, H. Yang, Y. Cheng, J. Xu, X. Xiang, L. Zhang, Y. Wang, *Journal of the European Ceramic Society* 2020, *40*, 2234-2238.
71. L. Protesescu, S. Yakunin, M. I. Bodnarchuk, F. Krieg, R. Caputo, C. H. Hendon, R. X. Yang, A. Walsh, M. V. Kovalenko, *Nano Lett.* 2015, *15*, 3692-3696.
72. M. Zhang, X. Wang, B. Yang, J. Zhu, G. Niu, H. Wu, L. Yin, X. Du, M. Niu, Y. Ge, Q. Xie, Y. Yan, J. Tang, *Adv. Funct. Mater.* 2020, *31*, 2007921.
73. M. Zhang, J. Zhu, B. Yang, G. Niu, H. Wu, X. Zhao, L. Yin, T. Jin, X. Liang, J. Tang, *Nano Lett.* 2021, *21*, 1392-1399.
74. J.-H. Wei, J.-F. Liao, X.-D. Wang, L. Zhou, Y. Jiang, D.-B. Kuang, *Matter* 2020, *3*, 892-903.
75. L. Zhou, J. F. Liao, D. B. Kuang, *Adv. Opt. Mater.* 2021, *9*, 2100544.
76. L.-J. Xu, X. Lin, Q. He, M. Worku, B. Ma, *Nat. Commun.* 2020, *11*, 4329.
77. J. Zhou, K. An, P. He, J. Yang, C. Zhou, Y. Luo, W. Kang, W. Hu, P. Feng, M. Zhou, X. Tang, *Adv. Opt. Mater.* 2021, *9*, 2002144.
78. E. A. Axtell Iii, J.-H. Liao, Z. Pikramenou, M. G. Kanatzidis, *Chemistry – A European Journal* 1996, *2*, 656-666.
79. J. Androulakis, S. C. Peter, H. Li, C. D. Malliakas, J. A. Peters, Z. Liu, B. W. Wessels, J.-H. Song, H. Jin, A. J. Freeman, M. G. Kanatzidis, *Adv. Mater.* 2011, *23*, 4163-4167.
80. H. Liang, S. Cui, R. Su, P. Guan, Y. He, L. Yang, L. Chen, Y. Zhang, Z. Mei, X. Du, *ACS Photon.* 2018, *6*, 351-359.
81. J. Chen, H. Tang, B. Liu, Z. Zhu, B. Gu, Z. Zhang, Q. Xu, J. Xu, L. Zhou, L. Chen, X. Ouyang, *ACS Appl. Mater. Interfaces.* 2021, *13*, 2879-2886.
82. F. A. Boroumand, M. Zhu, A. B. Dalton, J. L. Keddie, P. J. Sellin, J. J. Gutierrez, *Appl. Phys. Lett.* 2007, *91*, 033509.

83. A. Intaniwet, C. A. Mills, M. Shkunov, P. J. Sellin, J. L. Keddie, *Nanotechnology* 2012, *23*, 235502.
84. A. Ciavatti, T. Cramer, M. Carroli, L. Basiricò, R. Fuhrer, D. M. De Leeuw, B. Fraboni, *Appl. Phys. Lett.* 2017, *111*, 183301.
85. C. A. Mills, H. Al-Otaibi, A. Intaniwet, M. Shkunov, S. Pani, J. L. Keddie, P. J. Sellin, *Journal of Physics D: Applied Physics* 2013, *46*, 275102.
86. P. Büchele, M. Richter, S. F. Tedde, G. J. Matt, G. N. Ankah, R. Fischer, M. Biele, W. Metzger, S. Lilliu, O. Bikondoa, J. E. Macdonald, C. J. Brabec, T. Kraus, U. Lemmer, O. Schmidt, *Nat. Photon.* 2015, *9*, 843-848.
87. H. M. Thirimanne, K. D. G. I. Jayawardena, A. J. Parnell, R. M. I. Bandara, A. Karalasingam, S. Pani, J. E. Huerdler, D. G. Lidzey, S. F. Tedde, A. Nisbet, C. A. Mills, S. R. P. Silva, *Nat. Commun.* 2018, *9*, 2926.
88. Z. Xu, X. Liu, Y. Li, X. Liu, T. Yang, C. Ji, S. Han, Y. Xu, J. Luo, Z. Sun, *Angew. Chem. Int. Ed.* 2019, *58*, 15757-15761.
89. B. Xiao, Q. Sun, F. Wang, S. Wang, B. Zhang, J. Wang, W. Jie, P. Sellin, Y. Xu, *J. Mater. Chem. A* 2021.
90. Y. Shen, Y. Liu, H. Ye, Y. Zheng, Q. Wei, Y. Xia, Y. Chen, K. Zhao, W. Huang, S. Liu, *Angew. Chem. Int. Ed.* 2020, *59*, 14896-14902.
91. C. F. Wang, H. Li, M. G. Li, Y. Cui, X. Son, Q. W. Wang, J. Y. Jiang, M. M. Hua, Q. Xu, K. Zhao, H. Y. Ye, Y. Zhang, *Adv. Funct. Mater.* 2021, *31*, 2009457.
92. K. Tao, Y. Li, C. Ji, X. Liu, Z. Wu, S. Han, Z. Sun, J. Luo, *Chem. Mater.* 2019, *31*, 5927-5932.
93. B.-B. Zhang, X. Liu, B. Xiao, A. B. Hafsia, K. Gao, Y. Xu, J. Zhou, Y. Chen, *J. Phys. Chem. Lett.* 2020, *11*, 432-437.
94. J. Liu, B. Shabbir, C. Wang, T. Wan, Q. Ou, P. Yu, A. Tadich, X. Jiao, D. Chu, D. Qi, D. Li, R. Kan, Y. Huang, Y. Dong, J. Jasieniak, Y. Zhang, Q. Bao, *Adv. Mater.* 2019, *31*, 1901644.
95. X. Zheng, W. Zhao, P. Wang, H. Tan, M. I. Saidaminov, S. Tie, L. Chen, Y. Peng, J. Long, W.-H. Zhang, *J. Energy Chem.* 2020, *49*, 299-306.
96. Y. Liu, Z. Xu, Z. Yang, Y. Zhang, J. Cui, Y. He, H. Ye, K. Zhao, H. Sun, R. Lu, M. Liu, M. G. Kanatzidis, S. Liu, *Matter* 2020, *3*, 180-196.
97. S. Tie, W. Zhao, D. Xin, M. Zhang, J. Long, Q. Chen, X. Zheng, J. Zhu, W.-H. Zhang, *Adv. Mater.* 2020, *32*, 2001981.
98. D. Hao, D. Liu, Y. Shen, Q. Shi, J. Huang, *Adv. Funct. Mater.* 2021, *31*, 2100773.
99. Y. Xu, B. Jiao, T.-B. Song, C. C. Stoumpos, Y. He, I. Hadar, W. Lin, W. Jie, M. G. Kanatzidis, *ACS Photon.* 2018, *6*, 196-203.
100. J. Guo, Y. Xu, W. Yang, B. Xiao, Q. Sun, X. Zhang, B. Zhang, M. Zhu, W. Jie, *ACS Appl. Mater. Interfaces.* 2021, *13*, 23928–23935.
101. Y. Zhang, Y. Liu, Z. Xu, H. Ye, Z. Yang, J. You, M. Liu, Y. He, M. G. Kanatzidis, S. F. Liu, *Nat. Commun.* 2020, *11*, 2304.
102. M. Xia, J.-H. Yuan, G. Niu, X. Du, L. Yin, W. Pan, J. Luo, Z. Li, H. Zhao, K.-H. Xue, X. Miao, J. Tang, *Adv. Funct. Mater.* 2020, *30*, 1910648.
103. A. Xie, C. Hettiarachchi, F. Maddalena, M. E. Witkowski, M. Makowski, W. Drozdowski, A. Arramel, A. T. S. Wee, S. V. Springham, P. Q. Vuong, H. J. Kim, C. Dujardin, P. Coquet, M. D. Birowosuto, C. Dang, *Commun Mater* 2020, *1*, 37.
104. M. I. Saidaminov, A. L. Abdelhady, B. Murali, E. Alarousu, V. M. Burlakov, W. Peng, I. Dursun, L. Wang, Y. He, G. Maculan, A. Goriely, T. Wu, O. F. Mohammed, O. M. Bakr, *Nat. Commun.* 2015, *6*, 7586.
105. W. Wang, H. Meng, H. Qi, H. Xu, W. Du, Y. Yang, Y. Yi, S. Jing, S. Xu, F. Hong, J. Qin, J. Huang, Z. Xu, Y. Zhu, R. Xu, J. Lai, F. Xu, L. Wang, J. Zhu, *Adv. Mater.* 2020, *32*, 2001540.
106. L. Zhang, Y. Liu, X. Ye, Q. Han, C. Ge, S. Cui, Q. Guo, X. Zheng, Z. Zhai, X. Tao, *Cryst. Growth Des.* 2018, *18*, 6652-6660.
107. X. Wang, Y. Li, Y. Xu, Y. Pan, C. Zhu, D. Zhu, Y. Wu, G. Li, Q. Zhang, Q. Li, X. Zhang, J. Wu, J. Chen, W. Lei, *Chem. Mater.* 2020, *32*, 4973-4983.
108. C. Ma, F. Chen, X. Song, M. Chen, L. Gao, P. Wang, J. Wen, Z. Yang, Y. Tang, K. Zhao, S. Liu, *Adv. Funct. Mater.* 2021, *31*, 2100691.

109. W. Qian, X. Xu, J. Wang, Y. Xu, J. Chen, Y. Ge, J. Chen, S. Xiao, S. Yang, *Matter* 2021, *4*, 942-954.
110. X. Song, Q. Cui, Y. Liu, Z. Xu, H. Cohen, C. Ma, Y. Fan, Y. Zhang, H. Ye, Z. Peng, R. Li, Y. Chen, J. Wang, H. Sun, Z. Yang, Z. Liu, Z. Yang, W. Huang, G. Hodes, S. F. Liu, K. Zhao, *Adv. Mater.* 2020, *32*, 2003353.
111. X. Song, G. Hodes, K. Zhao, S. Liu, *Adv. Energy Mater.* 2021, *11*, 2003331.
112. S. Wu, C. Liang, J. Zhang, Z. Wu, X.-L. Wang, R. Zhou, Y. Wang, S. Wang, D.-S. Li, T. Wu, *Angew. Chem. Int. Ed.* 2020, *59*, 18605-18610.
113. J. Zhao, L. Zhao, Y. Deng, X. Xiao, Z. Ni, S. Xu, J. Huang, *Nat. Photon.* 2020, *14*, 612-617.
114. A. Ciavatti, R. Sorrentino, L. Basiricò, B. Passarella, M. Caironi, A. Petrozza, B. Fraboni, *Adv. Funct. Mater.* 2021, *31*, 2009072.
115. M. Du, Y. Dai, Z. Wang, S. Lv, G. Du, J. Li, Y. Qiu, J. Qiu, S. Zhou, *Adv. Mater. Technol.* 2020, *5*, 2000302.
116. B. Yang, W. Pan, H. Wu, G. Niu, J.-H. Yuan, K.-H. Xue, L. Yin, X. Du, X.-S. Miao, X. Yang, Q. Xie, J. Tang, *Nat. Commun.* 2019, *10*, 1989
117. Y. C. Kim, K. H. Kim, D.-Y. Son, D.-N. Jeong, J.-Y. Seo, Y. S. Choi, I. T. Han, S. Y. Lee, N.-G. Park, *Nature* 2017, *550*, 87-91.
118. L. Li, X. Liu, H. Zhang, B. Zhang, W. Jie, P. J. Sellin, C. Hu, G. Zeng, Y. Xu, *ACS Appl. Mater. Interfaces.* 2019, *11*, 7522-7528.
119. Y. Huang, L. Qiao, Y. Jiang, T. He, R. Long, F. Yang, L. Wang, X. Lei, M. Yuan, J. Chen, *Angew. Chem. Int. Ed.* 2019, *58*, 17834-17842.
120. Z. Li, F. Zhou, Q. Wang, L. Ding, Z. Jin, *Nano Energy* 2020, *71*, 104634.

Hybrid Imaging Detectors in X-Ray Phase-Contrast Applications

Luca Brombal and Luigi Rigon

1 X-Rays and Wave-Particle Duality

Traditionally, in the field of X-ray imaging, wave-particle duality is disregarded; X-ray's production, interaction, and detection are generally described using particle models, where the particles, the radiation quanta, are X-ray photons in the 10–100 keV energy range. At these energies, the interaction of X-rays with matter is dominated by photoelectric effect and Compton scattering. X-ray imaging relies on the attenuation (i.e., absorption or scattering) characteristics of the sample. Contrast is created when the sample introduces different attenuation in different regions; for instance, in medical X-ray imaging bones bring a much stronger attenuation of X-rays, compared to soft tissues, thereby producing a strong contrast.

Although the particle model provides a simple and effective description for the image formation, wave-particle duality calls for considering X-rays also as electromagnetic waves, with a wavelength in the range of 0.1–0.01 nm. Based on this principle, novel imaging techniques that take advantage of the wave nature of X-rays have been introduced and developed in the last two decades, initially at synchrotron radiation facilities, and later also at more portable X-ray sources. These techniques are commonly referred to as X-ray phase-contrast imaging (XPCI) [1, 2].

A simple wave model describing the X-ray interaction with the sample is introduced as follows [3]: X-rays are represented by plane waves traveling along the z direction, in a reference system where xy is the object plane. Due to the

L. Brombal (✉)
Division of Trieste, National Institute for Nuclear Physics (INFN), Trieste, Italy
e-mail: luca.brombal@ts.infn.it

L. Rigon
Division of Trieste, National Institute for Nuclear Physics (INFN), Trieste, Italy

Department of Physics, University of Trieste, Trieste, Italy

interaction with the object, the X-ray wavefield is modulated by the so-called object transmission function $T(x,y)$, a complex-valued function that can be written as:

$$T(x, y) = A(x, y)\, e^{i\phi(x,y)} \tag{1}$$

Here,

$$A(x, y) = e^{-\frac{2\pi}{\lambda} \int \beta(x,y,z)dz} \tag{2}$$

accounts for the wave amplitude reduction and

$$\phi(x, y) = -\frac{2\pi}{\lambda} \int \delta(x, y, z)\, dz \tag{3}$$

represents the phase shift introduced by the sample; λ is the X-ray wavelength; $\beta(x, y, z)$ and $\delta(x, y, z)$ are, respectively, the imaginary part and the real part decrement of the refractive index $n(x, y, z) = 1 - \delta(x, y, z) + i\beta(x, y, z)$, which describes the sample; finally, the integrals are carried out over the object extension along the z direction.

Equation (2) implies that the X-ray intensity is reduced by a factor $A^2(x, y) = e^{-\frac{4\pi}{\lambda} \int \beta(x,y,z)dz}$. This term accounts for X-ray attenuation, i.e., the physical process that leads to contrast formation in conventional X-ray imaging. In fact, Eq. (2) is equivalent to the well-known Beer–Lambert law:

$$\frac{I_{out}}{I_0} = e^{-\int \mu(x,y,z)dz}, \tag{4}$$

where I_0 is the intensity of the beam impinging on the object plane, I_{out} is the intensity of the outgoing beam, and $\mu(x, y, z) = 4\pi\beta(x, y, z)/\lambda$ is the linear attenuation coefficient of the object.

On the other hand, Eq. (3) introduces a phase shift $\phi(x, y)$, which has no counterpart in the X-ray particle model. This term is usually disregarded in conventional X-ray imaging because X-ray detectors are not sensitive to the phase of the radiation but to its intensity only. However, with suitable X-ray sources and tailored imaging techniques, the phase shift can be translated into detectable intensity modulation. This is the basic principle of XPCI, where the phase shift provides a supplementary pool of contrast, in addition to—and often much larger than—X-ray attenuation. In particular, XPCI can greatly enhance the visibility of structures featuring poor attenuation contrast such as, for instance, in biological soft tissues or plastic-based samples.

2 General Principles of XPCI

The simple model introduced in the previous section assumes a perfectly coherent source, both spatially (plane wave) and temporally (monochromatic X-rays) [3]. As a matter of fact, phase-contrast imaging in the X-ray regime was initially developed at highly coherent sources, such as synchrotrons, but a great degree of coherence is not always necessary. Indeed, different XPCI techniques require different degrees of coherence. Unfortunately, conventional X-ray tubes are poorly coherent, both spatially (short source-to-sample distance and/or large focal size) and temporally (broad energy spectrum), while micro-focus X-ray tubes reach high spatial coherence at the cost of low power. Since the implementation of XPCI at portable sources is highly desirable, recent research is pursuing two different courses. On the one hand, the aim is the construction of compact coherent X-ray sources, promising examples being liquid-metal-jet [4, 5] and inverse Compton scattering [6, 7] sources. On the other hand, the focus is on the development of XPCI techniques with low coherence requirements, as will be discussed in Sect. 3. Anyway, before reviewing the different techniques and their requirements, some common characteristics can be briefly highlighted.

To begin with, Eqs. (2) and (3) state that, while attenuation is related to the imaginary part β of the refractive index, the phase shift $\phi(x, y)$ is proportional to the real part decrement, δ. In medical imaging δ is much larger than β, and so are the local differences in δ compared with those in β [8]. For instance, Fig. 1 reports the values of δ and β in the 10–100 keV energy range for water, bone, and iodine. For ease of comparison, the three materials have been assumed to have the same density, $\rho = 1$ g/cm^3. It can be seen that in the case of water, δ overpowers β by two to three orders of magnitude. This holds also for soft tissues and plastic materials, implying that in phase contrast the signal can be much more significant than in attenuation: thus one can visualize details that are hardly detectable in conventional radiology.

Secondly, XPCI does not require, in principle, X-ray absorption, so XPCI could be performed with virtually no dose release. As a matter of practice, it is impossible to completely avoid absorption. However, XPCI can be performed at very low dose levels, compared to conventional X-ray imaging utilizing, for instance, X-ray energies higher than usual. In fact, while both δ and β decrease with decreasing X-ray wavelength (i.e., with increasing energy) δ is roughly proportional to λ^2 while β to λ^3. The faster decrease of β allows choosing suitably high X-ray energies, where δ is large enough to allow good phase contrast and β is small enough to warrant low dose delivery.

The last general remark goes as follows: in all imaging techniques, contrast stems from spatial variations of given physical quantities, related to the object. In XPCI techniques, fundamental quantities are the phase shift $\phi(x, y)$, but also its gradient and its Laplacian, measured in the object plane xy, namely $\nabla_{xy}\phi(x, y)$ and $\nabla^2_{xy}\phi(x, y)$. Thus, phase contrast can be particularly strong at the edges or at the interfaces of the sample, where the gradient and the Laplacian are typically large.

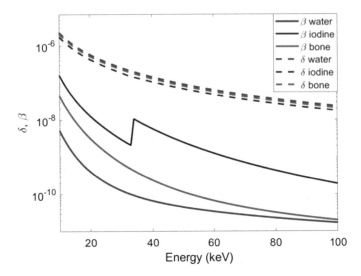

Fig. 1 δ (dashed lines) and β (solid lines) coefficients of water (blue), iodine (black), and bone (red) as a function of energy. All the materials are assumed to have density of 1 g/cm^3. Data are extracted from a publicly available database (http://ts-imaging.science.unimelb.edu.au/Services/Simple/)

XPCI techniques that are based on the gradient of the phase shift are sometimes referred to as differential phase-contrast imaging. Differential XPCI techniques highlight X-rays' refraction and ultra-small-angle-scattering (USAXS). In fact, the refraction angle α, between the incoming and the outgoing X-ray wave vectors, is proportional to the gradient of the phase shift: $\alpha \cong \frac{2\pi}{\lambda} \left| \nabla_{xy} \phi (x, y) \right|$. In biomedical samples, these deviations are generally in the order of 1–10 μrad, too small to be relevant in conventional X-ray imaging. Additionally, USAXS can be traced back to multiple refraction processes which occur at a scale smaller than the system's spatial resolution, and are thus perceived as a beam broadening. For this reason, the sensitivity to USAXS, sometimes obtained through dark-field imaging techniques, can reveal inhomogeneities on a subpixel scale [3].

3 X-Ray Phase-Contrast Imaging Techniques

XPCI is a relatively recent research field that arguably has not yet reached its full potential and will continue its development in the next decades. So far, several different techniques, with different characteristics and requirements, have been introduced. The following paragraphs aim at a short description of the most successful approaches, providing essential references for further reading.

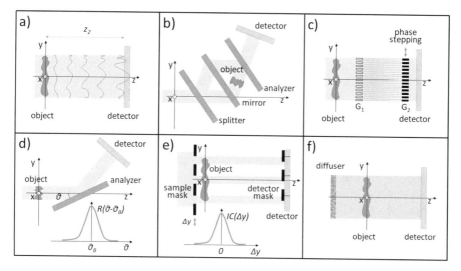

Fig. 2 Synoptic table sketching six different XPCI techniques: (**a**) Propagation-Based Imaging; (**b**) Interferometry; (**c**) Grating Interferometry; (**d**) Analyzer-Based Imaging; (**e**) Edge Illumination; (**f**) Speckle Imaging. For the sake of simplicity, in all panels, the X-ray beam (in yellow) is assumed parallel and directed from left to right, along the z-axis

3.1 Propagation-Based Imaging

Propagation-Based Imaging (PBI) can be straightforwardly implemented, provided that the X-ray source has a suitably high degree of spatial coherence and the detector has a conveniently high spatial resolution. When these two conditions are met, it is sufficient to choose appropriate values for the source-to-sample (z_1) and the sample-to-detector (z_2) distances. In particular, z_2, called propagation distance, allows the X-ray wave emerging from the object to evolve before impinging on the detector (Fig. 2a). Typically, PBI is performed in the near field regime, i.e., with $z_2 \ll \frac{a^2}{\lambda}$, a being the size of a detail of interest, compatible with the detector pixel size. Here, an edge enhancement proportional to $\nabla^2_{xy}\phi(x, y)$ is recorded [3, 9]. This edge effect is generally confined in a 1–100 μm region and it is characterized by a signal, whose intensity rapidly switches from a positive to a negative peak. If the spatial resolution of the imaging system is not adequate, the peaks are partially averaged and the signal is smeared out, strongly diminishing the edge-enhancement effect.

3.2 Interferometry

Early evidence of XPCI was provided by Bonse and Hart [10], who basically translated a well-known optical instrument, the Mach–Zehnder interferometer, from visible light to X-rays. Mirrors and splitters in the X-ray regime are obtained by means of three parallel crystals, which the beam traverses in Laue geometry. The first crystal (splitter) splits the beam into two branches traveling in different directions, the second crystal (mirror) reflects both branches, while the third crystal (analyzer) allows for beam recombination, creating an interference pattern. The sample is placed in one of the interferometer arms, altering the interference pattern (Fig. 2b). As typical with interferometric techniques, a very precise measure of the phase shift $\phi(x, y)$ can be obtained, and subtle phase-contrast effects can thus be revealed [11]. However, this technique has two significant drawbacks: it requires not only a high degree of coherence but also a nearly perfect crystal alignment at the atomic scale, due to the very small wavelength of X-rays. As a matter of practice, the latter requirement calls for extremely stable mechanics and it limits the field of view to few squared centimeters.

3.3   Grating Interferometry

Grating (or Talbot) interferometry (GI) is a different interferometric technique, with relaxed coherence and mechanical requirements compared to crystal interferometry. Two gratings, G_1 and G_2, usually dubbed phase and absorption grating, respectively, are placed between the sample and the detector (Fig. 2c). G_1 introduces a periodic phase shift, which in turn creates a strong interference fringe pattern with the same period at a given Talbot (or fractional) distance. Here the fringe pattern impinges on G_2, a highly absorbing grating with the same period of the fringe pattern (typically in the micrometer range); thus, the latter can be analyzed by performing a fine scan of G_2 with respect to G_1. This process, called phase stepping, requires the acquisition of multiple (usually 3–10) images [12] over a grating period.

As previously mentioned, the sample introduces attenuation, refraction, and ultra-small-angle scattering in the X-ray wavefield. In GI, these physical quantities are measured by comparing the periodic fringe pattern in the absence and in the presence of the sample; attenuation is described by the decrease in the mean value over the scan; refraction can be obtained from the shift of the signal; USAXS can be inferred from the decrease in the amplitude of the oscillations, i.e., from the reduction of the visibility of the fringe pattern [13].

While in principle GI requires a high degree of spatial coherence, it can be implemented at low coherence sources by introducing a third grating, G_0, close to the source (Talbot–Lau interferometer). G_0 is an absorbing mask with transmitting slits, which create an array of individually coherent, but mutually incoherent sources; under suitable geometric conditions, the latter contribute constructively to the image formation [14]. As far as the detector is concerned, GI does not necessarily require particularly high performances, since the phase effects are revealed by the gratings' system. However, very high-resolution detectors can in principle directly resolve the fringe pattern, thus avoiding the use of the analyzer grating G_2 [15].

3.4 Analyzer-Based Imaging

Analyzer-Based Imaging (ABI) takes advantage of an analyzer crystal, placed between the sample and the detector, which "analyzes" the X-ray wavefield emerging from the object. In practice, the analyzer crystal modulates the intensity that reaches the detector, according to the deviations that the X-rays have suffered while traversing the sample. The principle can be summarized as follows: in the absence of the object, the analyzer satisfies the diffraction condition at a given (Bragg) angle θ_B, efficiently reflecting the X-ray beam (Fig. 2d). As the crystal is detuned from θ_B, the reflectivity rapidly drops and vanishes; this defines the reflectivity curve, or rocking curve $R(\theta - \theta_B)$, as a function of the detuning angle. The tiny width of the rocking curve (typically 10–100 μrad) makes it an ideal tool to highlight X-ray deviations. In the presence of the sample, a local rocking curve can be considered on a pixel-per-pixel basis. Thus, ABI highlights refraction, by the shift of the local rocking curve; ultra-small-angle scattering (in the microradian range) due to the widening the rocking curve; and extinction (i.e., the rejection of small-angle scattering, in the milliradian scale), which adds to the X-ray attenuation in the sample to give the so-called apparent absorption [16].

Sampling the local rocking curve requires the acquisition of multiple (typically 3–10) images. Alternatively, two or three working points, i.e., different values of the detuning angle $(\theta - \theta_B)$ can be chosen, and the images therein obtained can be combined to yield parametric images describing apparent absorption, refraction [17], and USAXS [16]. Typical choice of the working points includes the peak, the flanks, and the toes of the rocking curve. At the peak of the rocking curve $(\theta = \theta_B)$, undeviated X-rays most likely reach the detector, while deviated photons are rejected or attenuated, thus maximizing extinction. At the flanks, where the slope of the rocking curve is the steepest, refraction is highlighted. At the toes, where undeviated X-rays are almost completely suppressed, ABI allows performing dark-field imaging, which reveals the contrast due to USAXS.

ABI does not have particular requirements in terms of detector performances, because the conversion of the phase effects into intensity modulation is performed by the analyzer crystal, which basically acts as a narrow angular band-pass filter. On the other hand, ABI requires high mechanical stability and a monochromatic and collimated X-ray beam, thus it is generally operated with synchrotron radiation.

3.5 Edge Illumination X-Ray Imaging

Edge Illumination (EI) was first introduced pursuing an alternative to the ABI technique with more relaxed requirements [18]. The principle of EI can be described as follows: a collimated X-ray beam, in the absence of the object, hits the detector at the very edge of its sensitive region, half in and half out of it. Refraction and USAXS introduced by the object move and widen the narrow beam, more in (or out of) the sensitive region of the detector, thus increasing (or decreasing) the recorded signal. This principle can be implemented by means of two absorbing masks with properly designed slits (or apertures); the first, placed before the object, creates an array of tiny beamlets; the second, located just before the detector (or directly on it) generates an array of insensitive regions so that each beamlet hits the edge between sensitive and insensitive regions (Fig. 2e). By changing the relative displacement (Δy) between the sample and the detector masks, the fraction of each beamlet captured by the sensitive regions of the detector can be modified, progressively ranging from null to full illumination. This defines the illumination curve, $IC(\Delta y)$, which is equivalent to the ABI's rocking curve. In fact, EI parallels ABI, both in its mathematical models and in the resulting images [16]. However, unlike ABI, EI can be easily implemented on a conventional X-ray tube, since it does not need monochromatic radiation.

Some similarities may be noted also between EI and GI, particularly due to the double mask arrangement in EI, which resembles the double grating system of GI. However, it is worth pointing out that the two techniques bear important differences, both conceptually and practically. Conceptually, unlike GI, EI belongs to the non-interferometric XPCI techniques, which means, for instance, that the EI signal can be observed also with a single beamlet [18]. Practically, the distinctive scale of the structures (few μm in GI versus several tens of μm in EI) makes EI less demanding in terms of mechanical stability, compared to GI. Moreover, while in GI the analyzer grating G_2 is somehow independent of the particular detector, in EI a close relation exists between the detector mask and the detector itself since the period of the detector mask must match the pixel pitch. In fact, the detector mask provides sharp edges, which redefine the pixel aperture and thus the detector point spread function (PSF). In case the detector itself features a sharp separation between neighboring pixels, the use of the detector mask can be avoided; a single mask set-up can then be implemented by means of a "skipped" sample mask, i.e., a sample mask with a period that corresponds to twice the pixel pitch, illuminating every other pixel edge when projected onto the detector. This arrangement also leads to a different shape of the illumination curve, from a peaked to a sigmoid function [19]. Alternatively, a different single mask set-up can be implemented with the so-called beam-tracking approach, whereby a high-resolution detector directly allows resolving the beamlets' footprint details [20].

3.6 Speckle Imaging

Speckle imaging is implemented by means of a diffuser, which generates a randomly distributed speckle pattern in the X-ray beam impinging on the object (Fig. 2f). Typically employed diffusers are common sandpaper or biological filter membranes with micron-sized structures [21]. The speckle pattern can be so strong that it overwhelms the object's X-ray image. However, the object introduces modulations in the speckle pattern in terms of refraction, transmission, and ultra-small-angle scattering. Thus, by quantitatively comparing the speckle pattern in the presence and in the absence of the object, the usual parametric phase-contrast images can be obtained. Despite its recent introduction, speckle imaging is very attractive due to the intrinsic ability to detect phase-effects in 2 dimensions, to the ease of the setup (as simple as adding a foil of sandpaper) and to the very mild requirements in terms of coherence, which make it suitable for conventional X-ray sources. However, the detector must have a sufficient spatial resolution to clearly resolve the speckles and to accurately track the changes in their pattern.

4 Post-processing and Phase Retrieval

All the different XPCI techniques introduced in the previous section share the ability to highlight phase contrast, i.e., contrast related to the phase shift term $\phi(x, y)$, alongside the conventional attenuation contrast. GI, ABI, EI, and speckle imaging yield parametric images, which are based on the attenuation, refraction, and USAXS characteristic of the sample. This is generally obtained by means of multiple exposures, which are recorded while scanning a particular element of the setup. Several post-processing algorithms have been devised, to retrieve the parametric images combining the single exposures, with different characteristics in terms of complexity and need of computing power [1–3, 21].

As already mentioned, regardless of the used phase-contrast technique (among GI, ABI, EI, and speckle imaging), the refraction image provides a map of the refraction angle, which is proportional to the gradient $\nabla_{xy}\phi(x, y)$. In general, the procedure of extracting the phase shift $\phi(x, y)$ from the refraction image is referred to as phase retrieval and belongs to the so-called inverse problems. The advantage of phase retrieval is twofold: first, it allows a quantitative measure, which can be a valuable tool for material characterization since $\phi(x, y)$ is related to the refractive index decrement, δ, which in turn is related to the tissue electron density. Secondly, switching from $\nabla_{xy}\phi(x, y)$ to $\phi(x, y)$ transforms an edge contrast into an area contrast, extending the phase signal from the boundary to the bulk of the detail, thus mitigating the influence of noise and improving the detail visibility; in fact, while the boundaries of the different details are greatly enhanced in the refraction image, the bulk signal could be very weak because $\phi(x, y)$ may vary slowly, regardless of possible important differences in the phase-shift absolute value, differences which are brought to light with phase retrieval. Additionally, it is worth mentioning that in their common implementation GI, ABI, and EI are sensitive to phase in one direction only (along the y axis in Fig. 2), thus the refraction image essentially depicts $\partial_y\phi(x, y)$. Hence, phase-retrieval algorithms, which are essentially based on image integration [22, 23], allow obtaining a parametric image representing the phase shift along y. On the other hand, sophisticated setups enabling phase sensitivity in two directions (omni-directional phase-contrast) have been recently introduced, an example of this will be provided in the next section.

Phase retrieval can be readily implemented also in interferometry [11] and is particularly important in PBI because PBI does not provide any parametric image, but yields, by default, an image where the phase contrast, proportional to $\nabla^2_{xy}\phi\,(x, y)$, is superimposed to the attenuation contrast. For this reason, in order to rigorously uncouple phase and attenuation contributions, PBI would require the acquisition of two (or more) images at different energies or propagation distances. The latter approach, referred to as holographic reconstruction [24], was the first to be developed soon after the first PBI images were realized. On the other hand, where constraints in terms of acquisition time or radiation exposure are present, retrieval algorithms requiring a single image, referred to as single shot, are preferred.

Various single-shot phase-retrieval algorithms have been proposed in PBI, many of which are based on the transport of intensity equation (TIE). Among these methods, the most widely used is arguably the one proposed by Paganin et al. [25]. Through this algorithm, it is possible to recover the phase shift and related quantities under the assumption of a monomorphous object, i.e., an object in which the ratio of δ and β is constant. This method was further extended by Beltran et al. [26], enabling the application of the algorithm to multi-material samples. Interestingly, in a similar theoretical framework with analogous assumptions, single-shot phase-retrieval algorithms have been developed also for ABI [27], EI [28], and GI [29]. From a mere signal processing perspective, all single-shot algorithms share the same characteristic that is they are low-pass filters, leading to output images with lower noise with respect to the input [30], hence accomplishing the "transfer" of visibility from edges to large areas. For comprehensive reviews on the subject, the reader is referred to dedicated publications [31, 32].

5   Hybrid Pixel Detectors and X-Ray Phase-Contrast Imaging

As mentioned in Sect. 3, each phase-sensitive technique has specific requirements on the components of the experimental setup, including the imaging detector. For instance, in the case of PBI, the detector must feature high spatial resolution to allow for effective detection of the high spatial frequencies associated with the phase-contrast signal. In the early days, this led to extensive use of X-ray films which provided a much higher spatial resolution when compared with digital pixel detectors available at the time [3]. Conversely, other phase-sensitive techniques have much less stringent requirements in terms of spatial resolution as phase-related signals are revealed through optical elements (e.g., crystals, gratings, masks). On the other hand, these techniques require processing after image collection to separate attenuation, refraction, and USAXS, and, for this reason, they have been implemented with digital detectors since their first appearance.

Independently from developments in phase-sensitive techniques, the last two decades have witnessed an extraordinary expansion in the field of hybrid pixel detectors. A hybrid pixel detector is a 2-dimensional matrix of microscopic sensitive elements, each connected to its own readout electronics. The sensor material and the readout electronics are processed on different substrates and are electrically connected via micro-bumps to form the imaging system [33]. The advantage of hybrid over monolithic devices, which are made by single chips where the readout electronics is formed on the surface and the bulk is used as the sensor, is the possibility to combine the readout chip with sensors of different materials [34] such as silicon, GaAs, CdTe, etc. Disruptive technological advancements in the manufacturing of both readout chips and sensors led to an ever-increasing number of applications spanning from fundamental research to industrial imaging. For instance, the availability of high-Z sensor materials produced with high purity opened up the possibility of performing imaging at relatively high energies (>30 keV) with efficiency close to 1, being extremely appealing for medical imaging applications, especially in computed tomography (CT) [35–37]. Individual pixels in readout chips are being equipped with increasingly complex electronic circuits enabling, for instance, to count individual photons (photon-counting) [38], to discriminate the incoming radiation based on its energy (spectral imaging) [39, 40] and to virtually increase the detector's spatial resolution by detecting the center of mass of individual events (super-resolution) [41, 42]. These novel features not only provide advantages over conventional charge integrating devices, but they pave the way for entirely new imaging applications.

The paths of hybrid detectors and phase-sensitive techniques have remained quite separated for a long time but, recently, many efforts towards the integration of these technologies have been made. A strong scientific interest in this field is evidenced by a number of publications, which have appeared in recent years, combining a great variety of hybrid devices with many of the available phase-sensitive techniques. Following this trend, this section is organized into three parts, each describing one peculiar feature of hybrid detectors with some of its specific applications in the field of phase-contrast imaging. While all the cited applications share elements of interest and innovation, it is worth underlining that this should not be regarded as a fully comprehensive review of the research in this field, which is rapidly evolving and constantly featuring new original approaches.

5.1 XPCI with Photon-Counting Detectors

Photon-counting detectors (PCDs) feature a readout architecture that is capable of processing each single X-ray photon. A signal is attributed to a photon, i.e., the photon is counted, only when the energy released in the sensor material exceeds a threshold that is set above the system noise. This results in complete suppression of random electronic noise (i.e., noise is Poisson dominated), also permitting long acquisition time measurements with full noise rejection. If multiple counters are available in a single pixel, PCDs can also be operated with a (virtually) zero-dead time [43], maximizing the radiation dose efficiency when multiple frame scans are required, such as in CT. The "standard" sensor material for PCDs is silicon, which can be produced with extremely high purity and at a relatively low cost. On the other hand, silicon has a low attenuation coefficient (3.3 cm^{-1} at 30 keV), therefore limiting its use to energies typically below 20 or 30 keV. The higher energy range is usually covered by high-Z sensor materials such as cadmium telluride (CdTe), cadmium zinc telluride (CZT), and gallium arsenide (GaAs). These materials possess a high attenuation coefficient (17 cm^{-1} at 80 keV for CdTe), hence ensuring a high absorption efficiency (>80% at 80 keV for 1 mm of CdTe) and are the preferred choice for many medical and biomedical imaging applications. On the other hand, the manufacturing of high-Z crystalline sensors is still challenging. In fact, these materials typically present impurities that ultimately lead to local artifacts (e.g., charge trapping) which must be compensated through processing procedures tailored to the detector type [44, 45]. The dimension of individual PCDs matrices usually does not exceed a few cm^2, thus requiring the tiling of multiple modules to reach large sensitive areas, while the size of individual pixels is in the range from 50 to 500 μm [33]. Importantly, unlike conventional scintillator-based charge integrating devices, where the spatial blurring due to the scintillation process is usually of two or more pixels, the response function of PCDs can be satisfactorily approximated with a box function corresponding to the pixel size [46]. High efficiency, low noise, and access to details below 100 μm make these devices ideal for PBI biomedical applications, where high spatial resolution is needed to detect edge enhancement effects while images must be collected at a low radiation dose level.

Arguably, one of the most relevant applications in this field is propagation-based breast CT. Breast CT is an emerging three-dimensional X-ray technique that aims at providing better diagnostic performance compared to conventional planar mammography in terms of breast cancer detection and characterization [47]. Due to the need for a high contrast-to-noise ratio to distinguish between structures with extremely similar attenuation properties and the strict constraints in terms of radiation dose, breast CT is an extremely challenging imaging task. So far, this has prevented its widespread use in clinical practice. Researchers soon realized that PBI tomography combined with the use of suitable detector and phase-retrieval filters could overcome many limitations of conventional breast CT, providing high detail visibility at clinically compatible doses (around 5 mGy) [48, 49]. The application of PBI in the context of breast imaging has its origins in the early 2000s when the SYRMEP collaboration built a clinical facility dedicated to phase-contrast mammography at Elettra, the Italian synchrotron in Trieste (Italy). The results of a three-year-long study encompassing more than 70 women proved that PBI outperforms conventional mammography yielding a higher diagnostic accuracy [50]. Following the completion of the first study, the experimental setup underwent a radical upgrade to enable the implementation of tomographic imaging. The key element of this upgrade was the large-area CdTe photon-counting detector (Pixirad-8) featuring a 60 μm pixel pitch, 650 μm thick sensor, and dead-time-free mode. The potential of this novel setup was demonstrated on a large number of test objects and excised breast tissue samples [51] and, when compared to one of the

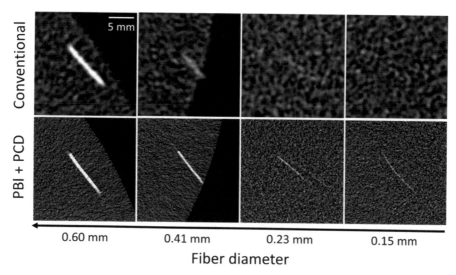

Fig. 3 Visibility comparison of fibroglandular details between a conventional breast CT scanner featuring a flat panel detector (top row) and the propagation-based setup featuring a photon-counting detector (bottom row) at the Elettra synchrotron (Trieste, Italy). Images are acquired at the same radiation (mean glandular) dose level of 6 mGy. The figure is adapted from Brombal et al. [52]

few commercially available breast CT scanners (an example of this comparison is shown in Fig. 3), the results were so promising that a new dedicated clinical study is now under development [52]. Contextually, another collaboration at the Australian Synchrotron Facility in Melbourne is developing a completely new clinical facility [53] entirely devoted to propagation-based breast CT.

Another interesting feature of the combined use of PCDs and PBI concerns the mitigation of the image noise dependence on the pixel size. In fact, due to obvious statistical reasons, noise magnitude in conventional X-ray imaging is strongly dependent on the detector pixel size. For detectors whose response function's width is mainly dictated by the pixel size, as in the case of PCDs, it can be demonstrated that the noise in tomographic images is inversely proportional to the square of the pixel size [55]. This implies that, when constraints in terms of radiation dose or scan time are present, such as in clinical or animal studies, high-resolution imaging at an acceptable noise level is not feasible. On the other hand, PBI requires small pixel sizes since phase-contrast effects are associated with high spatial frequencies. As previously described, from a signal processing perspective, the phase-retrieval algorithm is nothing more than a low-pass function that filters out the high spatial frequencies boosted by phase-contrast effects, bringing to a major reduction in the image noise [30]. For this reason, a larger high spatial frequency content in a PBI image (i.e., smaller pixel size) is associated with a more effective noise reduction (Fig. 4) due to the phase retrieval [56, 57]. This effect strongly mitigates the overall

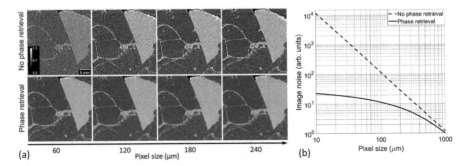

Fig. 4 Detail of a breast specimen PBI CT reconstruction obtained without (**a**, top row) and with (**a**, bottom row) phase retrieval at various pixel sizes, acquired at the SYRMEP beamline at Elettra (Trieste, Italy). In (**b**) the dependence of image noise on pixel size for images with (solid line) and without (dashed line) phase retrieval is shown. The images are adapted from Brombal [54]

image noise dependence on the pixel size, therefore enabling high-resolution low-dose CT imaging [54].

In addition to PBI, the use of low-noise hybrid detectors operated in photon-counting mode has been suggested in various studies with different phase-sensitive techniques such as, for instance, EI and GI. One illustrative example has been recently published in the context of GI tomography [58], where the researchers directly compared the performances of a conventional scintillation-based detector (Dexela 1512 by PerkinElmer, Waltham, MA), with a novel GaAs-based PCD prototype (SANTIS 0808 GaAs HR prototype by DECTRIS, Baden, Switzerland) (Fig. 5). Results indicated that the PCD can yield much sharper images at the same signal-to-noise ratio, being an extremely promising tool for a significant image quality improvement in GI. Similarly, Vila-Comamala et al. [59] described a high Talbot order GI setup that makes use of an EIGER 1M detector by Dectris, yielding exceptional values in terms of refraction sensitivity (<100 nrad). Concerning EI, Hagen et al. [60] suggested the potential of a small-animal scanner prototype featuring a CdTe-based PCD (Pixirad2-PixieIII, 650 μm sensor thickness, 62 μm pixel pitch), demonstrating that complete tomographic scans can be effectively collected in times below 15 min and radiation dose in the order of 300 mGy.

5.2 Spectral Radiology

Spectral radiography and tomography are used to separate and quantify different materials composing the investigated object, a process also known as material decomposition. This exploits the fact that the energy dependence of X-ray attenuation differs between elements (Fig. 1). Most of the material decomposition algorithms stem from the original work by Alvarez and Macovski [61]. These algorithms usually require as input two (or more) images acquired over different

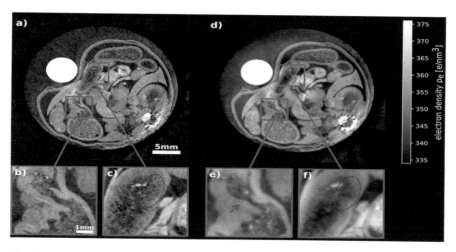

Fig. 5 Tomographic GI CT slices of the abdominal area of an ex vivo murine sample measured with the GaAs photon-counting SANTIS detector (**a**) and the CCD Dexela detector (**d**). Both measurements were conducted with the Mo 50 kVp spectrum filtered with the Talbot–Lau interferometer and a 4 cm water container. The images were recorded with an effective voxel size of 35.8 μm and are displayed in electron density values in units of e/nm^3. Subfigures (**b**) and (**e**) depict magnifications of the gastric area and subfigures (**c**) and (**f**) show a magnification of the intestine. In contrast to the Dexela images [(**e**) and (**f**)], features inside the gastric wall [(**b**), red arrow] and intestinal villi [(**c**), red ellipse] are resolved with the SANTIS prototype. Reproduced from Scholz et al. [58]

energy channels or spectra and the selection of two (or more) decomposition materials for which the energy dependence of the attenuation coefficient is known. As output they give a map of the projected thickness for each decomposition material, allowing to uncouple overlapping features in the input radiographs. A common example of this is the separation of bone and soft tissues in images of biological specimens [62]. The same concept applies to tomographic images where, starting from two (or more) tomograms acquired over different energies, density maps of given decomposition materials are computed [63]. The maximum number of different materials which can be decomposed is typically limited to the number of available energy channels.

During the last decade, spectral imaging technology has been successfully implemented with many different technical realizations. Scanners available on the market can feature two distinct sources with different X-ray spectra, a single source capable of fast voltage switching or dual-layer techniques where two detectors are superimposed, the second one receiving an X-ray spectrum hardened by the first one [64]. Concurrently, with the latest progress in X-ray detectors fabrication, the present research focus is rapidly shifting towards detector-based spectral imaging. Arguably, the main advantages of detector-based systems are the possibility of acquiring perfectly co-registered data over multiple energy channels in a single scan, and the simplification of the experimental setup, where only a single source and a single detector are needed. Spectral capabilities are usually obtained thanks to the presence of two or more energy thresholds per pixel, each with an associated counter. This scheme allows for energy binning, where each bin is defined by a pair of thresholds (or one, in the case of the highest energy bin) and its intensity is computed by subtracting the content of the higher energy threshold counter to the one of the lower energy threshold counter [65]. Alternatively, as in the case of Timepix3 [66] and Dosepix [67] chips, each event is digitized by means of the Time-over-Threshold (ToT) method, where the number of clock cycles in which the signal is above a given threshold is proportional to the energy released by the incoming X-ray. The latter method has the advantage of allowing for finer energy binning at the cost of a greatly reduced maximum count rate [33]. The energy resolution of spectral devices is in general dependent on the energy of the incoming radiation and the sensor material, and it can be as low as 1 keV at energies below 100 keV [33].

As mentioned in Sect. 2, for a given material the energy dependence of the attenuation coefficient, used in conventional spectral imaging, is in general quite different from that of the real part of the refractive index (δ, Fig. 1), which is responsible for phase effects. For this reason, the integration of phase-sensitive techniques in the context of spectral imaging can provide an additional channel, which can be used to improve or extend material decomposition algorithms. Specifically, the connection between spectral and phase-contrast imaging is established via the projected electron density that influences both real and imaginary parts of the refractive index [68]. Even though many research groups are presently focusing on the integration of these two techniques, spectral phase-contrast is still in its early days and the number of publications on this topic is still limited. Nonetheless, as first-line discrimination, it seems that different approaches to spectral phase contrast can be grouped in two categories: the first one uses the phase signal to increment the number of materials which can be decomposed in spectral acquisitions, while the second makes use of the phase signal to enforce consistency in material decomposition algorithms, therefore reducing the final image noise.

A rather enlightening example falling in the first category has been published by Ji et al. [69]. In their paper, the authors make use of a GI setup coupled with a CdTe-based photon-counting detector (model XC-Thor, XCounter AB, Sweden) with two energy thresholds per pixel and a rotating anode tube operated at 50 kV (G1582, Varian Medical Systems, Inc., Salt Lake City, UT). In conventional spectral imaging, this detector would allow for the decomposition of only two materials (one for each energy bin) but, by using the proportionality linking δ with the electronic density and extending the decomposition algorithm to a three materials scheme, the authors were able to successfully separate water (i.e., soft tissue), calcium (i.e., bone tissue), and iodine (i.e., contrast medium) (Fig. 6). The researchers also demonstrated that, compared to a three materials decomposition achievable with three energy bins, the spectral phase-contrast reconstructions yielded a significantly lower image noise. A similar example, again making use of GI, has been documented by Braig et al. [70]. In this case, the authors used monoenergetic (25 keV) images collected with photon-counting Pilatus-200 K detector (DECTRIS ltd., Baden, Switzerland), and exploited the attenuation and phase signals to generate spectral images of soft tissues and

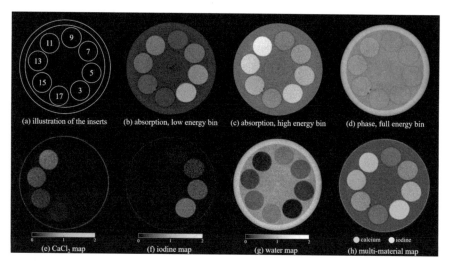

Fig. 6 Spectral GI CT images of a plastic phantom containing water, calcium-based, and iodine-based materials. In (**b**) and (**c**) the reconstructed absorption images in high and low energy bins, in (**d**) the phase image. Display range: (**b**)–(**c**): $[0.2, 0.6]$ cm^{-1}; (**d**): $[0.14, 0.32] \times 10^{-6}$. From (**e**) to (**g**) the decomposition coefficient maps, in (**h**) a multi-material map made by overlaying calcium iodine maps onto the full bin absorption image. In (**a**) a sketch of the phantom: the content of each insert can be found in the original paper. Reproduced from Ji et al. [69]

iodinated contrast medium from a single energy bin. Interestingly, in this case, the phase sensitivity enabled in a sense to achieve spectral images from a non-spectral detector.

On the other hand, an illustrative example of spectral phase-contrast imaging aiming at noise reduction is provided by Mechlem et al. [71]. This work, which is based on a numerical experiment on thorax radiography imaged through a GI setup, proposes a novel approach where the model of GI's signal is integrated within the spectral decomposition algorithm. In the study, both a realistic spectrum from a tungsten anode source at 120 kV and realistic energy response of a 2 mm thick cadmium telluride (CdTe) sensor with two energy thresholds are simulated. Results show that the combination of GI and spectral imaging benefits from the strengths of the individual methods; in fact, quantitatively accurate basis material images are obtained and the noise level is reduced when compared to conventional spectral X-ray imaging (Fig. 7). Schaff et al. [72] used a conceptually similar approach in the field of PBI. In this case, the authors show, based on experimental images on test objects and small-animal samples, that PBI spectral-phase contrast provides adequate material decomposition performance and reduced noise on the soft tissue component. Of note is that, despite the study having been conducted with monoenergetic synchrotron radiation (at 24 and 34 keV) and conventional charge integrating detector, the authors suggest its foreseeable extension to laboratory polychromatic sources combined with spectral detectors.

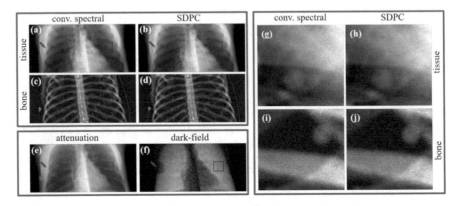

Fig. 7 Numerical simulation of human thorax radiography. The left part shows a comparison between conventional spectral material decomposition [(**a**), (**c**)], and spectral differential phase contrast (SDPC) material decomposition [(**b**), (**d**)] for both the tissue and bone images. The red rectangle in the images highlights the area where the artificial lung nodules are placed. Moreover, the conventional attenuation image (**e**), as well as the dark-field image (**f**) that was obtained by the proposed SDPC decomposition algorithm, is displayed. The red arrows highlight the location of a simulated pneumothorax. The right part of the figure shows a zoom of the area where the artificial lung nodules are placed. The zoomed images facilitate visual comparison of conventional spectral decomposition [(**g**), (**i**)] and SDPC decomposition [(**h**), (**j**)] regarding image quality and noise levels. Reproduced from Mechlem et al. [71]

We remark that the list of mentioned studies on spectral phase-contrast imaging is far from being complete: in this context, it is worth mentioning other significant works published in the field of EI [73], PBI [74, 75] and GI [76].

5.3   Subpixel Resolution

Subpixel or super-resolution detectors are devices in which the position of every single event can be determined with an uncertainty that is smaller than the physical pixel size. This allows the creation of fine virtual sampling grids, giving access to spatial details with dimensions below the pixel size. One of the best-known examples among super-resolution devices is the Timepix3 chip, where subpixel resolution is achieved by exploiting the charge sharing effect among neighboring pixels [41]. In fact, the sharing of the charge produced by a single photon event among detector pixels induces an electrical signal in multiple adjacent pixel electrodes. As mentioned in the previous section, when the signal is higher than a given triggering threshold, an event is registered and, by recording the signal's Time-over-Threshold (ToT), the energy deposited in that pixel can be found [34]. In addition to ToT, the Timepix3 chip is capable of reading Time of Arrival (ToA) for every single event. This enables the possibility to identify clusters of events belonging to the same photon interaction. For each cluster, the ToT measurements,

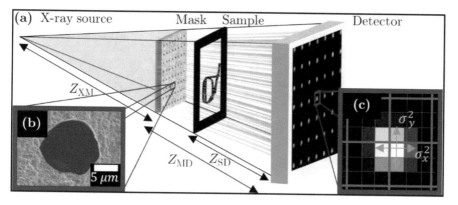

Fig. 8 Beam tracking experimental setup: (**a**) polychromatic X-rays are emitted from a micro-focus source and shaped into beamlets by the cylindrical apertures (**b**) of the absorption mask. The beamlets are attenuated, refracted, and scattered at the sample, and then detected with an Advapix detector with a 1-mm-thick silicon sensor and (**c**) a 55 μm pixel pitch (solid red lines) using a virtual subpixel (dashed red lines). Reproduced from Dreier et al. [77]

i.e., the amount of energy deposited in each pixel, can then be used to find the center of mass of the photon's interaction, i.e., its position [78]. Photons' positions are then binned into a finer matrix of virtual "subpixels," hence subpixel resolution is obtained. The same principle to achieve subpixel resolution can be found in the MONCH hybrid silicon detector [79, 80]. Despite being a charge integrating device, this detector, when operated with high frame rates and moderate photon fluxes, can discriminate single photons from the electronic noise (photon-counting performance). Thanks to its small pixel pitch (25 μm), the charge cloud generated by absorbed X-rays is shared between neighboring pixels for most of the photons. At low photon fluxes, interpolation algorithms can be applied to access extremely fine details at spatial scales down to 1 μm, as demonstrated by Cartier et al. [81].

In this context, many phase-contrast imaging applications making use of subpixel resolution devices have been documented, arguably the most interesting being in the fields of EI and GI. Specifically, a recent publication by Dreier et al. [77] showed that omni-directional phase sensitivity can be achieved in a single-shot thanks to the combined use of an absorbing mask with circular apertures and a super-resolution detector (Fig. 8). In this EI arrangement, also referred to as beam tracking [82], the apertures are aligned with the detector's pixel matrix such that each circular beamlet impinges at the intersection of 4 neighboring pixels, thus maximizing the charge sharing. As mentioned in Sect. 3, refraction and ultra-small-angle scattering effects act on the beamlets by shifting their position and broadening their width. Thanks to the subpixel resolution offered by the Timepix3-based Advapix detector (ADVACAM s.r.o., Prague, Czech Republic) the authors were able to detect and evaluate these effects in two dimensions, obtaining omni-directional refraction and dark-field images (Fig. 9). Of note is that the beam tracking technique was firstly implemented with conventional CCD detectors featuring a small pixel pitch. On

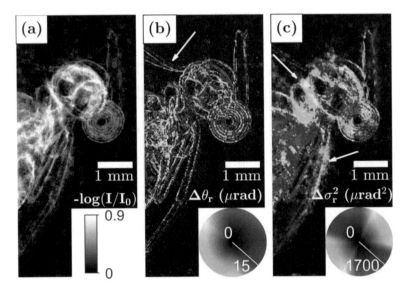

Fig. 9 Multi-modal images of a butterfly with attenuation (**a**), omni-directional refraction (**b**), and omni-directional dark-field contrast (**c**). Arrows indicate areas with increased contrast. Reproduced from Dreier et al. [77]

the other hand, such high-resolution detectors have rather limited efficiency and field of view, thus making the use of subpixel resolution hybrid detectors a more convenient choice. Contextually, very similar results have been obtained with an analogous setup by using the MONCH silicon detector [83]. The same detector has also been used in the context of GI, where Cartier et al. [81] demonstrated that the subpixel performance is sufficient for direct detection of the interference fringes without using the G_2 grating. Along with hybrid detectors, it is worth mentioning that applications of conventional devices also enabling subpixel resolution have been documented in the field of PBI by O'Connell et al. [84].

6   Conclusion

XPCI enables sensitivity to a new pool of contrast which has great potential in highlighting features that exhibit poor visibility in conventional imaging. Various techniques have been developed over the last two decades, each with specific characteristics that can be exploited in different applications as demonstrated by promising results in the context of biomedical imaging and material inspection. All the techniques share the common starting point that the sample-induced phase shift needs to be converted into an intensity modulation to be recorded on a pixelated detector, hence the integration with modern hybrid detectors appears to be a natural evolution of XPCI. This chapter is an attempt to outline the specific advantages that can be brought by these devices in the field of XPCI. An example is the integration of high-Z photon-counting detectors in PBI, where the high efficiency and sharp response are of great importance towards clinic implementations. Additionally unique features of modern hybrid detectors, such as spectral performances and super-resolution capabilities, are themselves enabling factors for the development of conceptually new XPCI solutions. For instance, this is the case of setups with omni-directional phase and USAXS sensitivity or spectral phase-contrast imaging. We believe that the integration of these research fields is only at the beginning and, as suggested by the growing number of "first time" articles appearing in the scientific literature, the best is yet to come.

References

1. Bravin, A., Coan, P., and Suortti, P. X-ray phase-contrast imaging: from pre-clinical applications towards clinics. *Physics in Medicine & Biology* 58.1 (2012): R1.
2. Endrizzi, M. "X-ray phase contrast imaging" *Nuclear Instruments and Methods in Physics Research Section A: Accelerators, Spectrometers, Detectors and Associated Equipment* 878 (2018) 88–98.
3. Rigon, L. X-ray imaging with coherent sources. *Comprehensive Biomedical Physics*, Elsevier, 2014. vol. 2, 193-220.
4. Hertz, H. M., et al. Electron-impact liquid-metal-jet hard x-ray sources. *Comprehensive Biomedical Physics*, Elsevier, 2014, vol. 8, 91-109.
5. Murrie, Rhiannon P., et al. Real-time in vivo imaging of regional lung function in a mouse model of cystic fibrosis on a laboratory X-ray source. *Scientific reports* 10.1 (2020): 1-8.
6. Carroll, F. E. et al. Pulsed tunable monochromatic X-ray beams from a compact source: new opportunities. *American journal of roentgenology* 181 (2003): 1197-1202.
7. Günther, B., et al. The versatile X-ray beamline of the Munich Compact Light Source: design, instrumentation and applications. *Journal of Synchrotron Radiation* 27.5 (2020).
8. Pelliccia, D., Kitchen, M. J., and Morgan, K. S.. Theory of X-ray Phase Contrast Imaging. *Handbook of X-ray Imaging: Physics and Technology*. CRC Press, 2018. 971-997.
9. Wilkins, S. W., et al. Phase-contrast imaging using polychromatic hard X-rays. *Nature* 384.6607 (1996): 335-338.
10. Bonse, U., and Hart, M. An X-ray interferometer. *Applied Physics Letters* 6.8 (1965): 155-156.
11. Momose, A. Phase-contrast X-ray imaging based on interferometry. *Journal of synchrotron radiation* 9.3 (2002): 136-142.
12. Weitkamp, T. et al. (2005). X-ray phase imaging with a grating interferometer. Opt. Express 13, 6296–6304.
13. Graetz, J., et al. Review and experimental verification of x-ray dark-field signal interpretations with respect to quantitative isotropic and anisotropic darkfield computed tomography. *Physics in Medicine & Biology* 65.23 (2020): 235017.
14. Pfeiffer, F., et al. Phase retrieval and differential phase-contrast imaging with low-brilliance X-ray sources. *Nature physics* 2.4 (2006): 258-261.
15. Takeda, Y., et al. X-ray phase imaging with single phase grating. *Japanese journal of applied physics* 46.1L (2007): L89.
16. Diemoz, P. C., et al. Non-Interferometric Techniques for X-ray Phase-Contrast Biomedical Imaging. *Handbook of X-ray Imaging*. CRC Press, 2017. 999-1024.
17. Chapman, D., et al. Diffraction enhanced x-ray imaging. *Physics in Medicine & Biology* 42.11 (1997): 2015.

18. Olivo, A., et al. An innovative digital imaging set-up allowing a low-dose approach to phase contrast applications in the medical field. *Medical physics* 28.8 (2001): 1610-1619.
19. Kallon, G. K., et al. Comparing signal intensity and refraction sensitivity of double and single mask edge illumination lab-based x-ray phase contrast imaging set-ups. *Journal of Physics D: Applied Physics* 50.41 (2017): 415401.
20. Vittoria, F. A., et al. Beam tracking approach for single-shot retrieval of absorption, refraction, and dark–field signals with laboratory x–ray sources. *Applied Physics Letters* 106.22 (2015): 224102.
21. Zdora, Marie-Christine. State of the art of x-ray speckle-based phase-contrast and dark-field imaging. *Journal of Imaging* 4.5 (2018): 60.
22. Massimi, L., et al. Fast, non-iterative algorithm for quantitative integration of X-ray differential phase-contrast images. *Optics Express* 28.26 (2020): 39677-39687.
23. Thüring, T., et al. Non-linear regularized phase retrieval for unidirectional X-ray differential phase contrast radiography. *Optics express* 19.25 (2011): 25545-25558.
24. Cloetens, P., et al. Holotomography: Quantitative phase tomography with micrometer resolution using hard synchrotron radiation x rays. *Applied physics letters* 75.19 (1999): 2912-2914.
25. Paganin, D., et al. Simultaneous phase and amplitude extraction from a single defocused image of a homogeneous object. *Journal of microscopy* 206.1 (2002): 33-40.
26. Beltran, M. A., et al. 2D and 3D X-ray phase retrieval of multi-material objects using a single defocus distance. *Optics Express* 18.7 (2010): 6423-6436.
27. Briedis, D., et al. Analyser-based mammography using single-image reconstruction. *Physics in Medicine & Biology* 50.15 (2005): 3599.
28. Diemoz, P. C., et al. Single-image phase retrieval using an edge illumination X-ray phase-contrast imaging setup. *Journal of synchrotron radiation* 22.4 (2015): 1072-1077.
29. Wang, X., et al. Single-shot phase retrieval method for synchrotron-based high-energy x-ray grating interferometry. *Medical physics* 46.3 (2019): 1317-1322.
30. Gureyev, T. E., et al. On the "unreasonable" effectiveness of transport of intensity imaging and optical deconvolution. *JOSA A* 34.12 (2017): 2251-2260.
31. Burvall, A., et al. Phase retrieval in X-ray phase-contrast imaging suitable for tomography. *Optics express* 19.11 (2011): 10359-10376.
32. Chen, R. C., Rigon, L. and Longo, R. Comparison of single distance phase retrieval algorithms by considering different object composition and the effect of statistical and structural noise. *Optics express* 21 (2013): 7384–7399.
33. Ballabriga, R., et al. Review of hybrid pixel detector readout ASICs for spectroscopic X-ray imaging. *Journal of Instrumentation* 11.01 (2016): P01007.
34. Jakůbek, J. Semiconductor pixel detectors and their applications in life sciences. *Journal of Instrumentation* 4.03 (2009): P03013.
35. Kalender, W. A., et al. Technical feasibility proof for high-resolution low-dose photon-counting CT of the breast. *European radiology* 27.3 (2017): 1081-1086.
36. Symons, R., et al. Feasibility of dose-reduced chest CT with photon-counting detectors: initial results in humans. *Radiology* 285.3 (2017): 980-989.
37. Willemink, M. J., et al. Photon-counting CT: technical principles and clinical prospects. *Radiology* 289.2 (2018): 293-312.
38. Llopart, X., et al. Medipix2, a 64k pixel read out chip with 55/spl mu/m square elements working in single photon counting mode. *2001 IEEE Nuclear Science Symposium Conference Record (Cat. No. 01CH37310)*. Vol. 3. IEEE, 2001.
39. Bellazzini, R., et al. Chromatic X-ray imaging with a fine pitch CdTe sensor coupled to a large area photon counting pixel ASIC. *Journal of Instrumentation* 8.02 (2013): C02028.
40. Frojdh, E., et al. Count rate linearity and spectral response of the Medipix3RX chip coupled to a 300μm silicon sensor under high flux conditions. *Journal of Instrumentation* 9.04 (2014): C04028.
41. Khalil, M., et al. Subpixel resolution in CdTe Timepix3 pixel detectors. *Journal of synchrotron radiation* 25.6 (2018): 1650-1657.

42. Schubert, A., et al. Micrometre resolution of a charge integrating microstrip detector with single photon sensitivity. *Journal of synchrotron radiation* 19.3 (2012): 359-365.
43. Ballabriga, R., et al. The Medipix3RX: a high resolution, zero dead-time pixel detector readout chip allowing spectroscopic imaging. *Journal of Instrumentation* 8.02 (2013): C02016.
44. Brombal, L., et al. Large-area single-photon-counting CdTe detector for synchrotron radiation computed tomography: a dedicated pre-processing procedure. *Journal of synchrotron radiation* 25.4 (2018a): 1068-1077.
45. Delogu, P., et al. Optimization of the equalization procedure for a single-photon counting CdTe detector used for CT. *Journal of Instrumentation* 12.11 (2017): C11014.
46. Di Trapani, V., et al. Characterization of the acquisition modes implemented in Pixirad-1/Pixie-III X-ray Detector: Effects of charge sharing correction on spectral resolution and image quality. *Nuclear Instruments and Methods in Physics Research Section A: Accelerators, Spectrometers, Detectors and Associated Equipment* 955 (2020): 163220.
47. Zhao, Y., et al. High-resolution, low-dose phase contrast X-ray tomography for 3D diagnosis of human breast cancers. *Proceedings of the National Academy of Sciences* 109.45 (2012): 18290-18294.
48. Gureyev, T. E., et al. Propagation-based x-ray phase-contrast tomography of mastectomy samples using synchrotron radiation. *Medical physics* 46.12 (2019): 5478-5487.
49. Longo, R., et al. Towards breast tomography with synchrotron radiation at Elettra: first images. *Physics in Medicine & Biology* 61.4 (2016): 1634.
50. Castelli, E., et al. Mammography with synchrotron radiation: first clinical experience with phase-detection technique. *Radiology* 259.3 (2011): 684-694.
51. Longo, R., et al. Advancements towards the implementation of clinical phase-contrast breast computed tomography at Elettra. *Journal of synchrotron radiation* 26.4 (2019): 1343-1353.
52. Brombal, L., et al. Image quality comparison between a phase-contrast synchrotron radiation breast CT and a clinical breast CT: a phantom based study. *Scientific reports* 9.1 (2019): 1-12.
53. Taba, S. T., et al. Toward improving breast cancer imaging: radiological assessment of propagation-based phase-contrast CT technology. *Academic radiology* 26.6 (2019): e79-e89.
54. Brombal, L. Effectiveness of X-ray phase-contrast tomography: effects of pixel size and magnification on image noise. *Journal of Instrumentation* 15.01 (2020): C01005.
55. Nesterets, Yakov I., Timur E. Gureyev, and Matthew R. Dimmock. Optimisation of a propagation-based x-ray phase-contrast micro-CT system. *Journal of Physics D: Applied Physics* 51.11 (2018): 115402.
56. Brombal, L., et al. Phase-contrast breast CT: the effect of propagation distance. *Physics in Medicine & Biology* 63(24) (2018b): 24NT03.
57. Kitchen, M. J., et al. CT dose reduction factors in the thousands using X-ray phase contrast. *Scientific reports* 7.1 (2017): 1-9.
58. Scholz, J., et al. Biomedical x-ray imaging with a GaAs photon-counting detector: A comparative study. *APL Photonics* 5.10 (2020): 106108.
59. Vila-Comamala, J., et al. High sensitivity X-ray phase contrast imaging by laboratory grating-based interferometry at high Talbot order geometry. *Optics Express* 29.2 (2021): 2049-2064.
60. Hagen, C. K., et al. A preliminary investigation into the use of edge illumination X-ray phase contrast micro-CT for preclinical imaging. *Molecular imaging and biology* 22.3 (2020): 539-548.
61. Alvarez, R. E., and Macovski, A. Energy-selective reconstructions in x-ray computerised tomography. *Physics in Medicine & Biology* 21.5 (1976): 733.
62. Carnibella, R. P., Fouras, A., and Kitchen, M. J. Single-exposure dual-energy-subtraction X-ray imaging using a synchrotron source. *Journal of synchrotron radiation* 19.6 (2012): 954-959.
63. Brooks, R. A. A quantitative theory of the Hounsfield unit and its application to dual energy scanning. *Journal of computer assisted tomography* 1.4 (1977): 487-493.
64. Johnson, T. R. C. Dual-energy CT: general principles. *American Journal of Roentgenology* 199(Suppl. 5) (2012): S3-S8.

65. Brun, F., et al. Single-shot K-edge subtraction x-ray discrete computed tomography with a polychromatic source and the Pixie-III detector. *Physics in Medicine & Biology* 65.5 (2020): 055016.
66. Poikela, T., et al. Timepix3: a 65K channel hybrid pixel readout chip with simultaneous ToA/ToT and sparse readout. *Journal of instrumentation* 9.05 (2014): C05013.
67. Wong, W. *A hybrid pixel detector ASIC with energy binning for real-time, spectroscopic dose measurements*. Diss. Mid Sweden University, 2012.
68. Mechlem, K., et al. A theoretical framework for comparing noise characteristics of spectral, differential phase-contrast and spectral differential phase-contrast x-ray imaging. *Physics in Medicine & Biology* 65.6 (2020): 065010.
69. Ji, X., et al. Dual Energy Differential Phase Contrast CT (DE-DPC-CT) Imaging. *IEEE transactions on medical imaging* 39.11 (2020): 3278-3289.
70. Braig, E., et al. Direct quantitative material decomposition employing grating-based X-ray phase-contrast CT. *Scientific reports* 8.1 (2018): 1-10.
71. Mechlem, K., et al. Spectral differential phase contrast x-ray radiography. *IEEE transactions on medical imaging* 39.3 (2019): 578-587.
72. Schaff, F., et al. Material decomposition using spectral propagation-based phase-contrast x-ray imaging. *IEEE Transactions on Medical Imaging* 39.12 (2020): 3891-3899.
73. Das, M., and Liang, Z. Spectral x-ray phase contrast imaging for single-shot retrieval of absorption, phase, and differential-phase imagery. *Optics letters* 39.21 (2014): 6343-6346.
74. Gureyev, T. E., et al. Quantitative analysis of two-component samples using in-line hard X-ray images. *Journal of synchrotron radiation* 9.3 (2002): 148-153.
75. Vazquez, I., et al. Quantitative phase retrieval with low photon counts using an energy resolving quantum detector. *JOSA A* 38.1 (2021): 71-79.
76. Epple, F. M., et al. Phase unwrapping in spectral X-ray differential phase-contrast imaging with an energy-resolving photon-counting pixel detector. *IEEE transactions on medical imaging* 34.3 (2014): 816-823.
77. Dreier, E. S., et al. Single-shot, omni-directional x-ray scattering imaging with a laboratory source and single-photon localization. *Optics letters* 45.4 (2020a): 1021-1024.
78. Dreier, E. S., et al. Virtual subpixel approach for single-mask phase-contrast imaging using Timepix3. *Journal of Instrumentation* 14.01 (2019): C01011.
79. Dinapoli, R., et al. MÖNCH, a small pitch, integrating hybrid pixel detector for X-ray applications. *Journal of Instrumentation* 9.05 (2014): C05015.
80. Ramilli, M., et al. Measurements with MÖNCH, a 25 μm pixel pitch hybrid pixel detector. *Journal of Instrumentation* 12.01 (2017): C01071.
81. Cartier, S., et al. Micrometer-resolution imaging using MÖNCH: towards G2-less grating interferometry. *Journal of synchrotron radiation* 23.6 (2016): 1462-1473.
82. Vittoria, F. A., et al. Multimodal phase-based X-ray microtomography with nonmicrofocal laboratory sources. *Physical Review Applied* 8.6 (2017): 064009.
83. Dreier, E. S., et al. Tracking based, high-resolution single-shot multimodal x-ray imaging in the laboratory enabled by the sub-pixel resolution capabilities of the MÖNCH detector. *Applied Physics Letters* 117.26 (2020b): 264101.
84. O'Connell, Dylan W., et al. Photon-counting, energy-resolving and super-resolution phase contrast X-ray imaging using an integrating detector. *Optics express* 28.5 (2020): 7080-7094.

Algorithm for Generating Effective Atomic Number, Soft-Tissue, and Bone Images Based on Analysis Using an Energy-Resolving Photon-Counting Detector

Natsumi Kimoto, Hiroaki Hayashi, Cheonghae Lee, Tatsuya Maeda, and Akitoshi Katsumata

1 Introduction

An X-ray image is essential for medical examination and non-invasive inspection. With the shift from analog detectors using X-ray film to digital detectors such as computed radiography (CR) and digital radiography (DR) systems, various techniques for image processing have been proposed to improve diagnosis [1, 2]. In recent years, new attempts to incorporate artificial intelligence into diagnostic technology are also attracting attention [3]. These techniques focus on how much useful information can be extracted from a generated X-ray image. We would also like to create a new process of image generation and investigate new possibilities for X-ray images.

Figure 1 shows the concept of the diagnostic system that we want to propose. In conventional diagnosis as shown in the right panel of Fig. 1, a medical doctor interprets a gray-scale image reflecting the absorbed X-ray energy based on anatomical knowledge. It means that a biological perspective plays an important role in this diagnosis. On the other hand, our goal is to carefully analyze the interaction of X-rays related to image generation; during the production of X-ray images, X-rays penetrating the human body which further interact with detector materials as shown in the left panel of Fig. 1. The analysis of the interaction in each process will lead to the generation of a novel concept X-ray image based on the perspective

N. Kimoto (✉) · H. Hayashi
College of Medical, Pharmaceutical and Health Sciences, Kanazawa University, Kanazawa, Ishikawa, Japan

C. Lee · T. Maeda
Graduate School of Medical Sciences, Kanazawa University, Kanazawa, Ishikawa, Japan

A. Katsumata
Department of Oral Radiology, Asahi University, Gifu, Japan

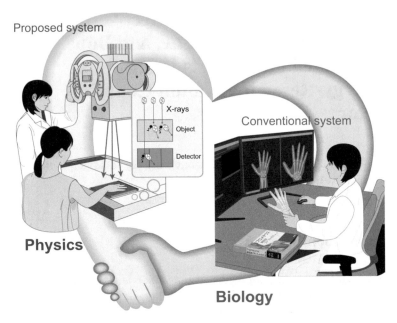

Fig. 1 Concept of the diagnostic system that we want to propose. The fusion of conventional qualitative diagnosis based on anatomical knowledge and quantitative diagnosis based on physics will lead to novel medical diagnosis

of physics. We expect that the fusion of conventional qualitative diagnosis based on biology and our proposed quantitative diagnosis based on physics will lead to innovations in diagnostic performance.

Currently, an energy-resolving photon-counting detector (ERPCD) which can generate images corresponding to the energy of each X-ray is being developed [4–6]. In our research, we focus on two-dimensional X-ray images and devise ways to make the best use of the X-ray attenuation information obtained from the ERPCD. In this chapter, we explain a novel method to identify materials using an ERPCD. First, we will explain the basic principle of material identification based on analysis of X-ray attenuation. Next, we will describe the issues which need to be considered when analyzing X-rays detected by ERPCD. Finally, we will present algorithms to derive an effective atomic number (Z_{eff}) image and to extract mass thickness (ρt) images related to soft-tissue and bone. Although the main topic of this chapter is to explain the novel method based on analysis of X-ray attenuation, the physics of X-ray attenuation is common for generating X-ray images using various systems such as plain X-ray and computed tomography (CT) examinations and we are confident that the chapter will be useful for future research using ERPCDs. If readers are interested in these contents, we recommend reading our papers [7–16] and books [17–19] for more detailed information.

2 Energy-Resolving Photon-Counting Detector to Realize Quantitative Analysis of an Object

Next-generation type ERPCD has been developed to realize novel analysis for such things as extracting properties of an object [4–6]. This section describes the processing procedure of the ERPCD, and the principle of material identification based on the analysis of X-ray attenuation.

To clearly understand the mechanism of an ERPCD, we need to compare it with an energy integrating detector (EID) [20, 21] which has been traditionally applied to medical and industrial X-ray examinations, as shown in Fig. 2. Figure 2(a) shows a schematic drawing of an EID. Assuming that four X-rays having different energies of E_1, E_2, E_3, and E_1 are incident to the pixel of the EID, corresponding charge clouds are generated in a pixel. Then, they are collected by an integrated circuit, and the corresponding totally absorbed energy ($\sum_i E_i$) values are digitalized and presented in a gray-scale image; this is the processing procedure for a conventional X-ray image. On the other hand, an ERPCD can analyze X-rays individually and discriminate X-ray energies corresponding to several energy thresholds [4–6]. Figure 2(b) shows a schematic drawing of an ERPCD. When setting threshold levels at $E_0{'}$, $E_1{'}$, $E_2{'}$, and $E_3{'}$, we can separate signals into three energy regions of $E_0{'} < E_i < E_1{'}$, $E_1{'} < E_i < E_2{'}$, $E_2{'} < E_i < E_3{'}$. In the case in which four X-rays having energies of E_1, E_2, E_3, and E_1 are incident to the pixel of ERPCD, E_1, E_2, and E_3 are classified into low, middle, and high energy bins, respectively. This is a main

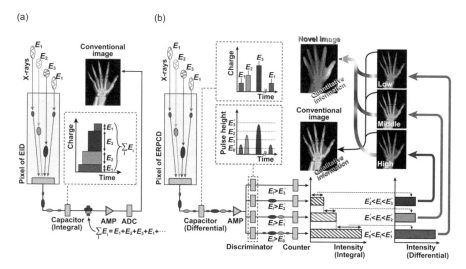

Fig. 2 Comparison between (**a**) energy integrating detector (EID) and (**b**) energy-resolving photon-counting detector (ERPCD). The EID generates images with information related to the sum of the energies. On the other hand, the ERPCD which is the focus of our research can analyze each X-ray energy

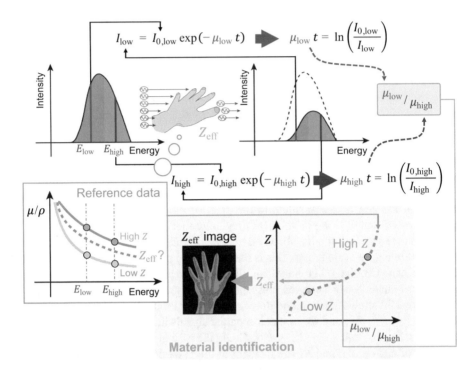

Fig. 3 Concept of material identification is based on the analysis of μ for determining differences in X-ray energies. We calculate μt values from the difference of intensities between the incident and penetrating X-rays and derive a $\frac{\mu_{\text{low}}}{\mu_{\text{high}}}$ value using $\mu_{\text{low}}t$ and $\mu_{\text{high}}t$ which correspond to low and high energies, respectively. Z_{eff} of an object can be determined from the experimentally derived $\frac{\mu_{\text{low}}}{\mu_{\text{high}}}$ value using a reference curve

difference from the EID, which is used to measure not the energy of an X-ray but the number of X-rays. In this case, two counts are measured in the low energy bin and one count is measured in each of the middle and high energy bins. We would like to create quantitative images that extract the properties of an object based on analysis using X-ray energy information. We can also produce a conventional X-ray image which provides qualitative information because the image related to the energy bin has information concerning intensity and effective energy.

Image generation using an ERPCD can quantitatively analyze the X-ray attenuation of an object. Here, we will introduce the method to extract the properties of an object from X-ray energy information. Figure 3 shows our concept for material identification. Assuming that an object having a certain Z_{eff} is measured, X-ray attenuation can be determined using the well-known formula [22]:

$$I = I_0 \times \exp\left(-\mu t\right), \tag{1}$$

where I_0 and I are intensities of the incident and penetrating X-rays, respectively, and μ and t are linear attenuation coefficient and material thickness, respectively. It is important to understand the fact that the probability of X-ray attenuation, μ, is determined by the atomic number and a function of the energy of the X-rays [22]. This means that if we can derive μ corresponding to a certain X-ray energy, we can determine the atomic number of an object. However, the physical quantity obtained from one piece of energy information is μt, and we cannot derive μ without using the information for t. To extract μ from μt, dual-energy information is required and ERPCD can be used to provide information corresponding to multiple energies with a single irradiation. In this study, we focus our attention on the low energy E_{low} and high energy E_{high} and calculate the corresponding μt values as $\mu_{\text{low}}t$ and $\mu_{\text{high}}t$. Then, taking the ratio of these values, we can derive the value $\frac{\mu_{\text{low}}}{\mu_{\text{high}}}$ ($=\frac{\mu_{\text{low}}t}{\mu_{\text{high}}t}$) which can be uniquely determined by atomic number without information for t [23]. In order to determine the Z_{eff} from $\frac{\mu_{\text{low}}}{\mu_{\text{high}}}$, a unique relationship between $\frac{\mu_{\text{low}}}{\mu_{\text{high}}}$ and Z_{eff} is available, and these data can be derived from the reference data μ/ρ [24]. When using the theoretical relationship, experimentally derived $\frac{\mu_{\text{low}}}{\mu_{\text{high}}}$ can be converted into Z_{eff}. This means that a novel image that reflects the Z_{eff} of an object can be produced by analyzing X-ray energy information generated from an ERPCD. We should note that this description is based on ideal energy absorption resulting from the interaction between incident X-rays and detector materials and focuses on monochromatic X-ray energy. However, actual cases are far from the ideal condition, and therefore the value measured with an ERPCD needs to be corrected to a value related to monochromatic X-ray.

3   Issues that Need to Be Considered to Achieve Accurate Material Identification

In the description in the former section, we explained that X-ray energy information needs to be analyzed accurately in order to derive the properties of an object. In general, medical and industrial applications, (1) polychromatic X-rays and (2) multi-pixel-type imaging detectors are widely applied, and therefore it is difficult to handle X-ray energy information ideally. In order to clarify the issues caused by these two restrictions, we show the differences in X-ray spectra obtained in each stage from incident to detection as shown in Fig. 4. The upper right panel shows the X-ray spectrum in which X-rays having various energies are generated from the X-ray equipment. When using an ERPCD having two energy bins, the low and high energy bins, presented by red and blue, respectively, are analyzed quantitatively. To perform the analysis, it is necessary to distinguish the energies, and we generally use the effective energies of each energy bin. In the spectrum obtained after penetrating an object, the degree of X-ray attenuation varies in the difference of Z_{eff} of an object, and as a result, the effective energy changes. This phenomenon is known

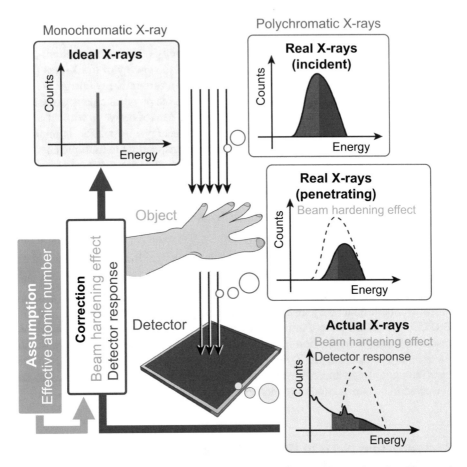

Fig. 4 Comparison of X-ray spectra obtained at each stage from incident to detection. The actual X-ray spectrum includes both the beam hardening effect and detector response. Our approach is to treat the measured polychromatic X-rays as if they are monochromatic X-rays by correcting for the beam hardening effect and detector response

as the "beam hardening effect" [25]. Furthermore, when X-rays are incident to the detector, the X-rays interact with detector materials, and the corresponding X-ray spectrum generated from the detector is much different than the incident X-ray spectrum. Although current commercial products use to take a practical approach for material decomposition based on the actual X-ray spectrum [26–29], our approach is much different. We apply the correction for the beam hardening effect and detector response to actual data by taking into consideration the Z_{eff} of an object [14–16, 19]. This method can process actual polychromatic X-rays as those equivalent to ideal monochromatic X-rays, which results in the accomplishment of what is considered "ideal" analysis based on monochromatic X-rays. In the following sections, we will carefully analyze the effects of the beam hardening effect and detector response and also introduce a novel method to correct these effects.

3.1 Beam Hardening Effect Depends on Effective Atomic Number

In this section, we will explain the beam hardening effect which is one of the issues to be considered for precise material identification. Figure 5 shows the relationship between the beam hardening effect and Z_{eff}. Figure 5(a) shows Z_{eff} values of the elements which compose the human body. The human body targeted for clinical examination is composed of various organs and tissue, and the Z_{eff} values of each organ and tissue are distributed at relatively low atomic number values. These Z_{eff} values can be calculated from the elemental composition of each organ [24] using a well-known formula [30]. As shown in Fig. 5(a), adipose tissue is 6.33, muscle is 7.45, and cortical bone is 13.23; it is obvious that there is a slight difference in Z_{eff} between the soft-tissues, and it can be roughly divided into soft-tissue at around $Z_{eff} = 7$ and bone around $Z_{eff} = 13$. Next, we present the beam hardening effect depending on the difference of Z_{eff}. Figure 5(b) shows the changes in the spectra and the effective energies ($\overline{E}$) corresponding to objects having various Z_{eff} values. The black line is the incident X-ray spectrum at 50 kV which is reproduced by a semi-empirical formula [31], and the $\overline{E}$ of the energy region 20–32 keV is 27.4 keV. We also present the X-ray spectra after penetrating objects of which Z_{eff} values and ρt value are set to be 5–15 and 1 g/cm^2, respectively. The distribution of X-rays is different depending on Z_{eff}, and it is important to especially focus on $Z_{eff} = 7$ (blue line) and $Z_{eff} = 13$ (red line). The corresponding $\overline{E}$ values are 27.8 keV and 29.5 keV, respectively. It is found that the difference between incident X-ray and penetrating X-ray shows the beam hardening effect, and the degree of effect differs depending on the Z_{eff}. Unless these effects are properly corrected based on the information of an object, it is not possible to perform an analysis in which polychromatic X-rays are regarded as monochromatic X-rays.

3.2 Response of Multi-Pixel-Type Energy-Resolving Photon-Counting Detector

Another issue to be considered for precise material identification is detector response. In this section, we will explain the concept of the response function of a multi-pixel-type ERPCD and the procedure to reproduce an X-ray spectrum using the response function. Figure 6 shows a concept of the various phenomena related to detector response when X-rays $\mathbf{\Phi}$ are incident to the detector. Figure 6(a) shows an ideal case where all of the energies of the incident X-rays are absorbed completely

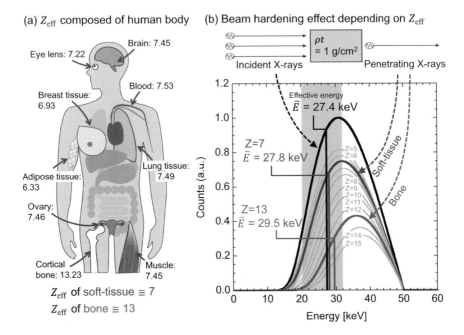

Fig. 5 Relationship between beam hardening effect and Z_{eff}. (**a**) shows the Z_{eff} values of various tissues which compose the human body. (**b**) shows the changes in the X-ray spectra and effective energies corresponding to objects having various Z_{eff} values

by the detector. Using the response function represented by an identity matrix **I**, the corresponding X-ray spectrum can be calculated by the matrix operation **IΦ** (= **Φ**). In actuality, the response function including phenomenon $\mathbf{R}^{(1)}$ which is related to radiation interactions and phenomenon $\mathbf{R}^{(2)}$ which is related to electric charge collecting processes should be considered to reproduce an X-ray spectrum measured with a multi-pixel-type ERPCD. Figure 6(b-1) shows $\mathbf{R}^{(1)}$ in which we take into consideration the transportation of the characteristics X-rays caused by the photoelectric effect and scattered X-rays due to the Compton scattering effect [7, 9, 17, 19, 32–36]. Figure 6(b-2) shows $\mathbf{R}^{(2)}$ in which the charge sharing effect [32–36] and energy resolution [32, 33, 35, 36] are considered. When considering both $\mathbf{R}^{(1)}$ and $\mathbf{R}^{(2)}$ operation of **Φ**, the X-ray spectrum under actual conditions can then be reproduced by matrix operation $\mathbf{R}^{(1)}\mathbf{R}^{(2)}\mathbf{Φ}$.

Here, we will look more closely at the response functions $\mathbf{R}^{(1)}$ and $\mathbf{R}^{(2)}$. First, the response function $\mathbf{R}^{(1)}$ is calculated by simulating the interaction between the incident X-rays and detector materials using Monte-Carlo simulation code EGS5 [37]. The detector materials are composed of cadmium zinc telluride (CZT) at a ratio of Cd:Zn:Te = 0.9:0.1:1.0 with a density of 5.8 g/cm³ and the outer size of the monolithic detector material is set as 10 mm × 10 mm with a thickness of 1.5 mm. The response function is defined as the spectra related to a center pixel of 200 μm × 200 μm. The 0–140 keV monochromatic X-rays of 10^6 photons

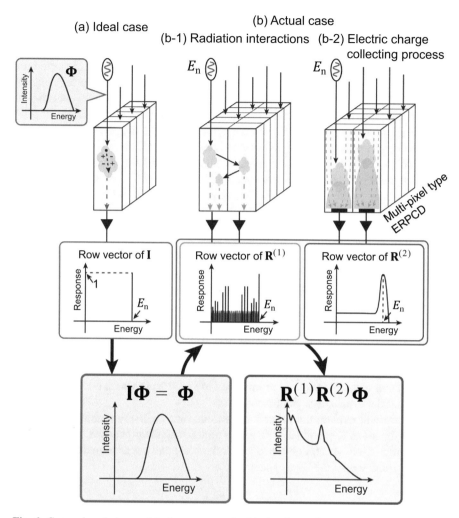

Fig. 6 Comparison between detector responses for ideal and actual cases. (**a**) In the ideal case, incident X-rays $\mathbf{\Phi}$ are totally absorbed within the pixel of the detector, and the detected X-ray spectrum ($\mathbf{I\Phi} = \mathbf{\Phi}$) is the same as the incident X-ray spectrum. (**b**) In actuality, incident X-rays are affected by two different phenomena: (**b-1**) radiation interactions and (**b-2**) electric charge collecting processes. Consequently, the detected X-ray spectrum $\mathbf{R}^{(1)}\mathbf{R}^{(2)}\mathbf{\Phi}$ is much different from that of the incident X-ray spectrum

are irradiated to the center pixel, and the total irradiation area is determined so as to establish equilibrium for the secondary produced radiation [9, 17, 19]. A two-dimensional matrix $\mathbf{R}^{(1)}$ is constructed from the elements of response functions corresponding to 0–140 keV monochromatic X-rays. Here, we demonstrate 80 keV monochromatic X-ray vector element for $\mathbf{R}^{(1)}$ as presented in Fig. 7(a). The red line represents the full energy peak (FEP) which appears when all of the incident X-ray energy is completely absorbed by the pixel of interest. On the other hand, the blue area relates to partially absorbed events, and some peaks are observed at 23–32 keV and 48–57 keV. It is important to understand these phenomena; when the X-rays are incident to the pixel, the photoelectric effect mainly occurs and then characteristic X-rays of Cd (23–27 keV) and Te (27–32 keV) are generated. When focusing on the pixel of interest, there is a possibility of the characteristic X-rays escaping to adjacent pixels, which results in the existence of escape peaks (EPs) at 48–57 keV. On the other hand, when focusing on the pixel adjacent to the pixel of interest, the characteristic X-rays generating at pixel of interest are incident, which results in the existence of peaks at 23–32 keV. It should be noted that the characteristic X-ray peaks are constant regardless of the energy of the incident X-rays, whereas the EPs vary according to the energy of the incident X-rays. Next, the response function $\mathbf{R}^{(2)}$ is reproduced taking into consideration charge sharing effect $\mathbf{r}^{(2,c)}$ and energy resolution $\mathbf{r}^{(2,e)}$. During charge collection process in a detector, electrons undergo a diffusion process and drift due to an electric field. Consequently, not all of the charges are collected by the pixel of interest and some charges spread into several adjacent pixels. This phenomenon is called the "charge sharing effect." One vector element in $\mathbf{r}^{(2,c)}$ is exemplified in Fig. 7(b-1). It consists of the peak component and the other part which is presented by a flat distribution. For our ERPCD, the respective proportions were optimized and the values were determined to be 35% and 65% so as to reproduce the X-ray spectrum measured with the ERPCD. In addition, as shown in Fig. 7(b-2), energy resolution $\mathbf{r}^{(2,e)}$ was determined as 5% at 80 keV so as to reproduce characteristic X-ray peaks. Using $\mathbf{r}^{(2,c)}$ and $\mathbf{r}^{(2,e)}$, the 80 keV vector element in $\mathbf{R}^{(2)}$ is exemplified in Fig. 7(b) in which the peaks around 80 keV and flat distribution can be observed.

Next, we will explain the procedure to reproduce an X-ray spectrum using the response functions $\mathbf{R}^{(1)}$ and $\mathbf{R}^{(2)}$. To easily understand the procedure, we reproduce two different X-ray spectra obtained under the ideal and actual situations as shown in Fig. 6. First, we set the X-ray spectrum,

$$\mathbf{\Phi} = \begin{pmatrix} \Phi_1 \\ \Phi_2 \\ \vdots \\ \Phi_j \\ \vdots \end{pmatrix}, \tag{2}$$

where the element of this vector is expressed as Φ_j. Figure 8(a) shows $\mathbf{\Phi}$ when setting the tube voltage at 50 kV.

In the ideal case, the response function can be expressed as the identity matrix:

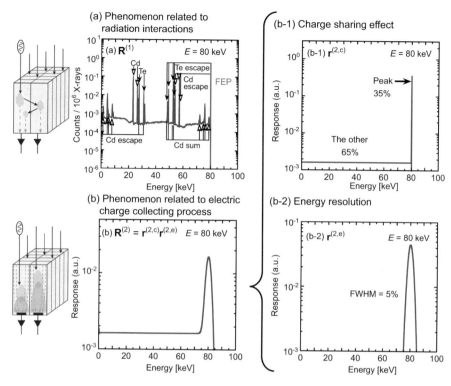

Fig. 7 Typical results of detector responses for 80 keV monochromatic X-ray. (**a**) is $\mathbf{R}^{(1)}$ including phenomenon related to radiation interaction; the red and blue lines show full energy peak and scattered X-rays, respectively. (**b**) is $\mathbf{R}^{(2)}$ including phenomenon related to the electric charge collection processes which are composed of (**b-1**) charge sharing effect $\mathbf{r}^{(2,c)}$ and (**b-2**) energy resolution $\mathbf{r}^{(2,e)}$

$$\mathbf{I} = \begin{pmatrix} 1 & 0 & \cdots & \cdots & 0 \\ 0 & 1 & \vdots & & \vdots \\ \vdots & \vdots & \ddots & \vdots & \vdots \\ \vdots & \vdots & \vdots & 1 & 0 \\ 0 & \cdots & \cdots & 0 & 1 \end{pmatrix}. \tag{3}$$

Figure 8(b-1) shows a two-dimensional color map presented by $\mathbf{I}$. It can be seen that unit values corresponding to FEPs appear in diagonal line.

In actual circumstances, the response function can be calculated by the matrix operation for $\mathbf{R}^{(1)}$ and $\mathbf{R}^{(2)}$,

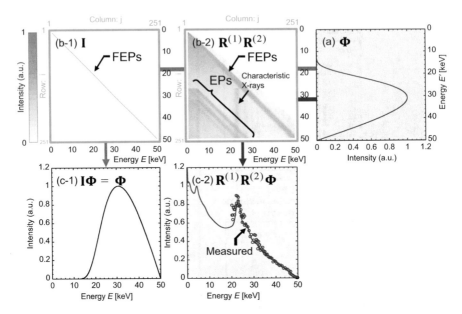

Fig. 8 The procedure for reproducing an X-ray spectrum taking into consideration detector response. (**a**) shows incident X-ray spectrum Φ. (**b-1**) and (**b-2**) show two-dimensional color maps of detector responses $\mathbf{I}$ and $\mathbf{R}^{(1)}\mathbf{R}^{(2)}$, respectively. (**c-1**) and (**c-2**) show reproduced X-ray spectra $\mathbf{I}\Phi = \Phi$ and $\mathbf{R}^{(1)}\mathbf{R}^{(2)}\Phi$, respectively. By performing the mathematical calculations, we can reproduce X-ray spectra measured with an ERPCD

$$\mathbf{R}^{(1)}\mathbf{R}^{(2)} = \begin{pmatrix} R_{1,1}^{(1)} & R_{1,2}^{(1)} & \cdots & \cdots\cdots \\ R_{2,1}^{(1)} & \ddots & \vdots & \vdots & \vdots \\ \vdots & \vdots & R_{i,k}^{(1)} & \vdots & \vdots \\ \vdots & \vdots & \vdots & \ddots & \vdots \\ \cdots & \cdots & \cdots\cdots & \cdots \end{pmatrix} \begin{pmatrix} R_{1,1}^{(2)} & R_{1,2}^{(2)} & \cdots & \cdots\cdots \\ R_{2,1}^{(2)} & \ddots & \vdots & \vdots & \vdots \\ \vdots & \vdots & R_{k,j}^{(2)} & \vdots & \vdots \\ \vdots & \vdots & \vdots & \ddots & \vdots \\ \cdots & \cdots & \cdots\cdots & \cdots \end{pmatrix}$$

$$= \begin{pmatrix} R_{1,1}^{(1,2)} & R_{1,2}^{(1,2)} & \cdots & \cdots\cdots \\ R_{2,1}^{(1,2)} & \ddots & \vdots & \vdots & \vdots \\ \vdots & \vdots & R_{i,j}^{(1,2)} & \vdots & \vdots \\ \vdots & \vdots & \vdots & \ddots & \vdots \\ \cdots & \cdots & \cdots\cdots & \cdots \end{pmatrix}, \tag{4}$$

where $R_{i,j}^{(1,2)}$ is described as

$$R_{i,j}^{(1,2)} = \sum_k R_{i,k}^{(1)} R_{k,j}^{(2)}.$$

$\mathbf{R}^{(1)}$ consists of the elements of $\mathbf{R}_{E'}^{(1)}$ corresponding to various incident X-ray energies E'. Then, $\mathbf{R}_{E'}^{(1)}$ is expressed as $\left\{ R_{i,1}^{(1)}, R_{i,2}^{(1)}, \ldots, R_{i,k}^{(1)}, \ldots \right\}$ where the corresponding response energy is the i-th element. In a similar way, $\mathbf{R}^{(2)}$ has elements of $\mathbf{R}_{E'}^{(2)}$. Then, the element $\mathbf{R}^{(1)}\mathbf{R}^{(2)}$ in the i-th row and j-th column is expressed as $R_{i,j}^{(1,2)}$. Figure 8(b-2) shows a two-dimensional color map of $\mathbf{R}^{(1)}\mathbf{R}^{(2)}$. The response corresponding to FEPs is shown with the diagonal line and the sharpness is affected by the energy resolution. The characteristic X-rays are observed at around 23 keV. EPs are represented by diagonal lines where the highest value is around 27 keV. The charge sharing effect causes an increase in intensities in the low energy region.

Finally, we will reproduce X-ray spectra using response functions $\mathbf{I}$ and $\mathbf{R}^{(1)}\mathbf{R}^{(2)}$. In ideal case, X-ray spectrum can be obtained by folding Φ with $\mathbf{I}$,

$$\mathbf{I}\Phi = \begin{pmatrix} 1 & 0 & \cdots & \cdots & 0 \\ 0 & 1 & \vdots & \vdots & \vdots \\ \vdots & \vdots & \ddots & \vdots & \vdots \\ \vdots & \vdots & \vdots & 1 & 0 \\ 0 & \cdots & \cdots & 0 & 1 \end{pmatrix} \begin{pmatrix} \Phi_1 \\ \Phi_2 \\ \vdots \\ \Phi_j \\ \vdots \end{pmatrix} = \begin{pmatrix} \Phi_1 \\ \Phi_2 \\ \vdots \\ \Phi_j \\ \vdots \end{pmatrix}. \tag{5}$$

Figure 8(c-1) shows $\mathbf{I}\Phi$. This is the same as incident X-ray spectrum Φ.

In the actual case, we can reproduce the X-ray spectrum by replacing $\mathbf{I}$ in Eq. (5) by $\mathbf{R}^{(1)}\mathbf{R}^{(2)}$:

$$\mathbf{R}^{(1)}\mathbf{R}^{(2)}\Phi = \begin{pmatrix} R_{1,1}^{(1,2)} & R_{1,2}^{(1,2)} & \cdots & \cdots & \cdots \\ R_{2,1}^{(1,2)} & \ddots & \vdots & \vdots & \vdots \\ \vdots & \vdots & R_{i,j}^{(1,2)} & \vdots & \vdots \\ \vdots & \vdots & \vdots & \ddots & \vdots \\ \cdots & \cdots & \cdots & \cdots & \cdots \end{pmatrix} \begin{pmatrix} \Phi_1 \\ \Phi_2 \\ \vdots \\ \Phi_j \\ \vdots \end{pmatrix} = \begin{pmatrix} \sum_j R_{1,j}^{(1,2)} \Phi_j \\ \sum_j R_{2,j}^{(1,2)} \Phi_j \\ \vdots \\ \sum_j R_{i,j}^{(1,2)} \Phi_j \\ \vdots \end{pmatrix}, \tag{6}$$

where each element is described as

$$\sum_j R_{i,j}^{(1,2)} \Phi_j = \sum_j \left(\sum_k R_{i,k}^{(1)} R_{k,j}^{(2)} \right) \Phi_j = \sum_{E'} \left(\sum_k R_{i,k}^{(1)} R_{k,E'}^{(2)} \right) \Phi\left(E'\right),$$

where j is the j-th element of Φ, for the sake of clarification, j is expressed as E'. Figure 8(c-2) shows the reproduced X-ray spectrum $\mathbf{R}^{(1)}\mathbf{R}^{(2)}\Phi$ plotted with the X-ray spectrum measured using our proto-type ERPCD [13, 15, 38]. $\mathbf{R}^{(1)}\mathbf{R}^{(2)}\mathbf{\Phi}$ is in good agreement with the experimental data. We can see that the $\mathbf{R}^{(1)}\mathbf{R}^{(2)}\mathbf{\Phi}$ has two major features: relatively large intensities in the low energy region and the presence of the characteristic X-ray peaks of Cd and Te. The $\mathbf{R}^{(1)}\mathbf{R}^{(2)}\mathbf{\Phi}$ is much different than that of $\mathbf{\Phi}$.

Here, in order to simplify the mathematical notation, $\mathbf{\Phi}$ is redefined as the following formations: $\mathbf{\Phi} = \{\Phi(0 \text{ keV}), \Phi(0.2 \text{ keV}), \cdots, \Phi(E_k), \cdots\}$ where E_k is X-ray energy at intervals of 0.2 keV (subscript of "k" means k-th vector element) for matrix formation and $\Phi(E)$ for continuous function. We also redefine $\mathbf{R}^{(1)}\mathbf{R}^{(2)}\mathbf{\Phi}$ as matrix formation: $\mathbf{R}^{(1)}\mathbf{R}^{(2)}\mathbf{\Phi} = \{R^{(1)}R^{(2)}\Phi(0 \text{ keV}), R^{(1)}R^{(2)}\Phi(0.2 \text{ keV}), \cdots, R^{(1)}R^{(2)}\Phi(E_k), \cdots\}$. These formation roles are applied to all equations in the following description.

To further investigate the effect of the detector response, we exemplify the analysis where μt is calculated using $\mathbf{\Phi}$ and $\mathbf{R}^{(1)}\mathbf{R}^{(2)}\mathbf{\Phi}$. Figure 9 demonstrates the analysis of μt when measuring aluminum ($Z_{\text{eff}} = 13$) having a $\rho t = 1.0$ g/cm^2 as the target object. The X-ray spectra being incident and penetrating object are described by $\mathbf{\Phi}_i$ and $\mathbf{\Phi}_p$, respectively. These are represented by black and red circles in Fig. 9(a-1). When these X-ray spectra are detected by the multi-pixel-type ERPCD, they are represented by $\mathbf{R}^{(1)}\mathbf{R}^{(2)}\mathbf{\Phi}_i$ and $\mathbf{R}^{(1)}\mathbf{R}^{(2)}\mathbf{\Phi}_p$ as shown in Fig. 9(a-2). Here, we calculate μt values in each case and compare them with the theoretical μt. Figure 9(b-1) shows the ideal case; the red circle shows $(\mu t)_{\text{meas}}$ calculated using the equation, $\ln\left(\frac{\Phi_i}{\Phi_p}\right)$, and the blue line shows the theoretical value calculated by multiplying ρt to μ/ρ. As a matter of course, the $(\mu t)_{\text{meas}}$ is in good agreement with the theoretical value. This indicates that the X-ray spectra retain the μt information related to the object, which means that we can extract information about the target object from the X-ray spectra. Figure 9(b-2) shows an actual case; the red circle is the $(\mu t)_{\text{meas}}$ calculated using the equation, $\ln\left(\frac{\mathbf{R}^{(1)}\mathbf{R}^{(2)}\mathbf{\Phi}_i}{\mathbf{R}^{(1)}\mathbf{R}^{(2)}\mathbf{\Phi}_p}\right)$, and the blue line is the theoretical value. The $(\mu t)_{\text{meas}}$ agrees with theoretical value in the energy region above approximately 32 keV. On the other hand, the difference from theoretical value increases in the lower energy region which is greatly affected by the cross-talk of characteristic X-rays and the charge sharing effect. This demonstration shows that regions above 32 keV have relatively pure information but lower energy regions, those below 32 keV have little information when setting 50 kV. It should be noted that the theoretical value cannot be sufficiently reproduced even when analyzing the regions above 32 keV.

3.3 Procedure to Correct for the Beam Hardening Effect and Detector Response

To accomplish highly accurate material identification, the biggest challenges that need solving are the issues of the beam hardening effect and detector response. In this section, we will introduce the correction procedure for the beam hardening

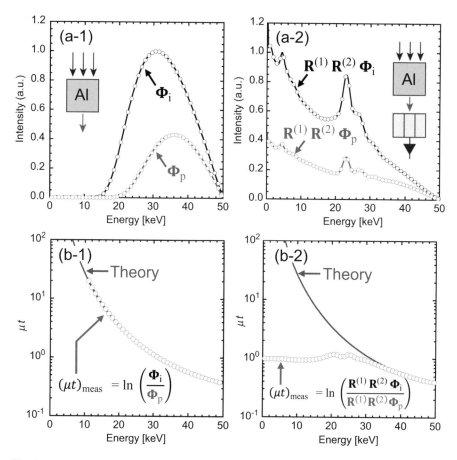

Fig. 9 Demonstration of how important it is to consider detector response. The results show μt analysis when measuring aluminum having $\rho t = 1.0$ g/cm^2. (**a-1**) and (**a-2**) show ideal X-ray spectra Φ and actual X-ray spectra $\mathbf{R}^{(1)}\mathbf{R}^{(2)}\Phi$, respectively. (**b-1**) and (**b-2**) are comparisons between μt values calculated from each X-ray spectrum and theoretical μt values

effect and detector response. Figure 10 shows a schematic drawing that explains the concept of the correction. The correction is to convert $(\mu t)_{\text{meas}}$ which is calculated from polychromatic X-rays detected by the ERPCD into $(\mu t)_{\text{cor}}$ calculated from monochromatic X-rays. To easily understand the correction of both effects, we focus on (a) monochromatic X-rays in Φ, (b) polychromatic X-rays in Φ, and (c) polychromatic X-rays in $\mathbf{R}^{(1)}\mathbf{R}^{(2)}\Phi$. (b) includes the beam hardening effect and (c) includes both beam hardening effect and detector response. As shown, common for conditions (a) through (c), we set the lower and upper energies of each energy bin as E_1 and E_2, respectively, and $\overline{E}$ is calculated. Assuming that an object with certain Z_{eff} and ρt values is measured, we simulate the X-ray spectra and calculate corresponding μt values related to conditions (a) through (c). Then, we derive

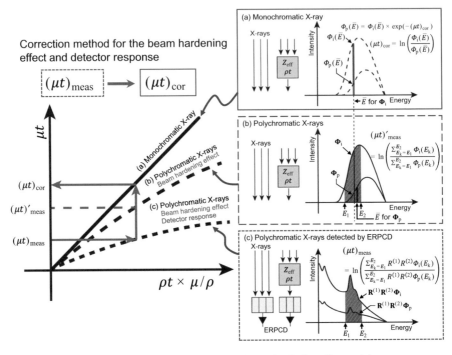

Fig. 10 Concept of the procedure to correct both the beam hardening effect and detector response. The correction is to convert $(\mu t)_{\text{meas}}$ calculated from polychromatic X-rays measured with an ERPCD to $(\mu t)_{\text{cor}}$ which is related to monochromatic X-rays. The former case consists of energy information disruption caused by the beam hardening effect and detector response, and the latter case is the retention of original energy information reflecting X-ray attenuation of the object. To clearly show the correction for both effects, we present each step of the correction: (**a**) $(\mu t)_{\text{cor}}$ calculated from monochromatic X-rays, (**b**) $(\mu t)'_{\text{meas}}$ calculated from polychromatic X-rays, and (**c**) $(\mu t)_{\text{meas}}$ calculated from polychromatic X-rays measured with the ERPCD

unique relationships between $\rho t \times \mu/\rho$ and μt which were proposed in a previous study [12, 14–16, 19].

First, we explain the relationship of (a) monochromatic X-ray energy $\overline{E}$. The $(\mu t)_{\text{cor}}$ is calculated from the equation, $\ln\left(\dfrac{\Phi_i(\overline{E})}{\Phi_p(\overline{E})}\right)$. Because the $\overline{E}$ is always constant even when ρt and/or Z_{eff} are varied, $(\mu t)_{\text{cor}}$ values can be plotted as a straight line $Y = X$; X and Y axes are $\rho t \times \mu/\rho$ and μt, respectively. Next, for (b) polychromatic X-rays, because the $\overline{E}$ of Φ_p, which is calculated from the equation, $\dfrac{\sum_{E_k=E_1}^{E_2} \Phi_p(E_k)E_k}{\sum_{E_k=E_1}^{E_2} \Phi_p(E_k)}$, is higher due to the beam hardening effect, the $(\mu t)'_{\text{meas}}$ calculated from the equation, $\ln\left(\dfrac{\sum_{E_k=E_1}^{E_2} \Phi_i(E_k)}{\sum_{E_k=E_1}^{E_2} \Phi_p(E_k)}\right)$, differs from the $(\mu t)_{\text{cor}}$; the $(\mu t)'_{\text{meas}}$ shows a curve which goes away from the $Y = X$ line. Finally, when using

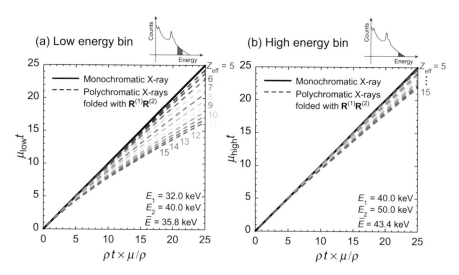

Fig. 11 Result of the correction curves for the beam hardening effect and detector response. (**a**) and (**b**) correspond to low and high energy bins, respectively. The solid and broken lines show monochromatic X-ray and polychromatic X-rays, respectively. The difference between monochromatic and polychromatic X-rays reflects the effects of beam hardening and detector response

(c) polychromatic X-rays detected using an ERPCD, $(\mu t)_{\text{meas}}$ calculated from the equation, $\ln \left(\dfrac{\sum_{E_k=E_1}^{E_2} R^{(1)} R^{(2)} \Phi_i(E_k)}{\sum_{E_k=E_1}^{E_2} R^{(1)} R^{(2)} \Phi_p(E_k)} \right)$, is much different from the Y = X line due to the beam hardening effect and detector response. Based on the relationship between (a) and (c), the $(\mu t)_{\text{meas}}$ measured with ERPCD can be converted into $(\mu t)_{\text{cor}}$ as shown by the direction of red arrows in the left panel. This correction means that both effects can be corrected simultaneously and polychromatic X-rays can be treated as monochromatic X-rays. This procedure was applied to virtual objects having $Z_{\text{eff}} = 5.0\text{–}15.0$ with $\rho t = 0\text{–}150$ g/cm^2 at intervals of 0.1 g/cm^2 to create the correction curves shown in Fig. 11. Figure 11(a) and 11(b) correspond to the low energy bin (32–40 keV) and high energy bin (40–50 keV), respectively. The solid line shows monochromatic X-rays, and the broken lines show polychromatic X-rays measured with an ERPCD for Z_{eff} between 5 and 15. The difference between the monochromatic X-ray and polychromatic X-ray shows the amount of the correction, and we can see a trend that the correction amount is larger for lower energy bin and higher Z_{eff}.

This curve can be used to simultaneously correct for the beam hardening effect and detector response taking into consideration Z_{eff}. Although we can apply the correction accurately under the assumption that the Z_{eff} of an object is known, it is a rare case that we know the Z_{eff} of an object in advance. To make the best use of the correction curve, we propose to determine Z_{eff} while estimating the amount of correction in later sections.

4   Material Identification Method Leading to Innovations in Imaging Technology

In this section, we would like to introduce our material identification method. The biggest advantage of our method is that the beam hardening effect and detector response can be corrected appropriately by taking into consideration the Z_{eff} of an object. This correction realizes the ideal analysis in which polychromatic X-rays measured with the ERPCD can be treated as those equivalent to monochromatic X-rays. As an application example showing the effectiveness of this analysis, we succeeded not only in identifying the Z_{eff} of an object but also in extracting ρt related to soft-tissue and bone. In the following section, we introduce the proposed method to derive Z_{eff} and ρt values of soft-tissue and bone. Then demonstrate the importance of correction for the beam hardening effect and detector response. Finally show the feasibility of our method visually using images produced with proto-type ERPCD.

4.1   Novel Method to Derive Effective Atomic Number Image and Extract Mass Thickness Images Related to Soft-Tissue and Bone

We will introduce a method to derive the Z_{eff} and extract the ρt values related to soft-tissue and bone. Figure 12 shows a diagram of the method. Assuming that a bilayer structure consisting of "a" and "b" is measured using an ERPCD having several energy bins, we perform the experiments without and with the presence of objects, as shown in the upper center panel. In the current system, we adopted to use two energy bins. Steps Fig. 12(a) through (f) are applied to derive Z_{eff} value and each ρt value. First, we prepare count values which are measured in the experiments without and with objects as shown in Fig. 12(a). The count values without and with objects correspond to mathematical expressions $\sum_{E_k=E_1}^{E_2} R^{(1)} R^{(2)} \Phi_i (E_k)$ and $\sum_{E_k=E_1}^{E_2} R^{(1)} R^{(2)} \Phi_p (E_k)$, respectively. Next, using these values, attenuation factor $(\mu t)_{meas}$ is calculated as

$$(\mu t)_{meas} = \ln \left(\frac{\sum_{E_k=E_1}^{E_2} R^{(1)} R^{(2)} \Phi_i (E_k)}{\sum_{E_k=E_1}^{E_2} R^{(1)} R^{(2)} \Phi_p (E_k)} \right). \tag{7}$$

This calculation is applied to low and high energy bins, and $(\mu_{low}t)_{meas}$ and $(\mu_{high}t)_{meas}$ are derived as shown in Fig. 12(b). Because $(\mu t)_{meas}$ is affected by the beam hardening effect and detector response, corrections related to various Z_{tent} values which are the values tentatively assigned as the Z_{eff} of an object are performed in each energy bin, namely $(\mu t)_{meas}$ is converted into $(\mu t)_{cor}$ values

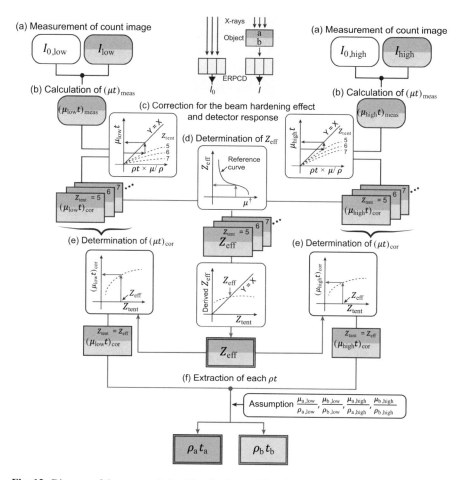

Fig. 12 Diagram of the proposed algorithm for determining the Z_{eff} and ρt values. Assuming that a bilayer structure consisting of elements "a" and "b" is measured using an ERPCD, we perform the experiments without and with the presence of the object. The six steps (a) through (f) are applied to measurement data from two energy bins to derive the Z_{eff} value and the ρt values corresponding to soft-tissue and bone

related to Z_{tent} values 5, 6, 7, . . ., 15, as shown in Fig. 12(c). Then, to extract μ using $(\mu_{low}t)_{cor}$ and $(\mu_{high}t)_{cor}$, the following calculation is performed:

$$\mu^\dagger \equiv \frac{\left(\mu_{high}t\right)_{cor}}{\sqrt{\left(\mu_{low}t\right)_{cor}{}^2 + \left(\mu_{high}t\right)_{cor}{}^2}} = \frac{\left(\mu_{high}\right)_{cor}}{\sqrt{\left(\mu_{low}\right)_{cor}{}^2 + \left(\mu_{high}\right)_{cor}{}^2}}, \tag{8}$$

where $\mu^\dagger$ is a normalized attenuation factor [8, 10–12, 14–16, 18, 19], and we obtain $\mu^\dagger$ values related to Z_{tent} values. In order to derive Z_{eff} from $\mu^\dagger$, a theoretical relationship between $\mu^{\dagger}$ and Z_{eff} is created using database μ/ρ [24]. Using this reference curve, experimentally derived $\mu^{\dagger}$ values can be converted into the corresponding Z_{eff} values as shown in Fig. 12(d). The derived Z_{eff} values are plotted as a function of Z_{tent}, and Z_{eff} of an object can be determined by searching for the intersectional point between the derived Z_{eff} curve and Y = X line. This algorithm to determine the Z_{eff} of an object takes advantage of the following trend. When Z_{tent} is the same as Z_{eff} of an object, the correction can perform appropriately, and the derived Z_{eff} becomes Z_{eff} of an object. On the other hand, when Z_{tent} is different from Z_{eff} of an object, the derived Z_{eff} is also different from Z_{eff} of an object due to improper correction. This means that the intersectional point of the Z_{tent} versus Z_{eff} plot indicates the Z_{eff} of an object, and we can determine Z_{eff} of the object using this algorithm.

Furthermore, taking advantage of the fact that the Z_{eff} of an object can be determined and simultaneously appropriate correction can be applied, we can perform further analysis and extract other properties of an object. We focus on the use of $(\mu t)_{cor}$ with the correction when obtaining the Z_{eff} of an object. Fortunately, when going back to Fig. 12(c), a database of $(\mu t)_{cor}$ values related to Z_{tent} values has been obtained. Now plotting $(\mu t)_{cor}$ as a function of Z_{tent}, the $(\mu t)_{cor}$ related to Z_{eff} can be determined as shown in Fig. 12(e). Because the $(\mu t)_{cor}$ related to Z_{eff} shows that the beam hardening effect and detector response can be corrected completely, ideal analysis based on monochromatic X-ray energy can be carried out. As one valuable application using $(\mu t)_{cor}$, we can extract ρt values of each element by solving simultaneous equations:

$$(\mu_{low} t)_{cor} = \frac{\mu_{a,low}}{\rho_a} \rho_a t_a + \frac{\mu_{b,low}}{\rho_b} \rho_b t_b, \tag{9a}$$

$$\left(\mu_{high} t\right)_{cor} = \frac{\mu_{a,high}}{\rho_a} \rho_a t_a + \frac{\mu_{b,high}}{\rho_b} \rho_b t_b, \tag{9b}$$

where $\frac{\mu_{a,low}}{\rho_a}$ and $\frac{\mu_{b,low}}{\rho_b}$ are μ/ρ values at low energy for "a" and "b," respectively, $\frac{\mu_{a,high}}{\rho_a}$ and $\frac{\mu_{b,high}}{\rho_b}$ are μ/ρ values at high energy for "a" and "b," respectively, additionally $\rho_a t_a$ and $\rho_b t_b$ are ρt values of "a" and "b," respectively. Because we have information related to monochromatic X-ray energy ($\overline{E}$ of Φ_i) under the assumption that bilayer structure consisting of elements "a" and "b" is measured, the μ/ρ values can be assigned by using a well-known database [24]. Consequently, the unknown values are $\rho_a t_a$ and $\rho_b t_b$, and therefore they can be derived by solving the simultaneous equations as shown in Fig. 12(f). This is one analysis that makes full use of our method in which monochromatic X-rays can be virtually obtained through the correction of beam hardening effect and detector response considering the identification of Z_{eff}. We are convinced that there are additional possibilities of discovering novel values using our method.

Next, to more clearly understand the procedure to determine Z_{eff} and ρt, we explain each process using typical results. Figure 13 shows typical results obtained when measuring a bilayer structure ($Z_{eff} = 10.5$) of acrylic ($\rho_{ac}t_{ac} = 2.5$ g/cm^2) and aluminum ($\rho_{Al}t_{Al} = 2.5$ g/cm^2). As shown in Fig. 13(a), we measure the count images obtained in the experiments without and with the presence of an object. Although we use count images from the low energy bin (32–40 keV) and high energy bin (40–50 keV) for analysis, for simplicity, we show the results related to a certain pixel of images from the low energy bin. The pixel values of interest in count images are 50,270 and 4,506 which correspond to the calculations $\sum_{E_k=E_1}^{E_2} R^{(1)} R^{(2)} \Phi_i (E_k)$ and $\sum_{E_k=E_1}^{E_2} R^{(1)} R^{(2)} \Phi_p (E_k)$, respectively. Substituting these values into Eq. (7), $(\mu_{low}t)_{meas}$ is calculated as 2.41 as shown in Fig. 13(b). Because the $(\mu_{low}t)_{meas}$ is affected by the beam hardening effect and detector response, correction is performed as shown in Fig. 13(c). Figure 13(c) shows the correction curves presented by the relationship between $\rho t \times \mu/\rho$ and $\mu_{low}t$. The solid and broken lines show monochromatic X-rays and polychromatic X-rays, respectively. In this graph, monochromatic X-rays are plotted on the Y = X line regardless of Z_{tent}, and the polychromatic X-ray trend depends on Z_{tent}. The difference between monochromatic X-rays and polychromatic X-rays indicates the amount of correction; it is seen that at higher Z_{tent}, the amount of correction is larger. By assuming the Z_{tent} values are between 5 and 15, the $(\mu_{low}t)_{meas}$ can be converted into $(\mu_{low}t)_{cor}$ values which correspond to Z_{tent} values. In this demonstration of a typical case shown in Fig. 13(c), when we set the $Z_{tent} = 15$, the $(\mu_{low}t)_{meas} = 2.41$ is converted into $(\mu_{low}t)_{cor} = 2.66$, as represented by the red arrows. These corrections are also applied to the high energy bin to obtain $(\mu_{high}t)_{cor}$ values. In order to determine Z_{eff}, $\mu^\dagger$ values are calculated by substituting $(\mu_{low}t)_{cor}$ and $(\mu_{high}t)_{cor}$ values to Eq. (8). The upper figure in Fig. 13(d) shows the relationship between $\mu^\dagger$ and Z_{eff}. The broken line shows the reference curve determined using a well-known database [24]. As shown by the arrows, the $\mu^\dagger = 0.57$ corresponding to $Z_{tent} = 15$ can be converted to $Z_{eff} = 11.5$. To determine Z_{eff} of an object, the derived Z_{eff} values are plotted as a function of Z_{tent} as shown in the lower figure in Fig. 13(d). By searching for the intersectional point between the derived Z_{eff} curve and Y = X line, the Z_{eff} of an object is determined to be 10.7. The determined Z_{eff} is in good agreement with $Z_{eff} = 10.5$ which is predetermined as a reference value using a well-known formula [30].

Furthermore, we want to determine ρt by taking advantage of the Z_{eff}. Because $(\mu_{low}t)_{cor}$ values related to Z_{tent} values are already obtained in the correction process as shown in Fig. 13(c), the $(\mu_{low}t)_{cor}$ is plotted as a function of Z_{tent} as shown in Fig. 13(e). Then, the $(\mu_{low}t)_{cor}$ related to an $Z_{eff} = 10.7$ can be determined as 2.57. Because the determined $(\mu_{low}t)_{cor}$ has completely corrected the beam hardening effect and detector response, we can perform the following analysis which can be carried out under the limited conditions of being able to use monochromatic X-rays. Substituting $(\mu_{low}t)_{cor}$ and $(\mu_{low}t)_{cor}$ values into Eqs. (9a) and (9b) under the assumption that the object consists of acrylic and aluminum, namely, μ/ρ values are known, the $\rho_{ac}t_{ac}$ and $\rho_{Al}t_{Al}$ can be determined to be 2.7 g/cm^2 and 2.6 g/cm^2,

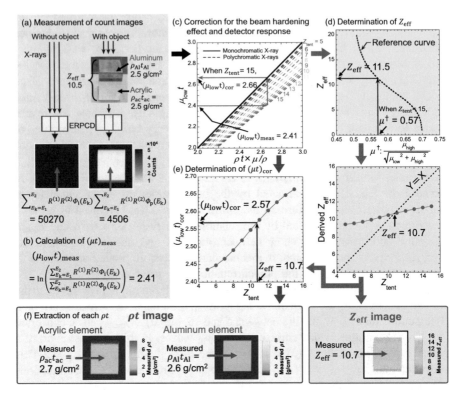

Fig. 13 Typical results to explain each process to derive Z_{eff} and ρt images. Using the data measured for a bilayer structure consisting of acrylic and aluminum, the Z_{eff} and ρt images can be derived by performing the six steps (a) through (f)

respectively, as shown in Fig. 13(f). The determined ρt values are in good agreement with reference ρt values of 2.5 g/cm². These procedures are applied pixel by pixel to create Z_{eff} and ρt images.

To confirm that our method can determine Z_{eff} and ρt values, we present images produced using our proto-type ERPCD. Figure 14 shows images of acrylic, aluminum, and bilayer structures of acrylic and aluminum having (a) $\rho t = 1.0$ g/cm² and (b) $\rho t = 5.0$ g/cm². As shown in the photograph, acrylic, aluminum, and bilayer structures are arranged in order from left to right. In our system, the conventional X-ray image, Z_{eff} image, and ρt image can be obtained. In conventional X-ray images which are equivalent to the image generated by EID, the image density varies depending on Z_{eff} and ρt. When the Z_{eff} is higher with the same ρt, the image density is smaller. Also, when the ρt is larger with the same Z_{eff}, the image density is smaller. It means that the property of an object cannot be identified because the image density changes depending on the two parameters, Z_{eff} and ρt. The Z_{eff} image is presented in color scale, and the region of interest (ROI) of 50×50 pixels is set in the center of each sample. The mean Z_{eff} value measured in ROI is presented below

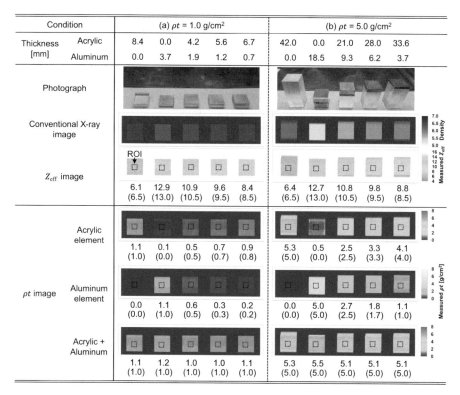

Fig. 14 The results for verifying our novel method to derive Z_{eff} and ρt images using a proto-type ERPCD system. The upper row of images shows photographs of samples having (**a**) $\rho t = 1.0$ g/cm^2 and (**b**) $\rho t = 5.0$ g/cm^2. The acrylic, aluminum, and bilayer structures of acrylic and aluminum are arranged in order left to right. The second row of images shows conventional X-ray images. The third row of images shows Z_{eff} images, and the ROI of 50 × 50 pixels is set in each image. The mean Z_{eff} value measured in ROI is presented below each image, and the theoretical Z_{eff} value is also presented in parentheses. The last three rows of images show ρt images for the acrylic element, the aluminum element, and both the acrylic and aluminum elements. The mean ρt value and theoretical value are presented below. In the Z_{eff} and ρt images, the measured values are in good agreement with theoretical values

each image, and theoretical Z_{eff} is provided in parentheses. The measured Z_{eff} value agrees with the theoretical Z_{eff} with an accuracy of $Z_{eff} \pm 0.4$. In the ρt images, elements for acrylic, aluminum, and acrylic and aluminum are presented. The ρt value for each sample is in good agreement with theoretical ρt with an accuracy of $\rho t \pm 0.5$ g/cm^2. Namely, these results indicate that our method can determine Z_{eff} and extract each ρt regardless of sample.

4.2   Importance of the Correction for the Beam Hardening
Effect and Detector Response

As explained in the above description, correction for both the beam hardening effect and detector response is the most important factor to pay attention for accurate material identification. In order to show the importance of the correction more clearly, we will compare the results with the applied correction and results of intentionally not applied correction. Since we have already described the algorithm including the correction using Fig. 12, we show another algorithm without applying the correction to the Z_{eff} and ρt determination, as shown in Fig. 15. The conditions are similar to those presented in Fig. 12, assuming that a bilayer structure consisting of "a" and "b" is measured as an object using an ERPCD, we analyze the experimental data without and with the presence of the object. As shown in Fig. 15(a) and (b), we measure count images from the low and high energy bins and calculate the corresponding $(\mu t)_{\text{meas}}$ values using Eq. (7). In the determination process of Z_{eff} as shown in Fig. 15(c), the $(\mu t)_{\text{meas}}$ values are substituted into Eq. (8), and the calculated $\mu^{\dagger}$ is converted into Z_{eff}. In the extraction process of ρt values as shown in Fig. 15(d), the $(\mu t)_{\text{meas}}$ values are substituted into Eqs. (9a) and (9b) under the assumption that μ/ρ values are known, and $\rho_a t_a$ and $\rho_b t_b$ can be determined. This algorithm is applied pixel by pixel to generate Z_{eff} and ρt images, and images with and without the corrections are compared.

Figure 16 shows the comparison between results with and without the correction for the beam hardening effect and detector response. The upper row shows a photograph of samples with a $\rho t = 5.0 \text{ g/cm}^2$; acrylic, aluminum, and three bilayer structures which are labeled as (a), (b), (c), (d), and (e), respectively. The left and right columns show the Z_{eff} and ρt images with and without corrections, respectively. It is obvious that the coloration is different in both images when comparing results with and without the correction applied. In order to investigate the difference quantitatively, we plot the mean value measured in the ROI of each image as shown in Fig. 17. The upper and lower figures correspond to the results of Z_{eff} and ρt, respectively. For each object (a) through (e), the reference value, measured value with correction, and measured value without correction are presented on the left, middle, and right, respectively. In the ρt results, the composition ratio of acrylic and aluminum is presented in yellow and green colors, respectively. In both Z_{eff} and ρt results with the correction applied, measured values are in good agreement with corresponding reference values for all samples. On the other hand, in the results without the correction, measured values are not in agreement with the corresponding reference values. This fact means that it is necessary to correct the beam hardening effect and detector response. Further analyzing the results without the correction, we can see that the trend of disagreement is especially noticeable in the samples with higher Z_{eff}. This can be explained by the physics of X-ray attenuation; the amount of correction becomes larger as an Z_{eff} is higher as shown in Figs. 11 and 13(c). A material having a higher Z_{eff} is strongly affected by the beam hardening effect and detector response, and therefore completely different values will be derived unless

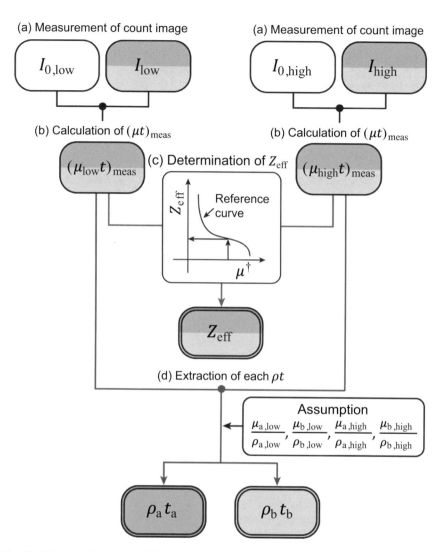

Fig. 15 Diagram of the material identification procedure without including correction of the beam hardening effect and detector response. Steps (a) through (d) are applied to derive Z_{eff} and ρt

the proper correction related to Z_{eff} is applied. In our method, we can perform the correction taking into consideration the Z_{eff} of an object and determine Z_{eff} and ρt values accurately. It was found that our method which takes into account the differences in Z_{eff} of an object plays an important role in achieving accurate material determination.

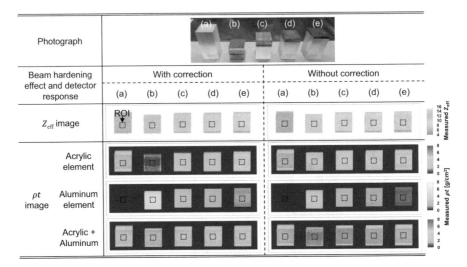

Fig. 16 Demonstration results to explain the importance of the correction for the beam hardening effect and detector response. The upper row shows a photograph of samples having $\rho t = 5.0$ g/cm^2. The acrylic, aluminum, bilayer structures are arranged in order from left to right and each object is label as (a), (b), (c), (d), and (e). The left and right columns show Z_{eff} and ρt images with and without corrections, respectively

4.3 Spillover Effect Expected to Clinical and Industrial Fields

In this section, we will discuss the usefulness of the images by visually presenting images measured with our proto-type ERPCD. Figure 18 shows images produced when measuring fish samples. Figure 18(a) shows photographs of freshwater and saltwater fish. Our system can generate four different images as shown in Fig. 18(b)–(e). Figure 18(b) shows the conventional X-ray image. We can clearly see the bone structure presented by white and the gas bladder shown in black. The X-ray path including bone has a large amount of X-ray attenuation, and therefore image density decreases and it is expressed by white color. On the other hand, the X-ray path including the gas bladder has a relatively small amount of attenuation when compared with that of the surrounding area filled with soft-tissue, and therefore the image density increases and expressed by black color. Although the value of image density does not directly represent the property of an object, we can identify the object by comparing the appearance presented by the contrast between the target and surrounding material with the appearance assumed in advance. This approach for image analysis is similar to that of medical diagnosis; normal/abnormal judgment can be performed qualitatively by analyzing the appearance aspect which can be read from the image based on anatomical knowledge. On the other hand, our system can produce quantitative images to express the functional aspect of an object. Figure 18(c) shows the Z_{eff} image. The bone structure can be seen as high Z_{eff} values of 9–11, with high Z_{eff} accumulations observed around the head, fins, and tail. On the

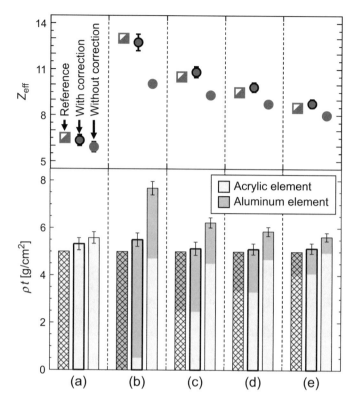

Fig. 17 Results of quantitative analysis of images with and without correction for the samples shown in Fig. 16. The mean values of the ROI set for each image are plotted. The upper and lower figures correspond to the results of Z_{eff} and ρt, respectively. The reference value, measured value with correction, and measured value without correction are compared. It is clearly seen that the correction process is important to obtain a quantitative value precisely

other hand, the gas bladder is not observed because the small X-ray attenuation is not reflected in the calculation of Z_{eff}. The coloration of the Z_{eff} image is different from that of a conventional X-ray image and reflects the physical phenomenon of X-ray attenuation. To extract various useful information from the Z_{eff} image, a new evaluation based on physical perspective is needed. Figure 18(d) shows a ρt image related to soft-tissue; in the analytical procedure, we assume that soft-tissue is equivalent to acrylic. By comparing the conventional X-ray image shown in Fig. 18(b), we can see that the structure of only soft-tissue is extracted clearly. The ρt increases in the thick area of the soft-tissue near the center and decreases in the area with the gas bladder; therefore, the bladder is easily identified as an area where soft-tissue does not exist. Figure 18(e) shows a ρt image related to bone which is assumed to be equivalent to aluminum. The structure of only bone can be extracted completely. By comparing to the Z_{eff} image shown in Fig. 18(c), it can be seen that the high Z_{eff} component is extracted more sharply. As you know, biological

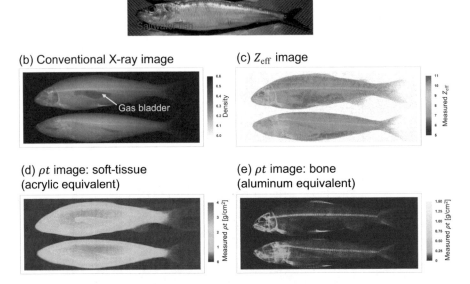

Fig. 18 Produced images of fish samples to demonstrate the feasibility of our method. (**a**) Shows a photograph of a freshwater fish and saltwater fish. In our system, we can obtain (**b**) conventional X-ray image, (**c**) Z_{eff} image, (**d**) ρt image for soft-tissue, and (**e**) ρt image for bone. We hope that these novel images lead to novel diagnosis techniques in the near future

tissue such as the human body, and food, etc., are often composed of soft-tissue and bone, and images corresponding to their separation will help achieve various diagnosis through examination. In the radiography of the lung area, images with bone removed are in great demand to confirm the presence of lesions [39–41]. In the mammography examination, identification of calcification buried in soft-tissue is also necessary [42]. Furthermore, our system can accurately measure the ρt of bone from a bone image. The ρt of bone is synonymous with bone mineral density (BMD) which is used for the diagnostic index of osteoporosis, which indicates that our system may be able to diagnose osteoporosis without using specific equipment (dual-energy X-ray absorptiometry) [43–45]. These are just a few examples of how images can be useful. We think there are various fields in which Z_{eff} and ρt images can be used effectively, so we would like you to consider the usefulness of the images.

Finally, we discuss the spillover effect of the proposed system using Fig. 19. We propose to extract various properties of an object by analyzing X-ray attenuation. Although we focused attention on plain X-ray examination which produces a two-dimensional image, the analysis based on the physics of X-ray attenuation is common to all examinations to produce X-ray images. In the case of photon-

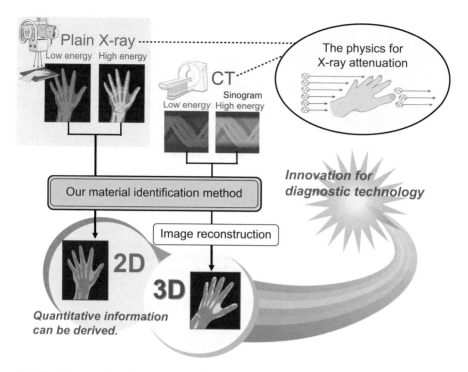

Fig. 19 Spillover effect of the proposed method. Our method based on the physics of X-ray attenuation which is common to plain X-ray and CT, therefore it has the potential to lead to innovations in diagnosis using novel quantitative images in various fields

counting CT, different energy-related sinograms are available. By applying the proposed algorithm to these sinograms and going through the process of image reconstruction, we can obtain a three-dimensional image with quantitative information. Photon-counting CT has been developed by many manufactures, and we are confident that this analysis technology will contribute to technical development. We hope that our system brings innovation to diagnostic technology.

5 Summary

In this chapter, an algorithm for extracting the properties of an object using an ERPCD was described. Based on the principle that the information of the object can be derived from the difference in the X-ray attenuation depending on X-ray energy, we analyzed polychromatic X-ray energies measured with an ERPCD. The most important issue of the analysis is that the measured energy information is distorted by the beam hardening effect and detector response. We described these effects from a viewpoint based on physics and presented a methodology to simultaneously correct for these effects. This correction can treat polychromatic X-rays as those equivalent to monochromatic X-rays, which results in the accomplishment of correct analysis in principle. In order to demonstrate our method, we succeeded in the derivation of Z_{eff} image and extraction of soft-tissue and bone images by applying our algorithm to each pixel of an image measured with a proto-type ERPCD. We hope that the content described in this chapter opens up possibilities for the use of photon-counting techniques in medical and industrial fields.

Acknowledgments The description in this chapter was partially supported by collaborative research between Kanazawa University and JOB CORPORATION (https://www.job-image.com/), Japan. The authors wish to express gratitude to members of JOB CORPORATION, Dr. Shuichiro Yamamoto, Mr. Masahiro Okada, Mr. Fumio Tsuchiya, Mr. Daisuke Hashimoto, Mr. Yasuhiro Kuramoto, and Mr. Masashi Yamasaki for their valuable contributions. We wish to thank Mr. Takumi Asakawa who belongs to GE Healthcare Japan for his important research results when he belonged to graduate school in Kanazawa University, Japan. We would like to thank Dr. Yuki Kanazawa who belongs to Tokushima University, Japan for discussing the feasibility of our research from a clinical point of view. We would also like to thank Dr. Yoshie Kodera and Dr. Shuji Koyama who belong to Nagoya University, Japan for discussing the clinical application of a photon-counting detector.

References

1. Pisano E. D., et al. 2000. Image Processing Algorithms for Digital Mammography: A Pictorial Essay. Radiographics. 20 (5), 1479–1491. doi: https://doi.org/10.1148/radiographics.20.5.g00se311479.
2. Prokop M, et al. 2003. Principles of Image Processing in Digital Chest Radiography. Journal of Thoracic Imaging. 18 (3) 148–164.
3. Kermany D. S., et al. 2018. Identifying Medical Diagnoses and Treatable Diseases by Image-Based Deep Learning. Cell. 172 (5), 1122–1131. doi: https://doi.org/10.1016/j.cell.2018.02.010.
4. Leng S., et al. 2019. Photon-counting detector CT: system design and clinical applications of an emerging technology. Radiographics. 39 (3), 729–743. doi: https://doi.org/10.1148/rg.2019180115.
5. Taguchi K. and Iwanczyk J.S. 2013. Vision 20/20: Single photon counting x-ray detectors in medical imaging. Med. Phys. 40 (10), 100901–1–19. doi: https://doi.org/10.1118/1.4820371.
6. Willemink M.J., et al. 2018. Photon-counting CT: technical principles and clinical prospects. Radiology. 289 (2), 293–312. doi: https://doi.org/10.1148/radiol.2018172656.
7. Asakawa T., et al. 2019. Importance of Considering the Response Function of Photon Counting Detectors with the Goal of Precise Material Identification. IEEE Nuclear Science Symposium and Medical Imaging Conference (NSS/MIC), 1–7. doi: https://doi.org/10.1109/NSS/MIC42101.2019.9059844.
8. Hayashi H., et al. 2015. A fundamental experiment for novel material identification method based on a photon counting technique: Using conventional X-ray equipment. IEEE Nuclear Science Symposium and Medical Imaging Conference (NSS/MIC), 1–4. doi: https://doi.org/10.1109/NSSMIC.2015.7582027.
9. Hayashi H., et al. 2017. Response functions of multi-pixel-type CdTe detector - Toward development of precise material identification on diagnostic X-ray images by means of photon counting-. Proc. SPIE 10132, Medical Imaging. 10132. 1–18. doi: https://doi.org/10.1117/12.2251185.

10. Kimoto N., et al. 2017. Precise material identification method based on a photon counting technique with correction of the beam hardening effect in X-ray spectra. Appl Radiat Isot. 124, 16–26. doi: https://doi.org/10.1016/j.apradiso.2017.01.049.

11. Kimoto N., et al. 2017. Development of a novel method based on a photon counting technique with the aim of precise material identification in clinical X-ray diagnosis. Proc. SPIE, Medical Imaging. 10132. 1–11. doi: https://doi.org/10.1117/12.2253564.

12. Kimoto N., et al. 2017. Novel material identification method using three energy bins of a photon counting detector taking into consideration Z-dependent beam hardening effect correction with the aim of producing an X-ray image with information of effective atomic number. IEEE Nuclear Science Symposium and Medical Imaging Conference (NSS/MIC), 1–4. doi: https://doi.org/10.1109/NSSMIC.2017.8533059.

13. Kimoto N., et al. 2018. Reproduction of response functions of a multi-pixel-type energy-resolved photon counting detector while taking into consideration interaction of X-rays, charge sharing and energy resolution. IEEE Nuclear Science Symposium and Medical Imaging Conference (NSS/MIC), 1–4. doi: https://doi.org/10.1109/NSSMIC.2018.8824417.

14. Kimoto N., et al. 2019. Feasibility study of photon counting detector for producing effective atomic number image. IEEE Nuclear Science Symposium and Medical Imaging Conference (NSS/MIC). 1–4. doi: https://doi.org/10.1109/NSS/MIC42101.2019.9059919.

15. Kimoto N., et al. 2021. Effective atomic number image determination with an energy-resolving photon-counting detector using polychromatic X-ray attenuation by correcting for the beam hardening effect and detector response. Applied Radiation and Isotopes. 170, 109617. doi: https://doi.org/10.1016/j.apradiso.2021.109617.

16. Kimoto N., et al. 2021. A novel algorithm for extracting soft-tissue and bone images measured using a photon-counting type X-ray imaging detector with the help of effective atomic number analysis. Applied Radiation and Isotopes. doi: https://doi.org/10.1016/j.apradiso.2021.109822

17. Reza., et al. 2017. Semiconductor radiation detectors: technology and applications. CRC Press, New York, 85–108.

18. Hayashi H., et al. 2019. Advances in Medicine and Biology. Nova Science Publishers, Inc., New York, 1–46.

19. Hayashi H., et al. 2021. Photon Counting Detectors for X-ray Imaging: Physics and Applications. Springer, Switzerland. 1–119. doi: https://doi.org/10.1007/978-3-030-62680-8.

20. Samei E. and Flynn J. M. 2003. An experimental comparison of detector performance for direct and indirect digital radiography systems. Med. Phys. 30 (4). 608–622. doi: https://doi.org/10.1118/1.1561285.

21. Spekowius G., Wendler T. 2006. Advances in Healthcare Technology: Shaping the Future of Medical Care. Springer, Netherlands. 49–64.

22. Knoll G.F., 2000. Radiation detection and measurement. John Wiley & Sons, New York, 1–802.

23. Heismann B. J., et al. 2003. Density and atomic number measurements with spectral x-ray attenuation method. Journal of Applied Physics. 94 (3), 2073–2079. doi: https://doi.org/10.1063/1.1586963.

24. Hubbell J.H., 1982. Photon mass attenuation and energy-absorption coefficients. Int J Appl Radiat Isot. 33 (11), 1269–1290. doi: https://doi.org/10.1016/0020-708X(82)90248-4.

25. Brooks R.A. and Di Chiro G. 1976. Beam hardening in x-ray reconstructive tomography. Phys Med Biol. 21 (3), 390–398. doi: https://doi.org/10.1088/0031-9155/21/3/004.

26. Good M. M., et al. 2011. Accuracies of the synthesized monochromatic CT numbers and effective atomic numbers obtained with a rapid kVp switching dual energy CT scanner. Med. Phys. 38 (4), 2222–2232. doi: https://doi.org/10.1118/1.3567509.

27. Johnson T. R. C. 2012. Dual-Energy CT: General Principles. American Journal of Roentgenology. 199 (5), S3–S8. doi: https://doi.org/10.2214/AJR.12.9116.

28. Tatsugami F., et al. 2014. Measurement of electron density and effective atomic number by dual-energy scan using a 320-detector computed tomography scanner with raw data-based analysis: a phantom study. J Comput Assist Tomogr. 38 (6), 824–827. doi: https://doi.org/10.1097/RCT.0000000000000129.

29. Wang X., et al. 2011. Material separation in x-ray CT with energy resolved photon-counting detectors. Med. Phys. 38 (3), 1534–1546. doi: https://doi.org/10.1118/1.3553401.
30. Spiers F.W. 1946. Effective atomic number and energy absorption in tissues. The British journal of radiology. 19 (218), 52–63. doi: https://doi.org/10.1259/0007-1285-19-218-52.
31. Birch R., Marshall M. 1979. Computation of bremsstrahlung X-ray spectra and comparison with spectra measured with a Ge(Li) detector. Phys Med Biol. 24 (3), 505–517. doi: https://doi.org/10.1088/0031-9155/24/3/002.
32. Hsieh S.S., et al. 2018. Spectral resolution and high-flux capability tradeoffs in CdTe detectors for clinical CT. Med. Phys. 45 (4), 1433–1443. doi: https://doi.org/10.1002/mp.12799.
33. Otfinowski P. 2018. Spatial resolution and detection efficiency of algorithms for charge sharing compensation in single photon counting hybrid pixel detectors. Nuclear Inst. and Methods in Physics Research, A. 882, 91–95. doi: https://doi.org/10.1016/j.nima.2017.10.092.
34. Taguchi K., et al. 2018. Spatio-energetic cross-talk in photon counting detectors: Numerical detector model (PcTK) and workflow for CT image quality assessment. Med. Phys. 45(5), 1985–1998. doi: https://doi.org/10.1002/mp.12863.
35. Trueb P., et al. 2017. Assessment of the spectral performance of hybrid photon counting x-ray detectors. Med. Phys. 44 (9), e207–e214. doi: https://doi.org/10.1002/mp.12323.
36. Zambon P., et al. 2018. Spectral response characterization of CdTe sensors of different pixel size with the IBEX ASIC. Nuclear Inst. and Methods in Physics Research, A. 892, 106–113. doi: https://doi.org/10.1016/j.nima.2018.03.006.
37. Hirayama H., et al. 2005. The EGS5 Code System. KEK Report. 2005-8 SLAC-R-730. 1–418.
38. Sasaki M, et al. 2019. A novel mammographic fusion imaging technique: the first results of tumor tissues detection from resected breast tissues using energy-resolved photon counting detector. Proc. SPIE. 1094864. doi: https://doi.org/10.1117/12.2512271.
39. Kuhlman J. E., et al. 2006. Dual-energy subtraction chest radiography: what to look for beyond calcified nodules. Radiographics. 26 (1), 79–92. doi: https://doi.org/10.1148/rg.261055034.
40. MacMahon H., et al. 2008. Dual energy subtraction and temporal subtraction chest radiography. J Thorac Imaging, 23 (2), 77–85. doi: https://doi.org/10.1097/RTI.0b013e318173dd38.
41. McAdams H.P., et al. 2006. Recent advances in chest radiography. Radiology. 241 (3), 663–683. doi: https://doi.org/10.1148/radiol.2413051535.
42. Kappadath S. C. and Shaw C. C. 2004. Quantitative evaluation of dual-energy digital mammography for calcification imaging. Phys Med Biol. 49 (12) 2563–2576. doi: https://doi.org/10.1088/0031-9155/49/12/007.
43. Blake M. G. and Fogelman I., 1997. Technical principles of dual energy X-ray absorptiometry. Seminars in Nuclear Medicine. 27 (3), 210–228. doi: https://doi.org/10.1016/S0001-2998(97)80025-6.
44. Cullum D. I., et al. 1989. X-ray dual-photon absorptiometry: a new method for the measurement of bone density. The British Journal of Radiology. 62 (739), 587–592. doi: https://doi.org/10.1259/0007-1285-62-739-587.
45. Theodorou J. D., et al. 2002. Dual energy X-ray absorptiometry in diagnosis of osteoporosis: basic principles, indications, and scan interpretation. Comprehensive Therapy. 28 (3), 190–200. doi: https://doi.org/10.1007/s12019-002-0028-6.

Absolute Density Determination and Compositional Analysis of Materials by Means of CdZnTe Spectroscopic Detectors

N. Zambelli, G. Benassi, and S. Zanettini

1 Introduction

X-ray technology has been used to inspect the internal part of objects since decades thanks to the property of X-rays of travel across matter. It has been firstly employed in health care for medical diagnostics, later in industry as a contactless technique for examining manufactures, and it has recently become essential for homeland security, especially for baggage inspection in airports or cargo containers check in seaports.

In this Chapter we will focus on the ultimate frontier of X-ray absorption technology, namely Hyperspectral X-ray Absorptiometry (HXA), as an instrument for the determination of absolute density and compositional analysis of materials. The term "absorptiometry" is usually associated with Dual-Energy X-ray Absorptiometry (DXA, or DEXA), which is the most widely known technique for bone densitometry; DXA is essential for screening of osteoporosis [1], and it is nowadays widely used for baggage scanners [2]. The adjective "hyperspectral" (a superlative of "spectral") refers to the spectroscopic ability of photon-counting detectors (PCDs) to discriminate the energy of every incident X-ray photon with extremely high accuracy. Indeed, the advent of direct conversion single photon-counting detectors has given the possibility to measure the X-ray attenuation of the inspected sample as a function of energy with outstanding energy resolution. Our attention is addressed to semiconductor detectors characterized by direct conversion of radiation into electric signal, such as silicon, CdTe and CdZnTe (CZT) sensors.

The acquisition of the transmitted X-ray spectrum (through the object) with a very fine binning gives access to specific information about the absorber traversed

N. Zambelli · G. Benassi · S. Zanettini (✉)
Due2lab srl, Scandiano (RE), Italy
e-mail: nicola.zambelli@due2lab.com; giacomo.benassi@due2lab.com;
silvia.zanettini@due2lab.com

107

by the beam, in particular its density and its elemental composition. Hyperspectral X-ray Absorptiometry basically relies on the accurate measurement of the energy dependence of the material attenuation coefficient, using a set-up composed by a polychromatic source in combination with a photon-counting detector, aligned in transmission configuration. The use of hyperspectral technology is a step forward with respect to multi-energy techniques, which subdivide the broadband spectrum in few energy bands, by using static or dynamic thresholds [3–6]. The aim of this Chapter is to explain how to use HXA for the direct measurement of the absolute density ρ (g/cm^3) of the material under investigation, and how to extract its refined elemental composition.

The determination of the absolute value of density of a certain material is not a trivial task. Non-destructive tests (NDTs) employing X-rays are already extensively used for several industrial applications, from internal inspection of food products [7] to defectiveness analysis of composites materials, for example, in the aircraft industry [8, 9]. However, traditional X-ray NDTs techniques generally employ X-ray detectors in the so-called integrating mode, which generates "grey-scale" radiographic images of the object. From such radiographs, only a relative value of the material density can be extracted. Consequently, a time-consuming calibration procedure with samples of known density and thickness is needed to retrieve the effective density of the object expressed in g/cm^3 units. Any variation in the material elemental composition which are not considered during the calibration procedure may cause a substantial mistake in the density value determination.

Hyperspectral X-ray Absorptiometry, instead, allows the direct measurement of the material density without the need of calibration, and will significantly boost the quantity of information obtainable with the same type of scan. This will lead to several advantages in monitoring the compliance of semi-finished product throughout the production line and the conformity of manufactures at the end. By measuring density in real-time, it is possible to give an early feedback of potential product defects or deviations from standard, and to immediately adjust production parameters and avoid material waste. For example, the tracing of density inhomogeneities in "green" (unfired) ceramic tiles after the pressing step is crucial, since such deviations strongly affect the robustness of the final product, especially for ceramic slabs which are few meters square large. Density non-uniformities should be detected before slabs firing, in order to adjust the applied pressure and possibly reuse raw powders before firing [10]. A contactless method to measure the absolute value of density allows to overcome the drawbacks of destructive analyses (such as mercury balance) that are nowadays employed. For these reasons, HXA is of strategical importance for real-time in-line monitoring of industrial production and will likely become a key technology for several industrial sectors, such as manufacturing of pressed wood panels production or ceramic tiles, where the density measurement of the manufacture is fundamental to inspect its mechanical properties.

Since X-ray mass attenuation coefficients are specific to each chemical element, Hyperspectral X-ray Absorptiometry also allows to evaluate the elemental composition of the sample. HXA is the only technique that allows the measurement of elemental composition of the whole volume of the sample, in contrast to X-ray fluorescence (XRF) [11] or Compton backscattering that inspect only the volume of material close to the exposed surface. K-edge absorption technique [12–14] is capable to inspect the sample internal volume, but it is better suited to trace high atomic number elements embedded in a low atomic number elements matrix, typically precious metals or rare earths inside ore samples, or contrast agents inside human body. K-edge technique is not usable when the material is mainly composed by light atoms such as carbon, oxygen, or hydrogen (typically organic materials), because the absorption edges of these elements are at very low energies (<1 keV), thus undetectable.

The experimental results presented in this Chapter have all been carried out by means of a single-pixel photon-counting CZT detector, even though this choice is quite unusual in the field of spectral X-ray attenuation, where, more often, the technology makes use of linear arrays or matrixes of pixels. Indeed, multi-pixel sensors facilitate and speed up imaging for those applications where an image of the object in three dimensions, especially human body in medical imaging applications, is needed. For this reason, they are usually employed for X-ray computed tomography (XCT or more commonly CT). CT scanners are also used in the wood sector to analyze tree logs before sawmill cutting, in order to localize wood knots or other internal defects for the optimization of wood boards cut [15]. Traditional CT systems provide 3D images that show the morphology of the absorber expressed as maps of arbitrary units, called Hounsfield Units (HU). Multi-energy technology is often used in combination with CT [5, 16, 17] and provides better material separation with respect to traditional CT. Recently the first attempts of spectral CT have been tested and published [18–21].

Nevertheless, the use of a point-like measurement configuration with collimation system is the most straightforward way to exploit HXA. In this configuration, the X-ray beam and the detector are aligned to each other on the two sides of the sample and collimated on a specific point of the inspected object; in this way, the local absolute density of that unit volume of the sample, also called voxel, will be determined. By moving the sample on an x–y plane perpendicular to the beam direction, it is possible to map its density in two dimensions. This "bottom-up" approach facilitates the correction of the distortions introduced by the detection system (which is real, and not ideal!) and brings experimental data closer to theory, resulting in a simplification of the problem and better understanding of HXA technology and its potential. For instance, an example of hyperspectral computed tomography with a CdTe point sensor for enhanced material identification has been demonstrated [22]. The final breakthrough for HXA would be the capability to record a 3D image in which each voxel is associated with the object absolute density value and chemical composition. The possibility to have this technology usable for medical imaging of patients will be a real step forward [20].

The X-ray energy range employed in HXA must be adapted to the object to be investigated, and the X-ray spectroscopic sensor must be adapted to that range. X-ray medical scanners, for example, obtain high-resolution images of soft tissue biological samples trying to minimize the radiation dose delivered to the patient; thus, they employ lowest possible X-ray energy (20 keV is the typical mammography energy range) and intensity. The low X-ray energy range is also suitable, for example, for the scanning of composite pieces for aircrafts, which are made of lightweight materials like carbon, paper, epoxy polymers. In the field of low energy X-ray transmission radiography, silicon detectors have already demonstrated their outstanding capabilities. This is due to their high detection efficiency in the $5 \div 20$ keV energy range and to the extremely low dark-current that provides exceptional sensitivity. However, the inspection of thick and dense objects requires instead the use of high energy X-rays and the use of high Z sensing materials, such as CdZnTe (CZT), is essential to detect such high energies. CZT (and CdTe as well) has a much larger radiation stopping power than silicon, thanks to its high density and to the high atomic number of its composing elements. For thick and heavy samples, low energy X-rays are completely darkened by the object self-absorption, thus it is necessary to work in the $30 \div 100$ keV range. A typical 300-μm thick silicon detector is almost transparent to 50 keV radiation (it absorbs below 25% of the total radiation above 20 keV, and 3% only at 50 keV), while 2 mm-thick CZT stops 100% of radiation up to 80 keV, and 85% at 100 keV. The use of higher X-ray energies leads to additional challenges related to the appearance of spectrum distortions due to physical phenomena happening inside the detector [23]. Since these features are strictly related to the detector response function and not to the inspected object, their correction is fundamental for ensuring a correct quantitative X-ray absorptiometry analysis, as it will be discussed in Sect. 4. The same correction is necessary for CdTe detectors when used for quantitative analysis [24].

The Chapter is structured as subsequently explained. Section 2 gives an overview of X-ray attenuation theory. Section 3 explains how to measure the area density (g/cm^2) and the absolute density (g/cm^3) by means of Hyperspectral X-ray Absorptiometry, while Sect. 4 illustrates the need to correct the CZT detector response function to ensure a correct quantitative analysis of blank and transmitted spectrum. A detailed description of laboratory set-up, and its adaptation to work in a real industrial environment, is given in Sect. 5. Several case studies showing the analysis of different materials are illustrated in the second part of the Chapter. The materials examined are divided in stoichiometric and non-stoichiometric compounds. Density determination in the non-stoichiometric case, when chemical composition of the material is not precisely known, is indeed more complex. Section 6 treats the analysis of stoichiometric plastic polymers such as polyethylene (PE), polycarbonate (PC), and polytetrafluoroethylene (PTFE), with an insight on the fast acquisition mode required for in-line industrial applications (Sect. 7). Section 8 discusses density estimation for a non-stoichiometric compound like wood. When chemical composition is only partially known, HXA can be used to better quantify the weight fractions of each chemical element, and potentially discover contaminants in the blend, in a process that we refer to as *elemental composition refinement*, which is described in Sect. 9.

2   X-Ray Attenuation Theory

When an incident X-ray beam of intensity I_0 is pointed towards an object, the measured intensity I behind that object for photons of energy E (acquired along the incident direction) is related to the linear attenuation coefficient μ of the material according to:

$$I(E) = I_0(E) \cdot e^{-\mu(E) \cdot x}, \tag{1}$$

which can be rewritten as:

$$\ln \frac{I_0(E)}{I(E)} = \mu(E) \cdot x \tag{2}$$

where x is the distance that beam travels through the material (cm), μ is the linear attenuation coefficient for a given material (cm^{-1}), which is a function of energy E, while I_0 and I are, respectively, the unattenuated incident intensity and the transmitted intensity, both measured as number of photon counts. If we consider a planar object and a set-up configuration in which the X-ray beam is perpendicular to the object surface, x corresponds to the material thickness.

The linear attenuation coefficient μ is an expression of the removal of X-ray photons from the incident beam by interactions of the photons with electrons of the material probed. These interactions can be absorption (removal from the beam) or scattering by the material (change of direction with associated energy reduction). Both phenomena affect the transmitted beam and are related to the material density of the object and the atomic number of the chemical elements composing the object. X-ray photons interact more with materials which are dense and made of elements with high atomic number Z. Indeed, the linear attenuation coefficient μ is proportional to material density ρ (g/cm^3) can be expressed as in Eq. (3) by defining mass attentuation coefficient μ_m:

$$\mu(E) = \left(\frac{\mu(E)}{\rho} \right) \cdot \rho = \mu_m(E) \cdot \rho \tag{3}$$

Mass attenuation coefficients μ_m (cm^2/g) strongly depend on photon energy. They are specific for each chemical element and known for all elements in a wide energy range. The National Institute of Standards and Technology (NIST) provides the values of μ_m for elements from $Z = 1$ to $Z = 92$ in the range 1 keV $\div$ 20 MeV [25].

For a fixed energy E' and a material composed by a single chemical element (graphite, for example), Eq. (2) is:

$$\ln \frac{I_0\left(E'\right)}{I\left(E'\right)} = \rho \cdot x \cdot \mu_m\left(E'\right), \tag{4}$$

For a fixed energy E' and a material composed by i chemical elements:

$$\ln \frac{I_0\left(E'\right)}{I\left(E'\right)} = \rho \cdot x \cdot \left[\sum_i w_i \cdot \mu_{m_i}\left(E'\right)\right], \qquad (5)$$

where w_i is the weight fraction of the i-th element inside the material. The mass attenuation coefficient of a chemical compound is the sum of the single element mass attenuation coefficients, each multiplied by the weight fraction of the i-th element. Density and thickness of the material are energy independent.

For variable energy, the photon attenuation of the sample is expressed by the following Equation:

$$\ln \frac{I_0(E)}{I(E)} = \rho \cdot x \cdot \left[\sum_i w_i \cdot \mu_{m_i}(E)\right]. \qquad (6)$$

The logarithm of the ratio $I_0(E)/I(E)$ in a certain energy range from E_1 to E_2 is known as *photon attenuation curve* and can be experimentally measured by using a polychromatic X-ray source and a spectroscopic energy-resolving detector.

According to Eq. (6), it is possible to determine the value of density ρ in units of g/cm^3 by measuring I and I_0 in a certain energy range from E_1 to E_2. This is possible if material thickness x and material elemental composition w_i are known, since mass attenuation coefficients μ_{mi} are tabulated and can be found in the NIST database [25]. The accuracy on density determination also depends on the precision of the measurement of thickness.

In the next paragraph we will detail how to measure the photon attenuation curve of a sample by means of Hyperspectral X-ray Absorptiometry and how to extract the absolute value of density.

3   Absolute Density Determination

The intensity of X-ray absorption is related to the material density ρ, which describes the mass distribution (dm/dV), and to the number of electrons of the atoms composing the material (atomic number, Z). The X-ray beam transmitted through the sample contains all the information about the absorber and a quantitative analysis of the X-ray photon attenuation allows to completely extract this information. Every single photon of both the blank spectrum and the transmitted spectrum must be counted and allocated to a specific energy bin, to correctly estimate how many photons of the incident beam have been taken away by the object, due to absorption or scattering. The number of photons absorbed by the object for each energy bin is given by the number of incident photons minus the number of transmitted photons. Therefore, a HXA analysis requires two measurement steps, as schematically illustrated in Fig. 1a. Step 1 represents the acquisition of the

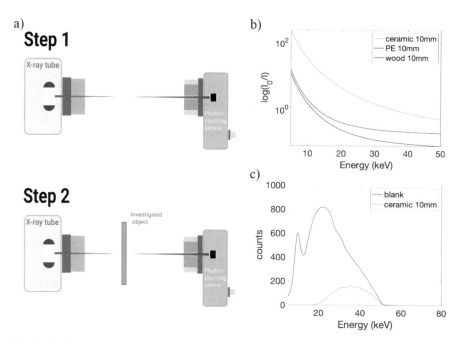

Fig. 1 (a) Hyperspectral X-ray Absorptiometry measurement concept: step 1 is the blank spectrum acquisition; step 2 is the transmitted spectrum acquisition. (**b**) Simulated photon attenuation curves in semilog scale for wood, polyethylene (PE), and ceramics of equal thickness. (**c**) Blank incident spectrum (gray) and transmitted spectrum (yellow) through 6 mm of ceramic material (simulated). (Modified from [34] with permission)

blank spectrum, $I_0(E)$, i.e., the spectrum of the X-ray polychromatic incident beam. Step 2 represents the spectrum acquisition of the beam transmitted through the sample, $I(E)$. The logarithm of the ratio $I_0(E)/I(E)$ as a function of energy, i.e., the photon attenuation curve (Eq. 6), depends on density, thickness, and elemental composition of the absorber. Density ρ and thickness x of the material are energy independent, while mass attenuation coefficients μ_m are strongly energy-dependent and determine the different trends of the photon attenuation curve for different compounds. To give an example, Fig. 1b illustrates the simulated photon attenuation curve in semilog scale for three different materials with equal thickness (10 mm): wood, polyethylene (PE), and ceramics. The average approximate density of these three materials is, respectively, 0.5 g/cm^3, 0.9 g/cm^3, and 2.2 g/cm^3. The most absorbing material is the one with highest ln(I_0/I) ratio; in this example, the most absorbing material is ceramics, and this is due to its significantly higher density, and to the fact that it contains silicon, aluminum, and oxygen, and other several elements with higher Z in smaller fraction. To appreciate the strong attenuation of a ceramic sample, due to its high density and to the presence of high Z elements, we report in Fig. 1c the transmitted spectrum of 10 mm of ceramic alongside with the blank spectrum used as incident X-ray radiation. Wood is the least attenuating material among the three. The photon attenuation curves of PE and wood, even though material densities are quite similar, differ in their shape because of diverse material composition: polyethilene is made of carbon and hydrogen (C_2H_4), while wood is made of carbon, hydrogen, and oxygen. The semilog scale emphasizes the different trend of each photon attenuation curve, which can be considered as a "fingerprint" of that specific material. The data reported in Fig. 1b, c are simulated.

The product of density and thickness $\rho \cdot x$ in Eq. (6) is called *area density* ρ_A and it is expressed in g/cm^2. The area density of a two-dimensional object corresponds to its mass per unit area. In the paper and textile industries, for example, it is called grammage and is expressed in grams per square meter. If the atomic weight fractions of the material w_i are known, Hyperspectral X-ray Absorptiometry can deliver a very precise value of area density ρ_A. In fact, Eq. (6) can be also expressed as:

$$\rho_A = \rho \cdot x = \frac{\ln \frac{I_0(E)}{I(E)}}{\left[\sum_i w_i \cdot \mu_{m_i}(E) \right]} \tag{7}$$

In Eq. (7) the experimental photon attenuation curve is divided by the sum of the mass attenuation coefficients, this results in a series of datapoints scattered around the average value of $\rho_A = \rho \cdot x$, which must be constant as a function of energy (slope equal to zero). A linear fit of this datapoints should return a straight line with intercept equal to ρ_A. A non-zero slope of the linear fit is an indication that the real sample chemical composition differs from the hypothesized composition.

To each point of the inspected object is associated a value of ρ_A, and this produces an x–y spatial distribution map of area density. In Fig. 2 we show an example of area density map. The sample is a 3D-printed disc made of ABS (Acrylonitrile Butadiene Styrene). ABS can be extruded or injection molded (for instance, it is the material the Lego bricks are made of) and it is one of the most common polymers for 3D printing. Nowadays 3D printing technique is widely used for prototyping, and it is important to check the prototype conformity and robustness at the end of printing process. The analyzed sample is an ABS disk with a diameter of 17 mm and a thickness of 4 mm, which has been printed on purpose with a special honeycomb internal structure (not visible from outside) to emulate differences in area density. For HXA analysis, the disk is mounted inside a sample holder with three small screws, as shown in the photograph (Fig. 2a). The x–y map has been recorded by moving the sample with a two-axis motor along the x and y-axis with 0.5 mm steps. Figure 2b reports the X-ray attenuation spatial map, as derived from integration of the transmitted spectrum of each single point. White zones such as the gap between ABS disk and sample holder are completely hollow, thus totally transparent to X-rays, while dark zones are strongly attenuating, e.g., the three metallic screws. Even though the internal honeycomb structure of the disk is visible in Fig. 2b, the integrated map of the transmitted beam expresses only relative variations of the X-ray absorption of different zones of the sample. Therefore, Fig. 2b is comparable with a typical gray-scale radiography obtained

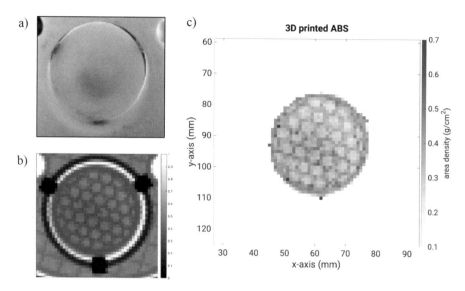

Fig. 2 Determination of the area density of a 3D-printed sample made of ABS. (**a**) Sample photograph; (**b**) Gray scale map of integrated transmitted intensity; (**c**) Area density map of a 3D-printed sample made of ABS

by traditional instruments (or with an image captured by a multi-energy system by integrating over a specific energy band).

On the contrary, the area density map in Fig. 2c clearly quantifies samples variations of area density, which ranges from 0.23 g/cm^2 at the center of the internal voids to 0.5 g/cm^2 in correspondence of honeycomb walls. An accurate blank spectrum acquisition has been necessary to measure the area density map. An approximate ratio of two exists between the area density of walls and voids. By supposing that the density of ABS printed polymer is uniform, we can infer that the area density variation is given by a twofold difference on thickness, which means that X-rays optical path inside ABS in correspondence of the honeycomb voids is roughly the half with respect to the disk total width, namely 2 mm instead of 4 mm (1 mm thick disk top surface and 1 mm thick disk bottom surface). This indeed corresponds to the 3D printer slice patterning. Moreover, the disk was affected by a defective area (visible in the photo as a brownish area close to the center of the disk) due to some inhomogeneity during 3D printing: interestingly this "burned" spot is observable in the area density map as a slightly denser area.

The value of absolute density ρ can be successively obtained dividing area density ρ_A by the object thickness x. The density of 3D printed ABS, as determined from the points corresponding to honeycomb walls, results approximately equal to 1.25 g/cm^3.

Thanks to this method the density of a flat object of known thickness (or, more generally, of an object of known geometry) can be measured locally point by point.

4 Correction of the Detector Response Function

The use of a spectroscopic detector operating in single photon-counting mode is a prerequisite for quantitative X-ray attenuation analysis. The photon-counting detector should ensure most importantly an optimal spectroscopic resolution to properly read the energy of the photons detected.

Besides that, the raw energy spectrum collected by CZT photon-counting sensors is usually distorted by physical interactions and limiting factors of the real detector and specific sensing material. The main spectrum distortions are listed in Table 1. Distortions must be identified and removed to avoid mistakes in the acquisition of the experimental photon attenuation curve.

CZT fluorescence is activated when incident energy is above 26 keV. Cd fluorescence peak appears in the spectrum at 26.7 keV, while Te fluorescence peak appears at 31.8 keV, due to a partial charge deposition of the incoming photon. Fluorescence is a non-negligible distortion, its yield can rise above 80% in CdTe [23]. Escape photons are the by-product of X-ray fluorescence. Escape peaks are generated when the energy of the incident photon is deposited in the material, after having created a fluorescence photon that leaves the detector volume. Thus, if X-ray source is monochromatic, escape peak appears at an energy equal to the incident photon energy minus the fluorescence binding energy. If X-ray source is polychromatic, counts related to escape photons are distributed over a large fraction of the spectrum. The generation of fluorescence and associated escape photons must be considered and corrected by using an algorithm that shifts the erroneous photon counts to the correct energy position [26, 27].

Pulse pile-up occurs when two or more photons imping the sensor very close in time and cannot be distinguished; when this happens, they are detected erroneously as a single photon with higher energy. Pile-up phenomena are very frequent when the application requires high X-ray fluxes, as it is often the case in NDTs for industry. At high flux rates, pile-up results in the distortion of the pulse amplitude measurement and in a subsequent loss of counts (also known as dead time losses). Its correction is currently one of the biggest challenges of Hyperspectral X-ray Absorptiometry. The resolution of pile-up issue is fundamental not only to ensure correct NDT measurements, but also for the development of spectral CT scanners for medical imaging [28, 29].

Insufficient charge collection is a reduction of the signal induced in the readout channel due to trapping of holes and electrons along their drift to the sensor

Table 1 Sources of X-ray spectrum distortions in CZT photon-counting detectors	
	• Fluorescence peaks
	• Escape photons
	• Pulse pile-up
	• Insufficient charge collection
	• Low angle scattering
	• Charge sharing (for multi-pixel detectors)

electrodes. Holes are more susceptible to trapping in CdTe and CZT material. For this reason, in this type of detectors, the "small-pixel" geometry is usually exploited to overcome this issue. Electron's losses are mostly linked to low quality of the CdZnTe material, for example, the presence of a high density of Te inclusions inside the sensing material. The effect of incomplete charge collection in the acquired spectrum is a low-energy shift of each photon detection. Incomplete Charge Collection (ICC) events are usually mitigated by reducing sensor thickness (at the cost of quantum efficiency) and increasing the electric field inside the material.

Low angle scattering is due to the radiation scattered by the sample and by the environment. The scattered photons enter the detector and fictitiously lower the attenuation coefficient of the sample. In our set-up, as it will be described in Sect. 5, we technically overcome this issue by collimating both the incident beam exiting the X-ray tube and the transmitted beam entering the detector.

Charge sharing also add up in case of multi-pixel arrays or pixel matrixes. Charge sharing results in a continuum of erroneous photon counts below the real energy of the incoming photon. Different techniques to correct it have been described in literature [30–32].

The correction of the distortions is fundamental for achieving an optimal quantitative analysis. Since these effects depend on energy and photon flux, different samples yield different distortions, and hence, their correction is one of the main challenges of the technique.

5 Laboratory Experimental Set-Up and DensXspect® System for Industry

Figure 3a represents a single-point laboratory set-up for Hyperspectral X-ray Absorptiometry. Polychromatic X-rays are generated by an X-ray tube and collimated on the sample. The aperture of the brass collimator placed on the X-ray source determines the investigated spot size: we typically use a hole with a diameter of 500 μm. The sample is located on an x–y motorized stage, which allows either to choose a specific point of investigation or to perform a 2-D scan of the sample. The CZT detector used by Due2lab has a thickness of 2.8 mm, full-area cathode, and a 2×2 mm^2 single-pixel electrode, surrounded by a guard-ring. The sensor is connected to a low-noise charge-sensing preamplification unit, powered by an ultralow ripple ±5 V power supply unit. The preamplified signal is processed by a 14-bit analog-to-digital converter (ADC, 100 Msps) and elaborated by a field-programmable gate array (FPGA) [33]. The system has a maximum number of energy bins equal to 2048, corresponding to a very small energy bin increment of 36 eV (0.036 keV) when used in the 5–80 keV, thus providing a digital spectroscopic resolution greatly higher than that of the CZT sensor itself. The CZT sensor fabrication, as well as front-end and digital pulse processing electronics

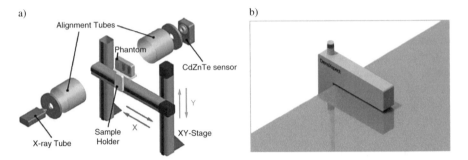

Fig. 3 Experimental set-up for the measurement of photon attenuation curve (**a**) in laboratory for off-line sample check and (**b**) along an industrial production chain for in-line real-time monitoring

development, is entirely realized by Due2lab. The typical energy resolution of this single-pixel detector is 3.4% at the 59.5 keV Am-241 photopeak [34] or lower. Source to detector distance is 30 cm, and the sample is placed in the middle (source to sample distance is 15 cm). This distance can be varied, and it should be reduced, if possible, to maximize the total number of photons detected in a certain time interval. Collimation is needed both exiting the X-ray tube and entering the sensor and perfect alignment of the two collimator apertures must be ensured. The transmitted beam is collimated on the center of the detector pixel, so that charge sharing with guard-ring does not lead to spectrum distortions. A very precise alignment of the system X-ray source/object/detector is fundamental to obtain a correct measurement of the absorbed photon intensity (this corresponds to match the conditions of "narrow beam" or "good geometry" measurement [35]).

When X-ray absorptiometry is used as non-destructive test for industry, the measurement configuration must be adapted to the production chain environment. Figure 3b shows DensXspect®, the prototype system developed by Due2lab for single-point measurement of the absolute density in industrial environments. X-ray beam passes vertically through a flat manufacture (wood panel, ceramic slabs, glass, etc.) which is sliding horizontally on the conveyor belt at a certain speed and has approximately constant thickness. The source-detector alignment is ensured by a mechanical bracket, which holds the X-ray tube and the high voltage generator on the bottom arm, and detector and readout electronics in the upper arm. DensXspect® is customizable according to application and user's needs.

X-ray absorptiometry imposes two requirements that are not always easily satisfied on a production chain and can sometimes hinder the possibility to exploit the technology for certain in-line applications. Since X-ray absorptiometry is a two-side measurement, it is necessary to place DensXspect® in correspondence of an interruption of the conveyor belt to ensure the measurement of the sample only. If the carrier cannot be interrupted, as for example for tubes with liquids flowing inside, it is necessary to know exactly the carrier X-ray attenuation to subtract its contribution. Secondary, periodic monitoring of the blank spectrum is crucial to exclude X-ray tube fluctuations as possible sources of photon attenuation curve artifacts. One single blank spectrum acquisition at the beginning of the measurement would not be sufficient. Thus, a direct path between the X-ray source and the detector, which means periodical interruptions of the scanned material sliding under the X-ray beam, is needed to register the blank at regular time intervals.

6 Measurement of Stoichiometric Materials

In this section we will give some examples illustrating the use of Hyperspectral X-ray Absorptiometry for the determination of absolute density of a stoichiometric compound. A stoichiometric compound is a material that presents an exact and fixed composition with a small integer ratio among its atomic components. Salt crystals (NaCl, $CaCO_3$) and mono-atomic compounds (graphite) are some examples of stoichiometric materials. Polyethylene (PE), polycarbonate (PC), and polytetrafluoroethylene (PTFE) are also stoichiometric compounds, since they are polymers whose elemental composition is, respectively, $(C_2H_4)_n$, $(C_6H_{14}O)_n$, and $(C_2F_4)_n$.

We have analyzed three samples of these polymers with our HXA laboratory set-up. Figure 4a shows the experimental photon attenuation curves of the three samples (solid lines). Since the samples had different thickness (2.0 mm for PTFE, 1.7 mm for PE, and 4.8 mm for PC), the photon attenuation has been normalized by the sample thickness to appreciate the influence of density and elemental composition on the trend of the curve. X-ray tube was operated at 25 kV.

The experimental photon attenuation curves are successively divided by the sum of the mass attenuation coefficients weighted according to the elemental composition of each specific polymer. The denominator of Eq. (7) can be calculated

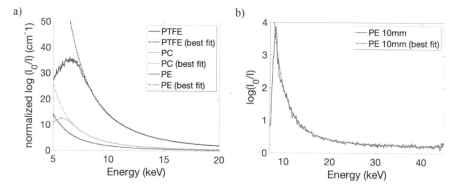

Fig. 4 (**a**) Experimental photon attenuation curves of polytetrafluoroethylene (PTFE), polycarbonate (PC), and polyethylene (PE). Each curve has been normalized by the thickness. Dashed lines represent the calculated photon attenuation curves obtained with extrapolated density values. (**b**) Experimental photon attenuation curve of a 10 mm thick PE sample best-fitted from 7 to 45 keV

Table 2 Area density and density measured for the polymer samples of PTFE, PC, and PE

Material	Elemental composition	Sample thickness (cm)	Area density ρ_A (g/cm^2)	Relative error on ρ_A (%)	Density ρ (g/cm^3)
PTFE	C_2F_4	0.20	0.440 ± 0.006	1.348	2.20 ± 0.03
PC	$C_6H_{14}O$	0.48	0.540 ± 0.010	1.178	1.12 ± 0.02
PE	C_2H_4	0.13	0.117 ± 0.003	2.222	0.90 ± 0.02

with precision because the chemical composition of the three samples is exactly known. PE molecular formula, for example, is $(C_2H_4)_n$, thus the relative weight fractions are 85.6% for carbon (0.856) and 14.4% (0.144) for hydrogen. The linear fit of Eq. (7) returns a specific area density value for each sample, as explained in Sect. 3. In this experiment the fit has been run on an energy interval ranging from 7 to 20 keV. The area density values determined for PTFE, PC, and PE samples are reported in Table 2. Area density includes the contribution of sample thickness; for this reason, it is higher for polycarbonate sample, which is the thickest of the three. By dividing area density values for samples thicknesses, we obtain absolute density values equal to 2.20 g/cm^3 for PTFE, 1.12 g/cm^3 for PC, and 0.90 g/cm^3 for PE. The ideal photon attenuation curves obtained with these density values are illustrated as dashed lines in Fig. 4a. Experimental and calculated curves fit perfectly, except for the lower energies where experimental dataset is affected by the noise of the detector and photon starvation due to complete absorption of the radiation.

We assume that the uncertainty on area density value is the standard deviation of the photon attenuation dataset points. If dataset points are highly scattered with respect to the intercept value found with the linear fit, relative error will be large. This is particularly true when measurement is done in a short acquisition time, thus photon statistics is poor. In our experimental conditions, with a total counts number equal to 25 MCounts for the blank spectrum over the whole energy range, the relative error is around 1–2%. By supposing that the error on thickness is ideally zero, the relative error on density reflects the error on area density. The density values experimentally obtained (Table 2) correspond to the expected values for the three polymers: PTFE is largely denser with respect to polycarbonate and polyethylene.

The correct choice of the energy region of interest (ROI) for the fit of the experimental dataset is a key step of density determination. Below a certain energy value E_1, the experimental photon attenuation curve deviates from expectation due to material self-absorption, resulting in photon shortage and poor statistics, as it is visible for PTFE and PC in Fig. 4a. Above energy value E_2, incident radiation is fully going through the sample, thus any useful information is present above this threshold. The two energy thresholds E_1 and E_2 are chosen by evaluating the experimental photon attenuation curve case by case. Fitting on short energy range leads to high relative error on area density estimation. It is necessary to select an energy ROI which only contains meaningful datapoints, but large enough to optimize density estimation. For samples with strong attenuation, the energy ROI will be shifted towards higher energies, because the low energy range is completely darkened by the object self-absorption. On the other hand, for lightweight materials such as polymers, the most suitable energy range is below 25 keV. In this last case, it is crucial to keep detector noise as low as possible, otherwise the low energy threshold E_1 shifts right, the relevant ROI shortens and consequently the relative error on density increases.

If the object to be investigated is thick, even if composed by lightweight chemical elements, it is necessary to operate at higher voltage to overcome sample self-absorption, as explained in the following experimental example.

To measure the absolute density of a 10 mm thick polyethylene sample, the X-ray tube voltage is set at 50 kV. 75 kCounts are collected for the blank spectrum. Figure 4b shows the photon attenuation curve of the 10-mm PE sample from 7 to 45 keV. The least-squares linear fit procedure has been run over the curve in the $10 \div 45$ keV energy range and returns a precise area density value. By dividing this value for PE sample thickness $x = (10.00 \pm 0.01)$ mm, we obtain an absolute density value of (0.94 ± 0.01), which correspond to a 1% error on density determination. The experimental acquisition and best-fit curve are shown in Fig. 4b. Measurements of other random points of the PE sample return the same density value, confirming that PE sample is homogeneous. This value extracted by means of Hyperspectral X-ray Absorptiometry is in a very good agreement with the value measured by measuring the weight/volume value for this specific sample, which is (0.95 ± 0.02) g/cm^3. The relative error is lower with respect to the 2.2% reported in Table 2 for thin PE sample. The larger available energetic range for best-fit minimization (up to 45 keV) resulted in a decrease of the relative error on density measurement.

To resume, in the case of the analysis of a stoichiometric compound, the relative chemical element weight fractions w_i are precisely known, thus the least-squares linear fit returns an area density value ρ_A extremely accurate, from which the absolute density value ρ can be determined if sample thickness is known.

7   Short Acquisition Time, High Photon Fluxes, and Experimental Error

With the aim of speed up density measurement, the 10-mm thick PE sample has been analyzed with a shorter acquisition time of 10 ms, which corresponds to a sampling rate of 100 Hz. X-ray tube current is increased at 600 μA. Under fast measurement conditions, the total number of photons counted in a 10-ms interval is roughly 4 kCounts over the whole energy range (against the 75 kCounts of the experiment described in Sect. 6), thus almost 20 times less photons than in slow measurement conditions. Despite the reduced statistics which scatters the experimental points of the photon attenuation of the photon attenuation curve (not shown), the S/N ratio still allows to determine the absolute density value by best-fit procedure. The experimental value determined is (0.94 ± 0.05) g/cm^3, which corresponds to a 5% error on density determination.

To lower the error on density determination at this very high sampling rate, it is necessary to maximize the X-ray tube fluence. The distance between X-ray tube and detector must be as short as possible to get maximum number of incident photons in the shortest measurement time interval. High photon fluxes are typically used for in-line real-time non-destructive tests to increase photon statistics and signal-to-noise ratio. To give an example, an X-ray flux of 0.8–1 MCps/mm^2 irradiating 4-mm thick wood panels is enough to enable a spectrum acquisition time of 100 ms, which corresponds to a sampling frequency of 10 Hz. These conditions lead to the determination of the panel area density with an error of 2%.

The minimum sampling frequency required is dictated by the travel speed of the product on the transport system. The use of high X-ray photon fluences requires sensor materials that are not sensible to self-polarization [36, 37] and advanced pile-up correction algorithms [29]. Pile-up is a distortion in the measurement of pulse amplitude due to the overlapping of two consecutive photons arrival, as briefly described in Sect. 4. At high flux, pulse pile-up distortion worsens, and only an efficient pile-up correction algorithm allows to avoid mistakes in the quantitative analysis.

8   Measurement of Non-stoichiometric Materials

In most cases the elemental composition of the material is not known a priori, either because the material is a non-stoichiometric compound or because it is a mixture of stoichiometric compounds whose relative percentage in the blend is not precisely defined.

A non-stoichiometric compound is a chemical compound having elemental composition whose proportions cannot be represented by a ratio of small natural numbers. Non-stoichiometric compounds are crystalline inorganic solids where a small percentage of atoms is missing (vacancies), or too many atoms are packed into an otherwise perfect lattice (interstitials atoms), or a certain fraction of atoms is substituted by another element. $Cd_{1-x}Zn_xTe$ itself, for example, is a crystalline non-stoichiometric compound. Many metal oxides and sulfides, such as $Fe_{1-x}O$ and $Fe_{1-x}S$, are metal-deficient owing to the presence of metal vacancies.

More generally, however, we refer to non-stoichiometric materials as those (non-crystalline) formed by a complex and variable mixture of stoichiometric compounds, for which an exact chemical formula cannot be defined. They can be organic (food, wood) or inorganic (ceramics, steel) materials.

Wood, for example, is a mixture of mainly two stoichiometric compounds, cellulose $(C_6H_{12}O_5)_n$ and lignin. Lignin is an organic polymer with high molecular weight, which can be composed by three different monomers: p-coumaryl alcohol, coniferyl alcohol, and sinapyl alcohol. The weight fraction of each monomer varies in relation to the vegetable family, the plant species, the organ and the biological tissue of the plant. Therefore, we know the elemental composition of wood only approximately: 50% carbon (C), 43% oxygen (O), 6% nitrogen (N), and 1% hydrogen (H).

The possibility to measure with precision the density of wood is fundamental since density has an influence on wood mechanical properties, such as strength and stiffness, which determine its performance for construction or other uses. Wood density in tree logs results from physiological processes related to cell growth, cell diameter, cell length, and cell wall thickness [11]. Several authors have used medical X-ray CT scanners to estimate the density of wood. However, they had to develop calibrated equations to relate CT numbers (CTNs, expressed in Hounsfield units) to density (g/cm^3 or kg/m^3) by comparing CT images with gravimetric densities for a wide number of wood specimens [38]. Hyperspectral X-ray Absorptiometry allows instead to extract the wood density directly, without the need of calibration.

We present the experimental density determination of two woody samples, respectively, fir wood and pine wood, both with thickness of (10.00 ± 0.02) mm (very low relative error on thickness: 0.2%). Considering that wood is less dense than several plastic polymers, with an average density that varies between 0.3 g/cm^3 (bamboo, for example) and 0.7–0.8 g/cm^3 (with the rare exception of ebony at 1.2 g/cm^3), and composed by low atomic number elements (N, C, O, and H), low energy X-rays are most suitable for inspecting these samples, even for a significant wood thickness. The possibility to use low energy X-rays for NDTs inspection systems is a great advantage because it significantly facilitates radioprotection authorization. According to Italian radioprotection laws, for example, legal authorization procedure is simplified for machines working below 30 keV. For this experiment, X-ray tube is polarized at 26 kV and operates at a current equal to 150 μA. Acquisition time is 1 s. Under these conditions, each acquisition of the blank spectrum collects roughly 20 kCounts over the whole energy range. Figure 5 reports the experimental photon attenuation curves for fir and pine wood samples between 10 and 25 keV.

The determination of absolute density for non-stoichiometric materials is more challenging in comparison to stoichiometric material because the denominator of Eq. (7) is not known a priori. In most cases, we only know an approximate elemental composition of the material, and eventually the list of possible chemical elements inside the material. The area density calculation is carried out by taking this unrefined composition as initial guess, and by running the linear fit minimization procedure considering the weight fractions w_i of each chemical element as variables. First, it is fundamental to start with an initial guess composition as close as possible to the truth, to maximize the accuracy on density value determination. The best elemental composition to be taken as initial guess (initial set of w_i coefficients) is the one that returns a straight line with slope very close to zero in Eq. (7), independently of the intercept with y-axis. The ideal slope is zero because, according to theory, area density ρ_A (the intercept) is a constant as a function of energy. Thus, w_i coefficients are left free to span within a certain range of possible values and the minimization process proceeds by sweeping all possible weight fractions combinations inside

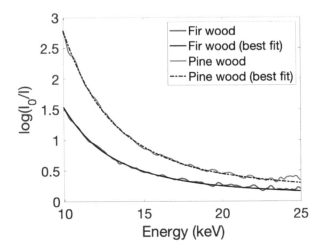

Fig. 5 Experimental photon attenuation curves of two different types of wood, fir (red) and pine (blue) and associated best-fit curves (black). Sample thickness is 10 mm. Resulting density is, respectively, (0.44 ± 0.01) g/cm^3 for fir and (0.86 ± 0.02) g/cm^3 for pine

that range. Obviously, the minimization process must be guided to avoid physical solutions that do not correspond to reality. The choice of correct boundaries to the possible weight fraction range of a certain chemical element ensures a higher probability to get to the most correct value of density. The output of this process is the area density value, and a refined set of w_i coefficients. If minimization goes as expected, the refined set of coefficients deviates only slightly from the initial guess.

Density values for fir and pine wood are the result of the linear fit minimization on the experimental photon attenuation curves of Fig. 5, carried out with an initial guess set of w_i coefficients that is left free to span over a possible range taken from literature [39]: C [48% ÷ 52%], O [42% ÷ 44%], H [5% ÷ 6%], N [0% ÷ 1%]. The resulting density values are 0.44 g/cm^3 for fir and 0.86 g/cm^3 for pine. The density of the same samples measured as weight/volume is, respectively, (0.43 ± 0.02) g/cm^3 for fir and (0.76 ± 0.02) g/cm^3 for pine. Fir wood density obtained from HXA is in perfect agreement with the weight/volume measurement. The discrepancy on pine density is likely related to the fact that natural wood samples are characterized by significant density spatial variations, visible to the eye and related to the presence of earlywood and latewood annual periodic growth. While weight/volume density measurement is an average of entire sample volume, density determination by means of X-ray attenuation is a sampling of very narrow wood spots of 0.2 mm^2 area. The density value of 0.86 g/cm^3 for pine is the average over three different spots of the sample. Density spatial variations are very accentuated in pine wood with respect to fir, and probably we probed three spots with density higher-than-average. A better overlap of the values derived from the two techniques can be obtained by sampling a higher number of wood spots.

Due2lab has developed the first in-line X-ray system prototype for real-time absolute density check of recycled wood pallets, in collaboration with IMAL-PAL group (Modena, Italy). The prototype has been developed in the framework of RIPALLET European project (http://www.ripallet.eu/index.php/en/results/) with the target to design a plant to produce a new generation of recycled, and recyclable, pressed wood pallets, produced from wood waste and manufactured minimizing the utilization of resins and chemical additives. The use of an innovative manufacturing process required a careful monitoring of the pallets area density to ensure their robustness and conformity. The wood pallets thickness is 4 mm and the sliding speed of the pallets on the conveyor belt is 4 m/s. The prototype is composed by five CZT photon-counting detectors, each coupled with an X-ray tube, which operate in parallel for real-time monitoring on five points of the pallet. The system acquisition time is as short as 100 ms (10 Hz sampling frequency) and the X-ray photon flux around 1 MCps/mm^2. Under these conditions, the inspection machine returns the wood panel area density value with an accuracy of 2%. Recycled wood panels are not affected by strong density spatial variations on millimetric scale as natural wood. This innovative NDT system is working in a real industrial environment since November 2018 [33].

9   Elemental Composition Refinement

In Sect. 8, we explain that the minimization procedure of the experimental photon attenuation curve to calculate the area density for a non-stoichiometric material is associated with the refinement of an initial guess of elemental weight fractions. In this Section we want to underline that the research process of correct atomic weight fraction is a powerful tool to get more precise information on material chemical composition. It can lead to discover the presence of contaminants, or the occurrence of small amounts of unexpected components in the blend. This technique can reveal the presence of a specific chemical element (above a certain threshold), or the presence of a particular layer material inside a multilayered structure. Since Hyperspectral X-ray Absorptiometry is not limited to the inspection of the sample surface, the tracing of a certain element applies to the whole internal volume of a solid piece of material. If a certain element is detected, the quantity measured is considered as equally distributed along the X-rays optical path. We refer to this specific use of HXA for getting more information on chemical composition as *elemental composition refinement*.

The first and most important step of this process is finding a correct initial guess of elemental composition, which is a set of w_i coefficients that returns a zero-slope straight line in Eq. (7) for ρ_A determination. A non-zero slope of this line is a strong indication that the assumed chemical composition is not completely correct.

The search for the proper initial guess is not a straightforward process. Fortunately, for certain materials the possible presence of an additional chemical element, not formerly considered in the blend, and its approximate concentration, is known

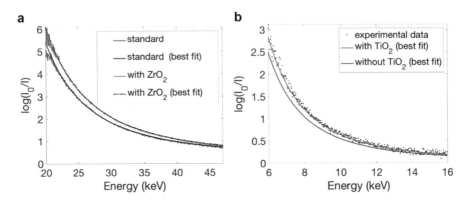

Fig. 6 Identification of low weight fraction components in non-stoichiometric materials: (**a**) Zirconium oxide (ZrO$_2$) in ceramic tiles and (**b**) Titanium oxide (TiO$_2$) in ABS

and this greatly helps the initial guess seeking. In this Section we present two experimental case studies that exploit this strategy: the detection of ZrO$_2$ inside ceramics tiles and the detection of TiO$_2$ in a plastic polymer for 3D printing.

Ceramics is a typical example of material blend. Ceramic tiles are composed for 90% of their weight by SiO$_2$ and Al$_2$O$_3$-based powders; a residual 10% comprises the following elements: K, Ca, Ti, Fe, Na, Mg, and P, which are present in the form of mineral powders. The exact weight fraction of the powder components differs depending on the soil provenance, and it is intentionally varied according to the final use. The mineral blend is pressed, dried, and annealed to obtain the tile, which is usually characterized by a bulk density above 2.0 g/cm^3. For the manufacturing of large ceramic slabs (few meters square), it is fundamental to monitor the density spatial homogeneity of the green ceramic before the firing step, to avoid cracks or defects on the final slab, and the density variations among different tiles. A relative density difference of 50 kg/m^3 (0.05 g/cm^3) leads to a shrinkage of 0.25% on the fired tile (i.e., 1.1 mm for a 45 × 45 cm^2 tile) [40].

We have analyzed two samples of different ceramic tiles, type A and type B, both with thickness of 6 mm and visual appearance almost identical. The measurements have been carried out under low X-ray flux and slow acquisition time conditions. Figure 6a shows the photon attenuation curves of the two ceramic samples: tile of type A (blue) is significantly more attenuating than type A (red). At a first sight, one can deduce that type A ceramic is denser than type B. The truth is that the difference is due to a small quantity of zirconium oxide (ZrO$_2$) that is added to ceramic powers for tile whitening, usually less than 1% in weight. HXA analysis confirms the presence of zirconium oxide and avoids blunders on density estimation. Indeed, if zirconia is added to the blend, a standard non-spectroscopic X-ray density inspection system, if not correctly recalibrated, would have overestimated the density of the tile containing ZrO$_2$ of 20%. The real absolute density values obtained from HXA are, respectively, 2.33 g/cm^3 for type A (with ZrO$_2$) and 2.39 g/cm^3 for type B ceramic. The measured density values are in good agreement with those measured by the tiles manufacturer with mercury balance method, which is still the standard technique for measuring density employed by ceramic industry.

The ability to correctly measure the density of a material composed by 11 chemical elements as ceramics by means of a non-destructive technique is a big achievement. In the ceramic tiles manufacturing, very small variations of density correspond to large differences in the force applied by the press; a density variation of 3% (e.g., from 1980 to 2040 kg/m^3) is obtained with a pressure variation of 40% (e.g., from 250 to 400 bars) [40]. This means that, to ensure an optimal feedback to the press, it is necessary to measure the density value of the green tile with a precision of 0.005 g/cm^3, that corresponds to an error of 0.2%. Improvements are still needed to obtain such measurement precision with HXA.

The presence of heavy elements among lighter ones is sometimes detected by means of K-edge absorption technique [41, 42]. Based on the position of elemental K-edges on the transmitted spectra, different elements can be identified inside the sample, and spatially located for 3-dimensional chemical imaging. However, we point out that in the present case study, the step-change in attenuation associated with zirconium K-edge absorption (located at 18 keV) is experimentally undetectable since completely darkened by strong sample self-absorption. Only the least-squares minimization of the entire photon attenuation curve from 20 to 47 keV allows to detect the tiny concentration of ZrO_2 and quantify it in the range 0.5–1%.

A second example of element composition refinement has been run on ABS case study, whose results are presented in Fig. 2. Initially, by considering a composition for ABS composed by different possible combinations of carbon, hydrogen, and nitrogen, which are the three chemical elements forming ABS monomers (acrylonitrile, butadiene, and styrene), the resulting mean area density was extremely high and inconsistent with the typical density of this copolymer. Having knowledge of the practice of adding small amounts of titanium dioxide inside polymers to whiten their color, we have run the minimization process starting from an initial elemental composition guess containing <1% amount of TiO_2. The least-squares minimization of the photon attenuation curve returns a reasonable ABS density value (around 1.25 g/cm^3) and a refined TiO_2 weight fraction of 0.4% added to ABS chemical composition. The presence of a very low amount of high atomic number atoms would have led to an important density overestimation. The presence of TiO_2 has been successively confirmed by XRF analysis, which revealed the presence of titanium characteristic X-ray fluorescence peaks at 4.5 and 4.9 keV (not reported here).

To conclude, Hyperspectral X-ray Absorptiometry has been capable, both for ceramics and plastics, to detect an anomaly with respect to the expected chemical composition of the material under examination. The precision of elemental composition refinement procedure and the accuracy on the determination of the exact single element weight fraction is still under study.

The ultimate goal would be the possibility to find the elemental composition initial guess without any assumption on the sample, which essentially means being capable to identify an unknown material from scratch. Even though, in principle, HXA provides all the necessary information to retrieve material chemical composition, the exploration of all the possible chemical elements combinations throughout the whole periodic table constitutes obviously an enormous calculation effort. The most promising way to perform direct material identification by HXA is exploiting machine learning techniques [43–45].

With respect to multi-energy techniques, Hyperspectral X-ray Absorptiometry enables a better material separation as the photon attenuation curves of very similar materials will show a different trend as a function of energy. Hyperspectral acquisition of X-ray photons allows to measure the exact trend of the photon attenuation curve as a function of energy but requires the ability to handle a huge quantity of data. The main challenge of using HXA is the downstream and the elaboration of the tremendous quantity of data generated by reading the incoming photon energy with a precision of few tens of eV.

10 Conclusions

This chapter illustrates how to use Hyperspectral X-ray Absorptiometry (HXA) to determine the absolute value of density of a specific object of known thickness or geometry. HXA technology works in transmission mode, with a polychromatic X-ray source irradiating the investigated object and a spectroscopic photon-counting detector acquiring the transmitted spectrum. The experiments presented in this chapter have been carried out by means of a CdZnTe detector. Other types of PCD sensors can be used, if they are suitable to the energy range that is the most appropriate for the analysis of the object (thicker/heavier sample requires higher X-ray energy). The outcome of HXA acquisition is the photon attenuation curve of a specific point of the object as a function of energy. If chemical composition is known with precision, as for stoichiometric materials, a least-squares minimization of this curve returns a density value extremely accurate. If material is non-stoichiometric, HXA allows to get better insight on its chemical (elemental) composition. Both absolute density measurement and elemental composition refinement are of great interest as non-destructive and contactless techniques that can be used in industry as analytical methods or non-conformity tests for quality control.

References

1. Lorente-Ramos R, Azpeitia-Armán J, Muñoz-Hernández A, et al (2011) Dual-Energy X-Ray Absorptiometry in the Diagnosis of Osteoporosis: A Practical Guide. American Journal of Roentgenology 196:897–904. https://doi.org/10.2214/AJR.10.5416
2. Devadithya S, Castañón D, University B, States U (2020) Material identification in presence of metal for baggage screening. Computational Imaging 9
3. Alvarez RE, Macovski A (1976) Energy-selective reconstructions in X-ray computerized tomography. Phys Med Biol 21:733–744. https://doi.org/10.1088/0031-9155/21/5/002

4. Le HQ, Molloi S (2011) Segmentation and quantification of materials with energy discriminating computed tomography: a phantom study. Med Phys 38:228–237. https://doi.org/10.1118/1.3525835
5. Wang X, Meier D, Taguchi K, et al (2011) Material separation in x-ray CT with energy resolved photon-counting detectors. Med Phys 38:1534–1546. https://doi.org/10.1118/1.3553401
6. Rebuffel V, Tartare M (2014) Multi-energy X-ray Techniques for NDT: a New Challenge. 10
7. Xiaobo Z, Xiaowei H, Povey M (2016) Non-invasive sensing for food reassurance. The Analyst 141:1587–1610. https://doi.org/10.1039/C5AN02152A
8. Jandejsek I, Jakubek J, Jakubek M, et al (2014) X-ray inspection of composite materials for aircraft structures using detectors of Medipix type. J Inst 9:C05062–C05062. https://doi.org/10.1088/1748-0221/9/05/C05062
9. Vavrik D, Jakubek J, Jandejsek I, et al (2015) Visualization of delamination in composite materials utilizing advanced X-ray imaging techniques. J Inst 10:C04012–C04012. https://doi.org/10.1088/1748-0221/10/04/C04012
10. Amorós JL, Cantavella V, Jarque JC, Felíu C (2008) Green strength testing of pressed compacts: An analysis of the different methods. Journal of the European Ceramic Society 28:701–710. https://doi.org/10.1016/j.jeurceramsoc.2007.09.040
11. Laforce B, Masschaele B, Boone MN, et al (2017) Integrated Three-Dimensional Microanalysis Combining X-Ray Microtomography and X-Ray Fluorescence Methodologies. Anal Chem 89:10617–10624. https://doi.org/10.1021/acs.analchem.7b03205
12. Meng B, Cong W, Xi Y, Wang G (2014) Image reconstruction for x-ray K-edge imaging with a photon counting detector. In: Developments in X-Ray Tomography IX. International Society for Optics and Photonics, p 921219
13. Panta RK, Bell ST, Healy JL, et al (2018) Element-specific spectral imaging of multiple contrast agents: a phantom study. J Inst 13:T02001–T02001. https://doi.org/10.1088/1748-0221/13/02/T02001
14. Egan CK, Jacques SDM, Wilson MD, et al (2015) 3D chemical imaging in the laboratory by hyperspectral X-ray computed tomography. Sci Rep 5:15979. https://doi.org/10.1038/srep15979
15. Beaulieu J, Dutilleul P (2019) Applications of computed tomography (CT) scanning technology in forest research: a timely update and review. Can J For Res 49:1173–1188. https://doi.org/10.1139/cjfr-2018-0537
16. Tomita Y, Shirayanagi Y, Matsui S, et al (2004) X-ray color scanner with multiple energy differentiate capability. In: IEEE Symposium Conference Record Nuclear Science 2004. IEEE, Rome, Italy, pp. 3733–3737
17. Alessio AM, MacDonald LR (2013) Quantitative material characterization from multi-energy photon counting CT. Med Phys 40:031108. https://doi.org/10.1118/1.4790692
18. Busi M, Mohan KA, Dooraghi AA, et al (2019) Method for system-independent material characterization from spectral X-ray CT. NDT & E International 107:102136. https://doi.org/10.1016/j.ndteint.2019.102136
19. Jumanazarov D, Koo J, Busi M, et al (2020) System-independent material classification through X-ray attenuation decomposition from spectral X-ray CT. NDT & E International 116:102336. https://doi.org/10.1016/j.ndteint.2020.102336
20. Garnett R (2020) A comprehensive review of dual-energy and multi-spectral computed tomography. Clinical Imaging 67:160–169. https://doi.org/10.1016/j.clinimag.2020.07.030
21. Danielsson M, Persson M, Sjölin M (2021) Photon-counting x-ray detectors for CT. Phys Med Biol 66:03TR01. https://doi.org/10.1088/1361-6560/abc5a5
22. Wu X, Wang Q, Ma J, et al (2017) A hyperspectral X-ray computed tomography system for enhanced material identification. Review of Scientific Instruments 88:083111. https://doi.org/10.1063/1.4998991
23. Ballabriga R, Alozy J, Campbell M, et al (2016) Review of hybrid pixel detector readout ASICs for spectroscopic X-ray imaging. J Inst 11:P01007. https://doi.org/10.1088/1748-0221/11/01/P01007

24. Redus RH, Pantazis JA, Pantazis TJ, et al (2009) Characterization of CdTe Detectors for Quantitative X-ray Spectroscopy. IEEE Trans Nucl Sci 56:2524–2532. https://doi.org/10.1109/TNS.2009.2024149
25. https://www.nist.gov/pml/xcom-photon-cross-sections-database
26. Miyajima S (2003) Thin CdTe detector in diagnostic x-ray spectroscopy. Med Phys 30:771–777. https://doi.org/10.1118/1.1566388
27. Plagnard J (2014) Comparison of measured and calculated spectra emitted by the X-ray tube used at the Gustave Roussy radiobiological service: Comparison of measured and calculated spectra emitted by an X-ray tube. X-Ray Spectrom 43:298–304. https://doi.org/10.1002/xrs.2554
28. Taguchi K, Zhang M, Frey EC, et al (2011) Modeling the performance of a photon counting x-ray detector for CT: energy response and pulse pileup effects. Med Phys 38:1089–1102. https://doi.org/10.1118/1.3539602
29. Hsieh SS, Iniewski K (2021) Improving Paralysis Compensation in Photon Counting Detectors. IEEE Transactions on Medical Imaging 40:3–11. https://doi.org/10.1109/TMI.2020.3019461
30. Abbene L, Gerardi G, Principato F, et al (2018) Dual-polarity pulse processing and analysis for charge-loss correction in cadmium–zinc–telluride pixel detectors. Journal of Synchrotron Radiation 25:1078–1092. https://doi.org/10.1107/S1600577518006422
31. Buttacavoli A, Principato F, Gerardi G, et al (2020) Room-temperature performance of 3 mm-thick cadmium–zinc–telluride pixel detectors with sub-millimetre pixelization. J Synchrotron Rad 27:1180–1189. https://doi.org/10.1107/S1600577520008942
32. Fey J, Procz S, Schütz MK, Fiederle M (2020) Investigations on performance and spectroscopic capabilities of a 3 mm CdTe Timepix detector. Nuclear Instruments and Methods in Physics Research Section A: Accelerators, Spectrometers, Detectors and Associated Equipment 977:164308. https://doi.org/10.1016/j.nima.2020.164308
33. Zambelli N, Zanettini S, Benassi G, et al (2018) High Performance CZT Detectors for In-Line Non-destructive X-Ray Based Density Measurements. In: 2018 IEEE Nuclear Science Symposium and Medical Imaging Conference Proceedings (NSS/MIC). pp. 1–4
34. Zambelli N, Zanettini S, Benassi G, et al (2020) CdZnTe-based X-ray Spectrometer for Absolute Density Determination. IEEE Transactions on Nuclear Science 1–1. https://doi.org/10.1109/TNS.2020.2996272
35. Knoll GF Radiation Detection and Measurements, 3rd ed
36. Thomas B, Veale MC, Wilson MD, et al (2017) Characterisation of Redlen high-flux CdZnTe. Journal of Instrumentation 12:C12045–C12045. https://doi.org/10.1088/1748-0221/12/12/C12045
37. Veale MC, Booker P, Cross S, et al (2020) Characterization of the Uniformity of High-Flux CdZnTe Material. Sensors 20:2747. https://doi.org/10.3390/s20102747
38. Osborne NL, Høibø ØA, Maguire DA (2016) Estimating the density of coast Douglas-fir wood samples at different moisture contents using medical X-ray computed tomography. Computers and Electronics in Agriculture 127:50–55. https://doi.org/10.1016/j.compag.2016.06.003
39. Hon DN-S, Shiraishi N (2001) Wood and cellulosic chemistry, 2nd ed., rev. expanded. Marcel Dekker, New York
40. Revel GM, Cavuto A, Pandarese G (2016) Laser ultrasonics for bulk-density distribution measurement on green ceramic tiles. Review of Scientific Instruments 87:102504. https://doi.org/10.1063/1.4964626
41. Schütz M, Procz S, Fey J, Fiederle M (2020) Element discrimination via spectroscopic X-ray imaging with a CdTe Medipix3RX detector. J Inst 15:P01006–P01006. https://doi.org/10.1088/1748-0221/15/01/P01006
42. Sittner J, Godinho JRA, Renno AD, et al (2020) Spectral X-ray computed micro tomography: 3-dimensional chemical imaging. X-Ray Spectrom xrs.3200. https://doi.org/10.1002/xrs.3200
43. Hu W, Chen S, Li Y, et al (2018) X-ray absorption spectrum combined with deep neural network for on-line detection of beverage preservatives. Review of Scientific Instruments 89:103108. https://doi.org/10.1063/1.5048281

44. Fang Z, Hu W, Wang R, Chen S (2019) Application of hyperspectral CT technology combined with machine learning in recognition of plastic components. NDT & E International 102:287–294. https://doi.org/10.1016/j.ndteint.2019.01.001
45. Fang Z, Wang R, Wang M, et al (2020) Effect of Reconstruction Algorithm on the Identification of 3D Printing Polymers Based on Hyperspectral CT Technology Combined with Artificial Neural Network. Materials 13:1963. https://doi.org/10.3390/ma13081963

A New Method of Estimating Incident X-Ray Spectra with Photon Counting Detectors Using a Limited Number of Energy Bins with Dedicated Clinical X-Ray Imaging Systems

Bahaa Ghammraoui and Stephen J. Glick

1   Introduction

For over a century, clinical X-ray imaging systems have operated in energy integrating mode. We are currently at the brink of a revolutionary change in clinical X-ray detector technology in which future detectors will be photon counting, allowing for energy discrimination. These photon counting detector (PCD) systems will have numerous benefits. Although PCDs have been used widely in nuclear imaging, their application in X-ray CT or radiography has been hampered due to the need for a high photon-count rate. Therefore, pixelated semiconductor-based PCDs equipped with fast readout electronics (ASICs) have been studied extensively to enable utilization in high-count rate situations. Unfortunately, these ASICs typically have a limited number of energy thresholds. Therefore, it is difficult to estimate an accurate full energy spectrum due to the need for sweeping the energy threshold, subsequently resulting in lengthy acquisitions and a higher dose to the patient. In addition, charge sharing, characteristic escape peaks, Compton scattering, and weighting potential cross talk can severely distort measured spectra in such pixelated detectors. A method to estimate an accurate full energy spectrum from these pixelated detectors would be desirable. In this chapter, we will discuss a novel method that uses an inverse problem-based approach for spectrum estimation and distortion correction with photon counting X-ray detectors using a limited number of energy bins.

B. Ghammraoui (✉) · S. J. Glick
Division of Imaging Diagnostics and Software Reliability, Center for Devices and Radiological Health, U.S. Food and Drug Administration, Silver Spring, MD, USA
e-mail: bahaa.ghammraoui@fda.hhs.gov

© The Author(s), under exclusive license to Springer Nature Switzerland AG 2023
K. (Kris) Iniewski (eds.), *Advanced X-Ray Radiation Detection*,
https://doi.org/10.1007/978-3-030-92989-3_6

2 Benefits of Estimating the Full Energy Spectrum in Pixelated X-Ray Photon Counting Detectors for Medical Imaging

2.1 Beam Quality

The selection of the incident X-ray spectrum for a given radiological procedure has a direct influence on many important parameters. Among these are patient exposure and transmission, image contrast and sharpness, and tube loading [13]. The characterization of X-ray spectra through half-value layer (HVL) measurements is well known and widely used for this purpose. An accurate HVL measurement with strong knowledge of the operating potential generally provides enough information about an X-ray spectrum for routine quality control purposes. Several techniques for measuring HVL have been found by using an ionization chamber (IC) with a series of measurements or radiochromic film with step-shaped aluminum (Al) filters [12]. Semiconductor solid state detectors have come into widespread use recently for quality control measurements, mainly because of their small size, ruggedness, and convenience of use. Some discrepancies between the methods, however, caused by differences in their radiation dosimeter properties (sensitivity, linearity, energy dependence, and directional dependence) have been observed. In addition, HVL provides limited information when using a filter material with a K-absorption edge. The most commonly used filters in radiological imaging preferentially remove lower-energy photons, but in some circumstances, filtering out high-energy photons is useful. In the latter case, the characterization of such X-ray spectra through HVL measurements could be imprecise; therefore, direct spectroscopic measurements of the X-ray spectra using multi-channel pulse height analyzer (MCA) systems might be beneficial and suitable. Unfortunately, MCA systems have the disadvantage of not being able to measure high-count rates and therefore cannot work under realistic operating conditions of an X-ray machine but instead require unsuitable geometries for pile-up mitigation [9].

In general, all the previously mentioned methods require several lengthy acquisitions in order to calculate the spectra over a 2D area. The 2D spectroscopic measurements are very important for characterizing X-ray spectra when using bowtie-shaped filters, as beam hardening and heel effects amplify the nonuniformities of the spectra shape across the radiation field. Therefore, accurate 2D spectroscopic measurements across the radiation field would be extremely useful for a beam quality study and dose optimization and calculation, especially when bowtie filters are used [2].

2.2 Energy Weighting

X-ray beams used in clinical imaging are polyenergetic and used to image objects with attenuation coefficients that are energy dependent and decrease with energy. Therefore, patient transmission increases with the X-ray energy, lowering the quantum noise in the acquired image, while image contrast decreases as X-ray energy increases, and vice versa.

The energy-integrated detector weights each detected photon proportionally to its energy, with higher energy photons having more weight. Therefore, image contrast decreases, since the contrast between materials generally depends on low-energy photons [19]. On the other hand, energy-resolved photon counting detectors (PCDs) can measure the energy deposited by each X-ray photon thereby allowing estimation of a full X-ray spectrum. By emphasizing the signal recorded from the low-energy photons, image contrast can be enhanced. This technique is called energy weighting, which basically means the weighting of each photon by an energy-dependent factor. The selection of the weighting factors most appropriate for a given imaging procedure is complicated and involves image contrast and noise considerations [26]. Several studies have shown the benefits of the energy weighting technique [11] using either the projection-based or image-based approaches when using photon counting with a limited number of energy bins. However, for a limited number of bins, the optimal arrangement of energy weighting that maximizes the benefits of the spectral information for tissue separation remains unclear, and higher numbers of energy bins would be more beneficial.

2.3 Material Decomposition

The technology of energy-resolved PCDs yields the possibility of decomposing conventional single-shot X-ray images into basis material images representing the density of various materials in the object being imaged. Applications and methods for material decomposition have been discussed extensively in the literature [22].

Spectral imaging methods are not limited to PCDs. For example, the Siemens SOMATON DRH in 1987 was the first available commercial dual-energy CT scanner using an energy-integrated detector. This system used the fast kV switching technique, in which the tube voltage was alternated between 85 and 125 kV from projection to projection [7, 15]. In general, the spectral imaging techniques with energy-integrated detectors (Dual Source, Fast kV switching, TwinBeam Dual Energ) are a compromise between the spectral separation and temporal coherence of the low- and high-energy data, dose efficiency, and the temporal and spatial resolution [7]. Furthermore, the maximum number of basis materials in material decomposition with most scanners that employ spectrally indiscriminate energy-integrating detectors is two.

Several studies have shown that the energy-resolving capabilities of PCDs can be used successfully for quantitative material decomposition [20] and K-edge imaging [23]. Furthermore, PCDs with more than three energy bins allow decomposition of more than three materials [14, 18].

However, the energy-resolving capability of PCDs is also limited, due to such phenomena as radiation scattering, beam hardening effects, limited detective quantum efficiency, detector dead time, inaccuracies with spectral models, and pulse pile-up and therefore could decrease the material decomposition accuracy [14]. On the other hand, an accurate model for the detector response or characteristics, or suitable correction techniques using calibration material measurements, could be used to correct for these degradation factors.

3 Methods for Accurate Measurement of the Full Energy Spectrum with Pixelated X-Ray Imaging Detectors

Several methods and techniques have been proposed for measuring the full energy spectrum with pixelated semiconductor-based PCDs. We will discuss some that are commonly used and their limitations. In addition, we will also discuss our recently proposed method that uses an inverse problem-based approach for spectrum estimation and distortion correction.

3.1 Concepts

A typical pixelated semiconductor-based PCD consists of a semiconductor diode, on which a large bias voltage is applied through pixelated electrodes. The conversion of the detected X-ray photon in the energy unit (keV) into an electrical voltage pulse (mV) involves several steps. An X-ray of energy E enters the detector and interacts with the semiconductor via the photoelectric, fluorescence, or Compton processes, to produce a cloud of electrons and holes with an amount proportional to the deposited energy. Generated charges are then drifted and attracted to the collection electrodes under the influence of an electric field caused by the applied voltage. The drifting process of the electron–hole pairs generates an electrical pulse with a height proportional to the energy deposited by the photon. Next, the pulse is processed by the connected ASIC electronics through charge-sensitive amplification and pulse shaping and then is sent to multiple pairs of voltage pulse height comparators and digital counters [24]. The latter compares each pulse to several preset threshold levels; thus, the detector can sort the incoming detector into a number of energy bins, depending on the energies of the upcoming photons.

3.2 Summary of the Conventional Methods

Following is a brief summary of the conventional methods commonly employed for measuring the full energy spectrum with pixelated semiconductor-based PCDs. One way to measure a full energy spectrum with such limited energy bin detectors is by sweeping the built-in thresholds and then differentiating the results. In that case, successful spectroscopic data highly relies on the uniformity or consistency of the exposure conditions during the whole energy threshold scanning process. In other words, the operational parameters of the source, such as tube voltage, current, focal spot size, and filtrations, must remain constant during the whole acquisition. For demonstration purposes, Fig. 1 shows an energy spectrum of ^{109}Cd by threshold sweep using the LAMBDA 60 K module planar detector (X-Spectrum GmbH, Germany) with a gallium arsenide (GaAs) sensor consisting of a GaAs semiconductor with a 256 × 256-pixel array 55 μm pixel size. The spectrum was obtained by sweeping the energy threshold from 5 to 30 keV with a decrement of 2 keV with charge summing mode activated, and the sweeping process was repeated several times to acquire the average and obtain better statistics. The spectrum was also averaged over 400 randomly selected pixels.

The energy threshold sweeping method requires long and repetitive acquisitions to reduce the noise effects that are amplified by the differentiating operator applied to the measured data. In addition, the measured spectra in that case represent the raw results from the detector; thus, these spectra suffer from all the spectral distortion discussed in Sect. 3.1. Spectral correction methods rely on prior knowledge of the spectral response of the detector, which can be applied to obtain the correct

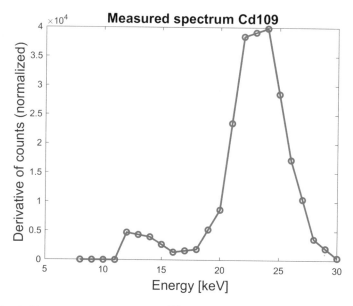

Fig. 1 Threshold sweeping measurement of a ^{109}Cd X-ray source

incident spectrum. The detector response function of a detector can be estimated accurately from a series of monochromatic beam measurements (e.g., synchrotron, fluorescence materials) [6], or less accurately from simulation that takes into account all the distortion effects [8, 25, 28], or, finally, from some semi-analytical method based on an analytical model with adjustable parameters tuned using some measured spectra from an a priori known source [21]. Full knowledge of the energy response functions of a detector is the most effective and reliable means of obtaining reliable and correct spectra. However, response functions change from pixel to pixel due to variations in the ASIC and crystal nonuniformity [24, 27]. Therefore, the energy response should be repeated for all the pixels to determine the pixel-by-pixel response variations. Although knowledge of the energy response functions provides unique information for spectral correction, it has prohibitive problems mainly caused by difficulties in measuring it, while most of the conventional imaging laboratories have limited accessibility monoenergetic X-ray emission with freely adjusted energy levels and very narrow energy bandwidth [24].

3.3 Full Spectrum Estimation as a Linear Inverse Problem

Recently, [9] demonstrated that a full energy spectrum can be reconstructed from a limited number of energy bins' data using a linear inverse problem approach. In this study, the incident X-ray spectrum was represented as a linear superposition of N_k basis known spectra I_k

$$I(E) = \sum_{k=1}^{N_k} a_k I_k(E), \tag{1}$$

where I has dimensions of number of emitted photons per keV, and a_k are expansion coefficients. With this assumption, the expected number of photons measured in the energy interval $E_i - \triangle E \leq E \leq E_i + \triangle E$ or the ith bin of the detector can be represented as

$$\langle m_i \rangle = \int_0^\infty I(E) S_i(E) D(E) dE = \sum_{k=1}^{N_k} a_k \int_0^\infty I_k(E) S_i(E) D(E) dE. \tag{2}$$

The terms $S_i(E)$ and $D(E)$ represent the energy bin sensitivity and the detector absorption, respectively. Therefore, Eq. (2) can be written as

$$\langle m_i \rangle = \sum_{k=1}^{N_k} a_k M_{ki}, \text{ with components } M_{ki} = \int_0^\infty I_k(E) S_i(E) D(E) dE. \tag{3}$$

$\langle m_i \rangle$ was measured as a calibration step from known basis spectra $I_k(E)$.

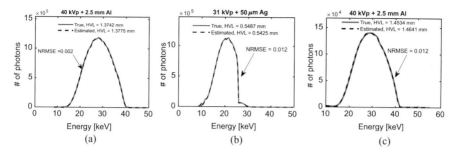

Fig. 2 Estimated spectra using the proposed algorithm: (**a**) and (**b**) were estimated from a simulation of an ideal photon counting detector with three energy bins. (**c**) was estimated from a one-pixel CdTe photon counting detector with three energy bins. Reprinted from Ghammraoui, B., Makeev, A., Glick, S.J., 2020. High-rate X-ray spectroscopy in mammography with photon counting detectors using a limited number of energy bins. Radiation Measurements 138, 106444, doi: https://doi.org/10.1016/j.radmeas.2020.106444. Copyright 2021 by the Copyright Clearance Center. Reprinted with permission

The equation was then solved for the $\{a_k\}$ using a penalized maximum-likelihood (PML) algorithm. After estimation of the $\{a_k\}$ for each known basis spectrum, Eq. 1 was used to generate or reconstruct the full incident spectrum. As shown here, the method is simple and can be done in most X-ray facilities with polychromatic source where there is no need for limited accessibility monoenergetic beams, and most importantly, the calibration for all pixels can be done all at once.

The authors evaluated the proposed algorithm through several numerical and experimental studies. For numerical validation, both hard continuous and soft spectra with K-edge from filter materials were estimated from an ideal PCD with three energy bins (Fig. 2a, b). In addition, experimental measurements of mammography X-ray spectra were estimated using two different PCDs, the one pixel cadmium telluride (CdTe) detector (Amptek XR-100T) operated with three energy bins and the high-count rate gallium arsenide (GaAs) 2D PCD (Lambda 60K detector) operated with four energy bins. Very good agreement can be observed between the estimated spectra, using the proposed algorithm, and true spectra and their corresponding HVL values (Figs. 2c and 3). Results from the GaAs detectors indicate the method's potential for estimating a full energy spectrum from a limited number of energy datasets, even in the presence of spectral distortion caused by pixels cross talk and nonuniform energy responses across the pixels.

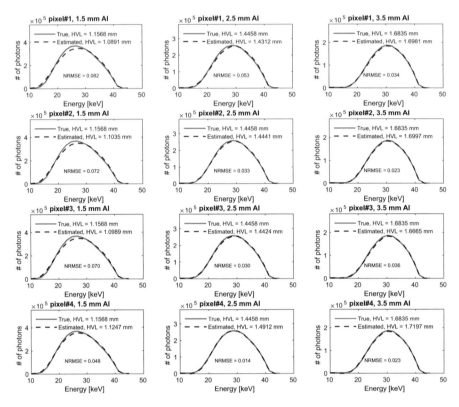

Fig. 3 Estimated spectra from four randomly selected spectra using the proposed algorithm from only three energy datasets. These are compared to the true spectra measured and corrected from the CdTe Amptek detector. The HVL values for all spectra are also presented. Spectra are for a filtered tungsten anode X-ray source with different thicknesses of added aluminum filtration. The thicknesses are shown next to the pixel number. Reprinted from Ghammraoui, B., Makeev, A., Glick, S.J., 2020. High-rate X-ray spectroscopy in mammography with photon counting detectors using a limited number of energy bins. Radiation Measurements 138, 106444, doi: https://doi.org/10.1016/j.radmeas.2020.106444. Copyright 2021 by the Copyright Clearance Center. Reprinted with permission

4 Examples of Full Spectrum Estimation Applications as a Linear Inverse Problem in Breast Imaging

4.1 Contrast-Enhanced Spectral Mammography

Contrast-enhanced spectral mammography (CEM) describes a mammogram that uses iodine contrast agent injected in an arm vein to highlight areas with suspicious lesions and unusual blood flow. Currently, CEM is performed through the dual-energy method, with low- and high-energy scans. An iodine-enhanced mammogram is then obtained by performing a weighted logarithmic subtraction of the low- and high-energy-acquired images for clearer visualization. The resulting image is supposed to show only where contrast has pooled, indicating areas that are potentially cancerous. The main drawback of the dual-energy method is the increased radiation dose from additional scans. Meanwhile, PCDs with multiple energy thresholds can also be used for contrast-enhanced spectral mammography with a single X-ray shot, reducing patient radiation exposure, as well as the scanning time and its associated patient motion artifacts [4, 10].

In this section, we will demonstrate through simulation how the full spectrum estimation method proposed in Sect. 3.3 can be used to improve image quality in photon counting contrast-enhanced spectral mammography. Photon Counting Toolkit (PcTK) software was used to simulate the photon counting CEM [29]. PcTK is a MATLAB program for a PCD model which takes into account spatio-energetic cross talk and correlation between PCD pixels. A cadmium telluride (CdTe)-based PCD with four energy thresholds was modeled in this study. The detector consisted of CdTe crystals with a size of $1080 \times 60 \, mm^2$, a pixel pitch of $100 \times 100 \, \mu m^2$, and a thickness of 0.75 m. We used the following parameters as inputs: a CdTe PCD with $d_{pix} = 100 \, \mu m$, a thickness of 0.75 mm, and a density of 5.85 g/cm^3. Shown in Fig. 4 are the spectral responses of the modeled detector with energy of 25, 35, and 45 keV, with flat-field irradiation.

The simulated phantom consisted of homogeneous breast tissue, which has the chemical composition of 50% adipose and 50% glandular tissues containing iodinated inserts of the same base material, to which an iodinated component is added, as illustrated in Fig. 5. The iodine concentrations were chosen to cover a typical clinical range of area densities of 0.25, 0.5, 1, and 2 mg/cm^2 [17]. Four

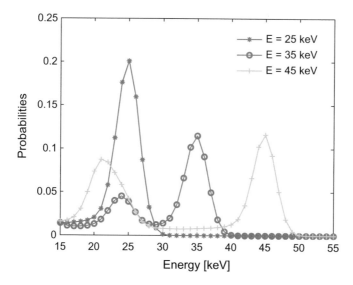

Fig. 4 The spectral response of the modeled CdTe detector with incident energy of 5, 35, and 45 keV, with flat-field irradiation

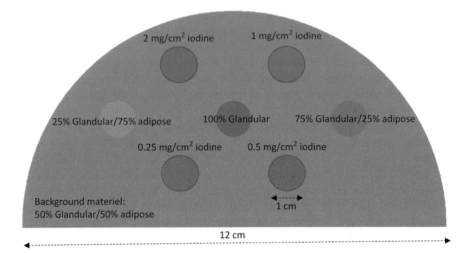

Fig. 5 The phantom corresponds to a 5 cm thick breast-equivalent material in a 50/50 ratio of glandular and adipose tissue containing iodinated inserts

other inserts, consisting of different adipose/glandular portions, were added for comparison. The eight inserts all were 10 mm high.

Four splitting energies of 15, 25, 31, and 39 keV were explored to separate incoming photons into three distinct datasets (bin #1: 15 keV to 25 keV, bin #2: 25 keV to 31 keV, bin #3: 31 keV to 39 keV, and bin #4: 39 keV and above). All of the X-ray spectra studied in this case were emitted from an X-ray tube with a tungsten target operating at 40 kVp and obtained from an Institute of Physics and Engineering in Medicine (IPEM) report [5]. The incident number of photons was chosen to obtain a mean glandular deposited dose equal to 1.5 mGy. We used six basis spectra in the calibration step, corresponding to the uniform incremental thickness of aluminum and an iodine aqueous solution of 20 mg/mL (from ($l_{al} = 1$ cm, $l_{id} = 0$ cm) to ($l_{al} = 4$ cm, $l_{id} = 2.5$ cm)).

As described in Sect. 3.3, we obtained system matrix elements M_{ki} from the output intensities of the four energy bins when measuring the incident basis spectra. Next, for each pixel in the phantom image, four energy bins' output $\{m_1, m_1, m_3, m_4\}$ was used to estimate its corresponding expansion coefficients $\{\alpha_k\}$. After maximum-likelihood estimation of expansion coefficients for each incident basis spectrum was completed, the unknown X-ray spectrum was estimated using Eq. (1). Furthermore, we used the estimated incident spectra at each pixel to estimate the propagation path length for the corresponding ray through iodine, adipose, and glandular tissues. We solved for the last parameters using a standard least-squares method. Figure 6 shows the iodine component image estimated after the full spectrum estimation process.

Fig. 6 Iodine component image estimated using the full spectrum estimation method described in Sect. 3.3

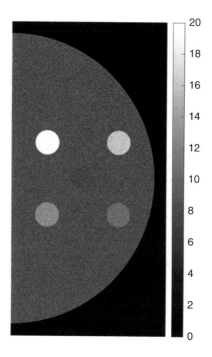

For comparison purposes, we show in Fig. 7 the iodine image using the standard dual-energy method and logarithmic subtraction [10]:

$$I_{low} = \log[\frac{n_{low}}{\overline{n}_{low}}]; \quad I_{high} = \log[\frac{n_{high}}{\overline{n}_{high}}] \tag{4}$$

$$\Rightarrow I_S = I_{HE} - w \times I_{LE}, \tag{5}$$

where n_{LE} and n_{HE} are the low- and high-energy raw images, w is the weight factor and $\overline{n}_{low}$ and $\overline{n}_{high}$ are the low- and high-energy flat field images. The optimum w was chosen to ensure a good cancellation of the contrast of the non-iodinated insert (100% glandular) with the background material of the phantom in the dual-energy image (I_S).

Although the standard dual-energy method using the raw images effectively removed the overlapping granularity from the image, the iodine inserts showed negative values. Quantitative information on iodine contents was completely lost due to the spectral distortion in the raw images caused mainly by the charge sharing and the Cd characteristics K-edge X-rays. However, the full spectrum estimation method was more efficient than the conventional logarithmic subtraction method in terms of iodine inserts visualization and could be used with more sophisticated material decomposition algorithms to provide improved quantitative estimates of iodine uptake.

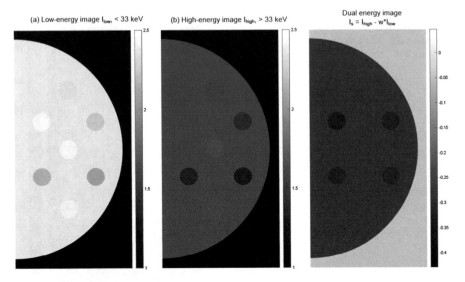

Fig. 7 Dual-energy images using the standard logarithmic subtraction method from the raw low- and high-energy images

4.2 Photon Counting Computed Tomography: Accurate Measurements of Attenuation Coefficients and Ring Artifacts Removal

Photon counting CT (PCCT) systems have many advantages over conventional energy-integrated CT systems, such as radiation exposure reduction, reconstructed images with higher spatial resolution, and reduction in beam-hardening artifacts, and they enable the use of multiple agents simultaneously. However, PCCT systems suffer from severe stripe artifacts in the projection domain, and concentric ring artifacts in the image domain, due to the lack of uniform detector sensitivity over the pixels that may compromise image quality and cause nonuniformity bias [1, 3]. Standard flat-field correction cannot correct for this nonuniformity, since part of it is caused by the thresholds' non-homogeneity over the pixels [3]. Many methods have been proposed to address this issue. They can be categorized into pre-processing in the projection space and post-processing in the image space, as discussed in the literature [1]. However, these methods usually come with penalties in terms of image resolution and contrast degradation caused by the smoothing filters used [3]. Accurate calibration methods based on measurements like the Signal-to-Equivalent Thickness Calibration (STC) approach [14] have been demonstrated to be successful in correcting for threshold non-homogeneity without compromising other image quality metrics. In this study, we showed by simulation an example of how the full spectrum estimation method could address this issue at its root to fix this

Fig. 8 The phantom consists of a 12 cm diameter 50% adipose and 50% glandular tissue-filled phantom containing two 20 mm diameter rods of 100% adipose and 100% glandular tissues

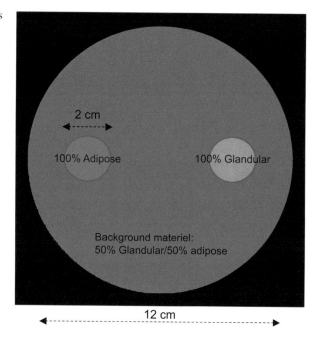

nonuniformity response between the pixels and thus avoid the need for additional correction methods.

We simulated a breast CT system that included a cone beam protruding from a 55 kV tungsten anode source, filtered by 1 mm of aluminum. The phantom was mounted on a rotating stage situated in front of the collimator (source–object distance = 100 cm and object–detector distance = 10 cm). The 12 cm diameter phantom (Fig. 8) consisted of 50% adipose and 50% glandular tissues, containing two 20 mm diameter rods of 100% adipose and 100% glandular tissues. The detector was modeled using PcTK software. It consisted of CdTe crystals and had a size of 130×130 mm^2, a pixel pitch of 100×100 μm^2, and a thickness of 0.75 m. We used the following parameters as inputs: a CdTe PCD with d_{pix}=100 μm, a thickness of 0.75 mm, and a density of 5.85 g/cm^3. Four splitting energies of 15, 25, 31, and 39 keV were explored. To simulate the thresholds' non-homogeneity, we added variance between the pixel thresholds consisting of ± 5 keV. The incident number of photons was chosen to obtain a mean glandular deposited dose equal to 1.5 mGy.

As in Sect. 4.1, the incident spectrum was estimated using the proposed full spectrum estimation method. Similar to the previous studies, six basis spectra were modeled after passing through 1, 2, 3, 4, 5, and 6 mm thick aluminum sheets. We used the estimated full energy spectrum at each pixel to obtain the monoenergetic sinograms. Next, we performed filtered back projection reconstruction using the publicly available MATLAB code developed by Kim [16] to reconstruct the monoenergetic images as shown in Fig. 9. For comparison purposes, we show in Fig. 10 the reconstructed images at different energy thresholds using raw images

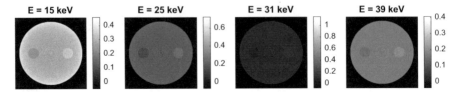

Fig. 9 The reconstructed images of the phantom in Fig. 8 using the raw recorded sinograms with standard flat-field correction

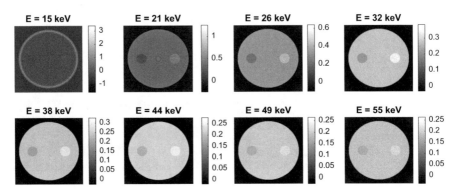

Fig. 10 The reconstructed monoenergetic images of the phantom in Fig. 8 using the monoenergetic sinograms obtained using the proposed full energy spectrum estimation method

corrected using the standard flat-field correction method. From these results, it can be seen that the proposed method was able to eliminate the ring artifacts without the need for any pre- or post-processing ring artifact correction methods. Further quantitative studies in terms of SNR and CNR are still needed to validate the conclusions.

References

1. An, K., Wang, J., Zhou, R., Liu, F., Wu, W., Aug 2020. Ring-artifacts removal for photon-counting CT. Opt. Express 28 (17), 25180–25193. URL http://www.opticsexpress.org/abstract.cfm?URI=oe-28-17-25180
2. Badal, A., 2016. Prototype adaptive bow-tie filter based on spatial exposure time modulation. In: Kontos, D., Flohr, T. G. (Eds.), Medical Imaging 2016: Physics of Medical Imaging. Vol. 9783. International Society for Optics and Photonics, SPIE, pp. 852–858.
3. Bader, A. M., Makeev, A., Glick, S. J., Ghammraoui, B., 2020. Development of a benchtop photon counting cone-beam CT system with a translate-rotate geometry. In: Medical Imaging 2020: Physics of Medical Imaging. Vol. 11312. International Society for Optics and Photonics, SPIE, pp. 1220–1226. URL https://doi.org/10.1117/12.2548380
4. Chen, B., Reiser, I., Wessel, J. C., Malakhov, N., Wawrzyniak, G., Hartsough, N. E., Gandhi, T., Chen, C.-T., Iwanczyk, J. S., Barber, W. C., 2015. Si-strip photon counting detectors for contrast-enhanced spectral mammography. In: Barber, H. B., Furenlid, L. R., Roehrig, H. N.

(Eds.), Medical Applications of Radiation Detectors V. Vol. 9594. International Society for Optics and Photonics, SPIE, pp. 26–33.

5. Cranley, K., Gilmore, B. J., Fogarty, G. W. A., Desponds, L., 1997. Catalogue of diagnostic x-ray spectra and other data. IPEM Report 78, The Institute of Physics and Engineering in Medicine, York, UK.

6. Ding, H., Cho, H.-M., Barber, W. C., Iwanczyk, J. S., Molloi, S., 2014. Characterization of energy response for photon-counting detectors using x-ray fluorescence. Medical Physics 41 (12), 121902.

7. Faby, S., Merz, J., Krauss, B., Kuemmel, F., 2018. Spectral Imaging with Dual Energy CT Comparison of different technologies and workflows.

8. Ghammraoui, B., Badal, A., Glick, S. J., 2018. Feasibility of estimating volumetric breast density from mammographic x-ray spectra using a cadmium telluride photon-counting detector. Med. Phys. 45 (8), 3604–3613.

9. Ghammraoui, B., Makeev, A., Glick, S. J., 2020. High-rate x-ray spectroscopy in mammography with photon counting detectors using a limited number of energy bins. Radiation Measurements 138, 106444.

10. Ghammraoui, B., Makeev, A., Zidan, A., Alayoubi, A., Glick, S. J., Under review. Characterization of a GaAs photon counting detector for mammography. Journal of Medical Imaging.

11. Glick, S. J., Ghammraoui, B., 2020. Imaging of the breast with photon-counting detectors. Spectral, Photon Counting Computed Tomography: Technology and Applications, 55.

12. Gotanda, T., Katsuda, T., Gotanda, R., Tabuchi, A., Yamamoto, K., Kuwano, T., Yatake, H., Kashiyama, K., Takeda, Y., 2008. Half-value layer measurement: simple process method using radiochromic film. Australas. Phys. Eng. Sci. Med. 32, 150–158.

13. Jennings, R. J., 1977. X-ray filters and beam quality. Quality assurance in diagnostic radiology, 105.

14. Juntunen, M. A. K., Inkinen, S. I., Ketola, J. H., Kotiaho, A., Kauppinen, M., Winkler, A., Nieminen, M. T., 2020. Framework for photon counting quantitative material decomposition. IEEE Transactions on Medical Imaging 39 (1), 35–47.

15. Kalender, W. A., Perman, W. H., Vetter, J. R., Klotz, E., 1986. Evaluation of a prototype dual-energy computed tomographic apparatus. i. phantom studies. Medical Physics 13 (3), 334–339.

16. Kim, K., 2012. 3d cone beam CT (CBCT) projection backprojection FDK, iterative reconstruction Matlab examples, Matlab central file exchange. retrieved December 16, 2019. Mathworks. March 10.

17. Klausz, R., Rouxel, M., Mancardi, X., Carton, A.-K.and Patoureaux, F. J., 2018. Introduction of a comprehensive phantom for the quality control of contrast enhanced spectral mammography. In: ECR 2018. ECR 2018.

18. Le, H. Q., Molloi, S., 2011. Segmentation and quantification of materials with energy discriminating computed tomography: A phantom study. Medical Physics 38 (1), 228–237.

19. Lee, S.-W., Choi, Y.-N., Cho, H.-M., Lee, Y.-J., Ryu, H.-J., Kim, H.-J., 2012. The effect of energy weighting on x-ray imaging based on photon counting detector: a Monte Carlo simulation. In: Pelc, N. J., Nishikawa, R. M., Whiting, B. R. (Eds.), Medical Imaging 2012: Physics of Medical Imaging. Vol. 8313. International Society for Optics and Photonics, SPIE, pp. 1584–1589.

20. Leng, S., Zhou, W., Yu, Z., Halaweish, A., Krauss, B., Schmidt, B., Yu, L., Kappler, S., McCollough, C., Aug 2017. Spectral performance of a whole-body research photon counting detector CT: quantitative accuracy in derived image sets. Physics in Medicine & Biology 62 (17), 7216–7232.
URL https://doi.org/10.1088/1361-6560/aa8103

21. Liu, X., Lee, H. K., 2014. Detector response function of an energy-resolved CdTe single photon counting detector. Journal of X-Ray Science and Technology 22 (6), 735–744.

22. McCollough, C. H., Leng, S., Yu, L., Fletcher, J. G., 2015. Dual- and multi-energy CT: Principles, technical approaches, and clinical applications. Radiology 276 (3), 637–653.

23. Persson, M., Huber, B., Karlsson, S., Liu, X., Chen, H., Xu, C., Yveborg, M., Bornefalk, H., Danielsson, M., 2014. Energy-resolved CT imaging with a photon-counting silicon-strip detector. Physics in Medicine and Biology 59 (22), 6709–6727.
24. Ren, L., Zheng, B., Liu, H., 2018. Ring-artifacts removal for photon-counting CT. Journal of X-Ray Science and Technology 28 (1), 1–28.
25. Schlomka, J. P., Roessl, E., Dorscheid, R., Dill, S., Martens, G., Istel, T., Baumer, C., Herrmann, C., Steadman, R., Zeitler, G., Livne, A., Proksa, R., 2008. Experimental feasibility of multi-energy photon-counting k-edge imaging in pre-clinical computed tomography. Physics in Medicine and Biology 53 (15), 4031–4047.
26. Shikhaliev, P. M., 2008. Computed tomography with energy-resolved detection: a feasibility study. Physics in Medicine and Biology 53 (5), 1475–1495.
27. Spartiotis, K., LeppÃnen, A., Pantsar, T., PyyhtiÃ, J., Laukka, P., Muukkonen, K., MÃnnistÃ, O., Kinnari, J., Schulman, T., 2005. A photon counting CdTe gamma- and x-ray camera. Nuclear Instruments and Methods in Physics Research Section A: Accelerators, Spectrometers, Detectors and Associated Equipment 550 (1), 267–277.
28. Taguchi, K., 2020. Photon Counting Detector Simulator: Photon Counting Toolkit (PcTK). (In) Spectral, Photon Counting Computed Tomography: Technology and Applications (Devices, Circuits, and Systems). Edited by Taguchi K, Blevis I, Iniewski K. CRC Press, Taylor & Francis Books, Inc., Ch. 18.
29. Taguchi, K., Stierstorfer, K., Polster, C., Lee, O., Kappler, S., 2018. Spatio-energetic cross-talk in photon counting detectors: Numerical detector model (PcTK) and workflow for CT image quality assessment. Medical Physics 45 (5), 1985–1998.

Optical Properties Modulation: A New Direction for the Fast Detection of Ionizing Radiation in PET

Yuli Wang and Shiva Abbaszadeh

1 Introduction

The inherent temporal variance generated during the scintillation process limits the time resolution in positron emission tomography (PET) utilizing scintillators. A coincidence time resolution around 100 ps can be realized using conventional scintillators. On the other hand, modulation of the optical properties of a material can be orders of magnitude faster enabling femtosecond or sub-picosecond time scale. In this chapter, we will review the recent developments of utilizing the optics pump-probe measurement to investigate the optical properties modulation method as a novel approach for radiation detection in PET.

The chapter starts with an explanation of the limitations on the coincidence time resolution of conventional scintillation-based PET detector and the impact of improved detector timing resolution on PET system performance. The concept of the optics pump-probe measurement for radiation detection is then introduced. Next, work related to the development of optical properties modulation method is summarized, including: (1) Theoretical simulation work demonstrating whether single annihilation photon can be detected due to ionization-induced optical property modulations (e.g., refractive index), (2) Proof-of-concept experimental arrangement demonstrating optical property modulation (e.g., small changes in refractive index) can be used to detect ionizing photons, and (3) Efforts to amplify the ionization-induced optical property modulation to improve the detection sensitivity of experimental setup to achieve single annihilation photon detection in PET.

Y. Wang · S. Abbaszadeh (✉)
UCSC, Santa Cruz, CA, USA
e-mail: ywang812@usc.edu; sabbasza@ucsc.edu

© The Author(s), under exclusive license to Springer Nature Switzerland AG 2023
K. (Kris) Iniewski (eds.), *Advanced X-Ray Radiation Detection*,
https://doi.org/10.1007/978-3-030-92989-3_7

1.1 Benefits of Better Time Resolution for PET Detector

In the past years, a lot of research has focused on improving the characteristics of scintillation crystals and other factors in PET systems to improve coincidence detection time-of-flight (ToF) capability, known as the 10-ps TOF challenge [1]. Researchers have been looking for crystal materials with faster rise time, shorter decay time, and higher light output that are more suitable for ToF-PET systems [2]. Crystals including lanthanum bromide ($LaBr_3$) and cerium-doped lutetium orthosilicate (LSO) are good candidates [3, 4]. Different crystal geometries and surface treatments have been studied to improve the PET time resolution [5]. Photodetectors including silicon photomultiplier tubes (SiPM) have been developed to detect scintillating light with higher quantum efficiency and faster response time [6]. The combinations of different scintillators [7] and the methods of utilizing different sizes of SiPMs in combination with scintillators [8, 9] have been studied. Other electronic components required in the ToF-PET system have also been improved to achieve smaller time jitter [10].

Significant improvement of 511 keV photon coincidence time resolution will bring a lot of signal amplification compared with existing systems [11]. This increased signal-to-noise ratio (SNR) in reconstructed PET images enables the advanced ability to visualize and quantify a smaller number of diseased cells in the presence of diffuse background signals common in any PET research [12]. Alternatively, patient injection dose and patient scan time (the two main limitations of clinical PET systems) can be reduced with better TOF performance [13]. If the time resolution is less than 20 ps (<20 ps), the image reconstruction time will also be greatly reduced [14, 15] enabling PET to be used for real-time imaging applications, including guided diagnostic interventions and surgery to treat the disease.

1.2 Coincidence Time Resolution Limitation of Conventional Scintillation Detectors

Even though a dramatically improved time resolution (~10 ps) is desired as discussed in Sect. 1.1, the coincidence time resolution of conventional PET detectors utilizing scintillation detection is largely limited by the scintillation process, which is essentially a form of "spontaneous emission." In [16, 17], the evolving process of the charge carriers created by ionizing radiation photons in scintillation crystals is reviewed and can be summarized as in Table 1.

The production of primary charge carriers happens at the first stage after the interaction between ionizing radiation photons and scintillation crystals. High kinetic energy is transferred to the primary charge carriers, and they immediately further excite a large number of secondary charge carriers through inelastic electron–electron scattering and Auger processes. This multiplication of charge carriers occurs within the femtosecond range. At the second stage within the sub-picosecond

Table 1 The evolving process of the charge carriers excited by high energy radiation photons in scintillation crystals [16, 17]

Time scale (s)	Process
10^{-16} to 10^{-14}	Production of primary charge carriers from interactions with ionizing radiation photons, production of secondary charge carriers through inelastic electron–electron scattering and Auger processes
10^{-14} to 10^{-12}	Thermalization of charge carriers, along with the production of phonons
10^{-12} to 10^{-10}	Localization of charge carriers through interactions with material defects and impurities, leading to the production of polarons
10^{-10} to 10^{-8}	Migration of charge carriers, radiative and/or non-radiative carrier recombination, production of scintillation light
10^{-10} to 10^{-9}	Typical optical transport time

time scale, the electron multiplication process stops and the thermalization of charge carriers starts with the production of phonons. The energy of charge carriers decreases below the threshold for inelastic electron–electron scattering. The third stage features the localization of charge carriers through their interactions with stable defects and impurities in the scintillation crystal. This stage is at the picosecond time scale. At the last stage within nanosecond time scale, charge carriers migrate to the luminescent centers and scintillation light is produced through radiative carrier recombination. Non-radiative decay happens simultaneously. A typical optical transport time for scintillation light to reach photodetectors has been added in Table 1 as a reference.

The production and emission of scintillation light occur only at the last stage. The stochastic nature of the processes occurring before scintillation emission leads to significant statistical fluctuations for the generation of the first scintillation photons. Consequently, the coincidence time resolution achievable by a scintillation based PET detector has an intrinsic limit which is estimated to be on the order of 100 ps in [17].

Apart from the intrinsic scintillation processes, other influencing factors on the timing properties of scintillation based PET detectors have also been extensively investigated, including the intrinsic light yield, decay time, scintillation light transit time variations (a function of crystal element length and photon interaction depth), surface treatment, the photon detection efficiency and time jitter of the photodetector, and the properties (bandwidth and time jitter) of readout electronics [18–21]. These factors determine the limitation for the coincidence time resolution of a realistic scintillation based PET detector. The current performance of "clinically relevant" PET scintillation detectors (with crystal length larger than or equal to 20 mm) is greater than 100 ps [7, 20, 22].

However, an order of magnitude improvement (e.g., 10 ps or better coincidence time resolution) would enable substantial gains in reconstructed image quality and accuracy using ToF PET methodology. It will also lead to many other benefits as discussed in Sect. 1.1. To dramatically improve the coincidence time resolution in PET and achieve a performance far below 100 ps, the intrinsic limitations placed by the scintillation process need to be overcome. Therefore, in this chapter, we review a novel ionizing photons detection mechanism utilizing the optics pump-probe measurement.

1.3 Concept of Optics Pump-Probe Measurement for Radiation Detection

One way to overcome the limitation for coincidence time resolution placed by the scintillation mechanism is to utilize the ultrafast charge carrier transient phenomena occurring before scintillation (as stated in Table 1) to generate ultrafast time stamps for radiation detection. The optics pump-probe method utilizing the mechanisms of optical property modulation, as widely exploited in the optical telecommunications industry [23, 24], shows great potential for the measurement of these ultrafast processes. First, unlike standard ionizing radiation detection methods, the optics pump-probe measurement introduces no extra time delay or variation. It is an "in place" and "instantaneous" detection method that has been widely used to observe femtosecond scale dynamical processes [25, 26]. Second, for the optical modulation material reviewed in this chapter (CdTe) with an electron–hole pair creation energy of approximately 5 eV, it can be estimated that the total number of charge carriers generated by one 511 keV photon is around 10^5. This relatively small number of carriers induces a very subtle modulation signal in the detector material. In optics pump-probe measurement, it is possible to make use of an optical resonant cavity with a high quality factor (high-Q) that would enable the observation of very weak signals. For example, single molecule detection has been achieved with high-Q optical microcavities [27, 28]. Therefore, it is possible to borrow the idea of optics pump-probe measurement to explore a totally new optics based method for 511 keV photon detection in PET utilizing the modulation of optical properties due to ionization.

A basic comparison of a conventional scintillation based PET detection method and the proposed optics based detection method is shown in Fig. 1. In a scintillation based PET detector, the ionization charges created by 511 keV photons are converted to scintillation light which is detected by a photodetector. The time resolution is limited by the scintillation mechanism and other factors as discussed in Sect. 1.2. In the proposed method, a probe laser is used for the detection of a small changes in the local electric field resulting from the ionization charges created from high energy (e.g., 511 keV) photon interactions. Significant changes in optical properties (such as refractive index, absorption coefficient, reflectivity, nonlinear susceptibility, etc.) that are induced by the electric field changes will modulate the characteristics (such as amplitude and phase) of the probe laser light. Therefore, by monitoring the probe laser state with a photodetector, we can detect the interactions of ionizing photons.

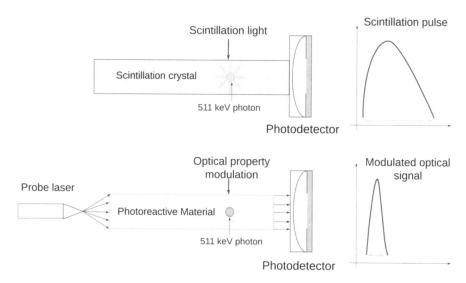

Fig. 1 Upper: Conventional scintillation based PET annihilation photon detection method; Lower: optics pump-probe based detection method

2 Simulation Work Summary

To investigate whether the optical property modulation effect induced by a single 511 keV photon interaction is detectable using modern optics techniques, and to guide the development of a more sensitive detection setup, in this section, we discuss simulation workflow to answer this question.

2.1 Mechanism of Ionizing Charge Induced Optical Property Modulation

During the interaction between a 511 keV photon and the detector material, the 511 keV photon first knocks out an inner-shell or outer-shell electron from the material and leaves this primary electron with high kinetic energy. This primary ejected electron then continues to interact with the material and deposit its energy. These ionization charges will modulate the optical properties of the detector material primarily in two ways, modulating the optical absorption coefficient and modulating the refractive index. For both properties, modulation is induced by either a change in the charge carrier density or a disturbance in the local electric field [29]. In this chapter, we mainly review on the simulation work to estimate the strength of refractive index modulation by electric field disturbance induced Pockels effect.

2.2 *Simulation Setup and Workflows*

Cadmium Telluride (CdTe) is used as the target detector crystal. CdTe is a semiconductor with high effective Z number (50) and high density (5.85 g/cm^3) [30, 31], which could improve the detector crystal's interaction probability with ionizing photons.

In Fig. 2 on the left, a schematic drawing of one potential design of a PET detector module is presented, using the optical property modulation-based detection method. Optical fibers are used to couple the probe laser into the detector crystals. Each detector crystal element is designed to have a small cross-section of, for example, 200 μm × 200 μm for better probe laser confinement, and sufficient length (e.g., 20 mm) to improve the annihilation photon detection efficiency. Multiple detector crystal elements are combined together to make a detector module. The photodetectors coupled to the crystals monitor modulations in the intensity or phase of the probe laser, in order to detect ionizing interactions.

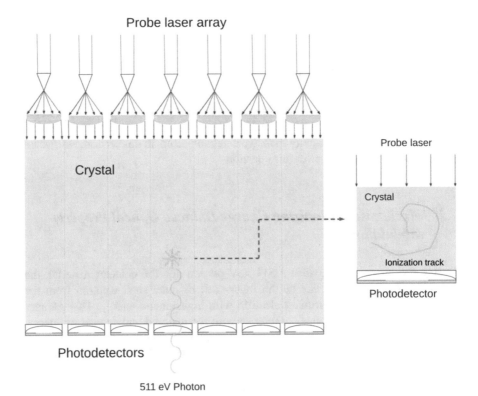

Fig. 2 (Left) One potential design of a PET detector module using the optical property modulation-based detection method. (Right) Zoom-in view of a small crystal element where a 511 keV photon has interacted and created an ionization track; This is the simulation volume studied in this work, which is adapted from [32]

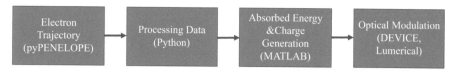

Fig. 3 Simulation flow to calculate the strength of induced optical property modulation. pyPENE-LOPE and Lumerical are two simulation tools. Python and MATLAB scripts are used to process and convert the data. (Adapted from [33])

The right of Fig. 2 illustrates a zoom-in view of a small portion of the detector crystal where a 511 keV photon has interacted with the crystal element and created an ionization track. This is the volume used to perform the simulations. A simulation volume of 200 μm × 200 μm × 200 μm is chosen and we focused on cases when the energy of the 511 keV photon is completely absorbed in a detector crystal element.

The simulation flow is shown in Fig. 3. To study the strength of the ionization charge induced optical property modulation effects, the energy deposition process during 511 keV photon interactions with detector materials was first simulated using pyPENELOPE. Considering Compton scattering is the most possible interaction event, the kinetic energy of electron depends on the scattering angle which ranges from around 350 keV to zero. Therefore, 350 keV is chosen as the representative energy for the incident electron to perform the simulation. All the energy of the ejected electron is assumed to deposit into the detector material. Other energies can be studied following the same procedure. The output of pyPENELOPE provides information on the spatial distribution of the deposited energy per injected electron per step, which is parsed by python and MATLAB scripts and ultimately is converted to a charge generation rate file.

Since the Pockels effect depends on electric field, the electric field distribution is then simulated. This simulation is done with Lumerical. For CdTe material, a simulation volume of 200 μm × 200 μm × 200 μm with bias voltages was used during the simulation. The electric field distribution at 10 ps after the electron is injected is calculated. Ten picosecond is chosen since it is the interesting detecting ultrafast processes after ionization charges are created.

With the simulated results for the 3D distribution of the electric field, the values could be averaged across the entire simulation volume to estimate the average carrier density change and electric field disturbance that can be observed by a probe laser (Probe technique was used for the modulation detection and the probe laser tends to detect an average effect across its beam area). With these single electron tracks spatially averaged charge density and electric field, the induced optical refractive index modulations from Pockels effect could be calculated. These results will determine a best case condition to evaluate if the resulting modulations can be detected with modern optics experiments.

2.3 Simulation Performance

The converted 3D distribution of the charge generation rate (in MATLAB) is shown in Fig. 4. The location of each sphere in Fig. 4 represents a collision event where the electron loses its energy to the crystal. The size of the sphere is proportional to how much energy is deposited. The red arrow marks the incident point of the electron. We could note that larger density of energy is deposited towards the end of the tracks, which is represented by larger spheres.

With this charge generation profile, the charge density and electric field distribution in the target material could be simulated. Based on the estimated electric field disturbance, the induced modulation in optical photo-refraction can be calculated. The main effect is Pockels effects, which lead to modulations in refractive index for the target crystal. The results are shown in Table 2. Since the Pockels effect induced modulation strength varies with the probe laser wavelength, the target wavelength is also marked in Table 2.

Based on the calculation and simulation results [32, 33], the largest refractive index modulation induced by a 511 keV photon is around 4.90E−04 (in CdTe). With modern optics experiments, refraction measurement sensitivity has been reported as better than 4.10E−04 [34]. Therefore, the simulation results show that with appropriate experiment design, it is possible to detect individual 511 keV photons depositing energy equal or higher than 350 keV with the proposed optical modulation-based detection method.

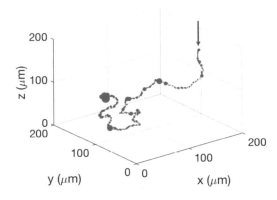

Fig. 4 3D energy deposition trajectory for a single 350 keV ejected electron in CdTe. Red arrow marks the incident point of the electron

Table 2 Estimation on modulation strength of photo-refractive [33]

Modulation mechanism	Material	Target wavelength	Δn
Pockels effect	CdTe	10.6 μm	4.90E−04

3 Experimental Work Summary

3.1 Light Transmission Based Method

An optical setup based on two crossed polarizers (light transmission based method) was proposed to detect ionizing radiation [35, 36], which leveraged an optics pump-probe measurement. A schematic of the arrangement of the optical setup is shown in Fig. 5. A tunable laser beam is first collimated, then the laser light transmits through the first polarizer. The linear polarized laser light from the first polarizer illuminates in front of the detector crystal and exits the crystal from the back side. The exiting light goes into the second crossed polarizer, followed by a focusing lens and a high sensitivity photodiode. Transmitted light intensity is the signal being monitored during the experiment. When the ionizing radiation source is not present, nominally the cross-polarizers cancel the transmitted signal. When ionization occurs within the detector crystal, this transient perturbs the linear polarization of the light entering the crystal, allowing some non-polarized light to transmit through the second polarizer and generate a signal in the photodetector. The detector crystal is biased and the whole setup is placed in a light-tight box during the experiment.

In the proposed method, the Pockels effect is used to measure the optical property modulation induced by radiation, which can modify the refractive index of each detector crystal due to the interaction between the crystal and the ionizing radiation photons. Detector crystals are regarded as Pockels cells and the induced refractive index of each Pockels cell [37] can be expressed by Eq. (1):

$$n(E) \approx n_0 - \frac{1}{2}\gamma n_0^3 E \tag{1}$$

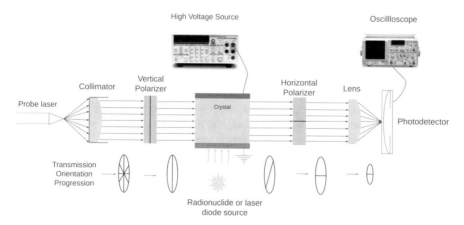

Fig. 5 Diagram of setup based on two crossed polarizers

where $n(E)$ is the refractive index of a detector crystal with an applied electric field E, E is the applied bias voltage, including the external voltage from a high voltage power supply and the internal voltage induced by the interaction between the detector crystal and ionizing radiation photons, γ is the coefficient of the Pockels effect, and n_0 is the refractive index of the detector crystal in the absence of a bias voltage.

The dependence of transmitted light intensity on applied bias voltage is expressed as Eq. (2) [38]:

$$I = I_0 \text{Sin}^2 \left(\frac{\pi n_0 \gamma d E}{\lambda} \right) \tag{2}$$

where I is the intensity of transmitted light through two crossed polarizers, I_0 is the maximum intensity of transmitted light through two crossed polarizers, d is the path length of light through the crystal, which approximately equals the thickness of the detector crystal, λ is the wavelength of the probe laser, and $n(E)$, γ, E have the same meanings as in Eq. (1).

Charge carriers are generated when the ionizing radiation photon interacts with the detector crystal. After generation, charge carriers immediately drift toward two opposite electrodes under the external applied bias voltage, producing an internal electrical field opposite to the external electrical field. In other words, local charge carrier separation will result in a disturbance in the distribution of the electric field within the crystal. According to Eq. (1), after an ionizing radiation interaction, the refractive index of the detector crystal would be modulated as a result. According to Eq. (2), the change of refractive index of the crystal would modify the polarization state of transmitted light through the crystal and then change the intensity of transmitted light through the second polarizer, which would ultimately modify the output of the photodetector. Therefore, monitoring the output of photodetector could achieve the goal to monitor the interaction of ionizing radiation.

3.2   Light Interference Based Method

Alternatively, a two-beam interference setup was proposed to achieve the optics pump-probe measurement [39]. The schematic for the optical setup is shown in Fig. 6. A tunable C-band (with central wavelength around 1550 nm) probe laser first goes through a linear polarizer and then illuminates the CdTe crystal. The light reflected back from the front and rear surfaces of the crystal interferes with each other and generates a stable interference pattern (shown in the inset). The interference pattern is expanded with a lens. An optical iris is placed after the lens to allow only a fixed portion of the interference pattern to pass through (shown by the black arrows in the inset). The light transmitted through the iris is further focused by another lens and

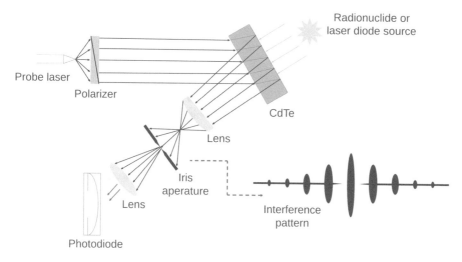

Fig. 6 Schematic of optical setup. The "signal" is represented by a spatial shift in the interference pattern caused by the creation of charge carriers in the crystal

detected with a photodiode detector. The detected light intensity is the signal being monitored during the experiment.

3.3 Experiment Performance Evaluation

A Na-22 radionuclide source induced modulation signal is observed. A 100 μCi Na-22 source [40] is used in experiments. The source is repeatedly placed 15 mm from the CdTe crystals and then removed. The change of signal level between the presence and absence of the source is observed and recorded in order to see how a radionuclide source modulates the signal level. Since the signal does not change immediately during the transition between the delivery of source to the detector crystal and the removal of the source, the selected data point for each measurement is the stable signal level observed at 5 min after placing or removing the source. After the settling time, the signal level remains stable at a fixed value. The value corresponding to the peak of the histogram is used as the magnitude of the stable signal level. All data points are taken consecutively in time.

The experimental result of the light transmission based method is shown in Fig. 7, from which we can observe that the output optical signal from CdTe crystals can be modulated by the presence of the Na-22 radionuclide source, due to a change in the refractive index of both detector crystals.

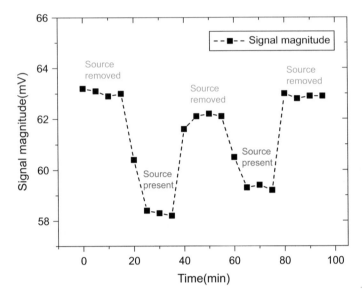

Fig. 7 Modulation signals for CdTe induced by an Na-22 source. Red bars indicate the magnitude of the modulation signal when the Na-22 source is placed nearby detector crystals, whereas blue bars represent the magnitude of the signal when the source is removed

3.4 Methods to Improve the Detection Sensitivity

Previous simulation and experimental work demonstrated that such an optical setup could be utilized in radiation detection, though its detection sensitivity needed improvement. Therefore, two methods are proposed to understand and improve the detection sensitivity of an optical polarization modulation-based method, including the angle between polarizers and electric field distribution within the detection crystal [41, 42]. Cadmium telluride (CdTe) was studied as the detector crystal.

The "magic" angle (i.e., optimal working angle) of the two crossed polarizers based optical setup with CdTe is explored theoretically and experimentally. The experimental results show that the detection sensitivity could be improved by around 10% by determining the appropriate "magic" angle. For the dependence of detection sensitivity on electric field distribution as well as on the bias voltage across the detector crystal using CdTe crystals, the experimental results show that a smaller electrode on the detector crystal, or a more concentrated electric field distribution could improve detection sensitivity. For CdTe, a detector crystal sample with 2.5 mm × 2.5 mm square electrode has twice the detection sensitivity of a detector crystal with 5 mm × 5 mm square electrode. Increasing the bias voltage before saturation for CdTe could further enhance the modulation strength and thus the sensitivity. These investigations demonstrated that by determining the proper working angle of polarizers and bias electrical distribution to the detector, we could improve the sensitivity of the proposed optical setup.

4 Potential Detection Structure with Proposed Detection Setup for Future PET Module

The ultimate goal would be to establish a setup (and develop an appropriate readout method) that is sensitive enough to measure the modulation signal from the ionization charges created by one individual 511 keV photon with measured response time and variance both on the order of picoseconds. When this goal is achieved, we will take steps towards designing a practical PET detector module that employs the proposed optics pump-probe measurement based detection concept as depicted in Fig. 1 lower figure. The exact structure of the detector module will depend on the final choice of the experimental setup, but it should have a low profile to enable assembly of detector modules with high crystal packing fraction relative to incoming photons. Two possible structures are shown in Figs. 8 and 9 to give the readers a basic concept.

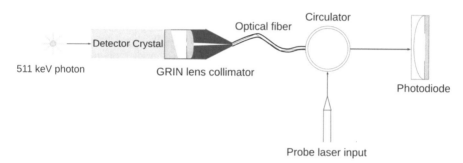

Fig. 8 Possible structure of a practical PET detector module built with the proposed optical property modulation-based detection concept

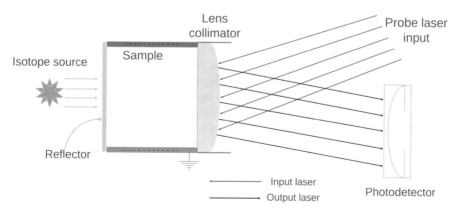

Fig. 9 Potential practical detector module using the proposed optical polarization modulation-based detection mechanism

For the structure shown in Fig. 8, a detector crystal with electro-optic modulation effect is used to sense the interactions of 511 keV photons. The crystal's optical properties will be changed by the ionization charges created from 511 keV photon interactions. A small GRIN lens and an optical fiber are used to collimate the probe laser light and to collect the reflected laser beam from the detector crystal. The intensity of the reflected laser light will be changed as a result of the optical property modulation of the detector crystal. A small optical circulator is used to separate the input probe laser from the reflected light and to guide the reflected laser beam into a photodiode detector that monitors the reflected light intensity change. This structure could be used similarly as conventional scintillation detector elements and be integrated to form a practical compact PET detector module.

For Fig. 9, in the proposed structure of the detector module, a detector crystal with Pockels effect will be used to monitor the induced signal by the 511 keV photon interactions. The input probe laser and the reflected output laser light will be collimated and collected by a lens collimator. The input probe laser beam is composed of an ensemble of sub-laser beams with 200 μm diameter, consistent with [32] which demonstrated a 200 μm diameter laser beam could improve the detection sensitivity of optical properties modulation method. The polarization changes the intensity or the phase of the reflected output laser will be modulated as long as the photons interact with the detector materials. The output laser will be detected directly by the photodetector.

5   Summary

In summary, this chapter discussed the recent developments of using the optics pump-probe measurement to attempt the optical properties modulation method for PET. The theoretical simulation work on the strength of the optical modulation signal induced by the annihilation ionizing photons demonstrates the detection feasibility of the optics pump-probe method. Two designs of optical setups and their experimental results were used to show the proof-of-concept that the optical property modulation effect can be used to detect ionizing photons. Moreover, both amplifying the modulation signal (different bias voltages) and improving the detection sensitivity of the setup (by exploring magic angle) were investigated as different efforts towards the goal to achieve single annihilation photon detection. In the end, two potential detection structures with proposed detection setup for future PET module are presented.

References

1. Lecoq, Paul, et al. "Roadmap toward the 10 ps time-of-flight PET challenge." Physics in Medicine & Biology 65.21 (2020): 21RM01.

2. Yeom, Jung Yeol, et al. "First performance results of Ce: GAGG scintillation crystals with silicon photomultipliers." IEEE transactions on nuclear science 60.2 (2013): 988-992.
3. Melcher, C. L., and J. S. Schweitzer. "Cerium-doped lutetium oxyorthosilicate: a fast, efficient new scintillator." IEEE Transactions on Nuclear Science 39.4 (1992): 502-505.
4. Li, Mohan, Yuli Wang, and Shiva Abbaszadeh. "Development and initial characterization of a high-resolution PET detector module with DOI." Biomedical Physics & Engineering Express 6.6 (2020): 065020.
5. Kronberger, Matthias, Etiennette Auffray, and Paul Lecoq. "Probing the concepts of photonic crystals on scintillating materials." IEEE Transactions on nuclear science 55.3 (2008): 1102-1106.
6. Yeom, Jung Yeol, et al. "Fast timing silicon photomultipliers for scintillation detectors." IEEE Photonics Technology Letters 25.14 (2013): 1309-1312.
7. Cates, Joshua W., and Craig S. Levin. "Advances in coincidence time resolution for PET." Physics in Medicine & Biology 61.6 (2016): 2255.
8. Schaart, Dennis R., et al. "LaBr3: Ce and SiPMs for time-of-flight PET: achieving 100 ps coincidence resolving time." Physics in Medicine & Biology 55.7 (2010): N179.
9. Romanchek, Gregory, et al. "Performance of Optical Coupling Materials in Scintillation Detectors Post Temperature Exposure." Sensors 20.21 (2020): 6092.
10. Conti, Maurizio. "State of the art and challenges of time-of-flight PET." Physica Medica 25.1 (2009): 1-11.
11. Karp, Joel S., et al. "Benefit of time-of-flight in PET: experimental and clinical results." Journal of Nuclear Medicine 49.3 (2008): 462-470.
12. Surti, Suleman, et al. "Investigation of time-of-flight benefit for fully 3-DPET." IEEE transactions on medical imaging 25.5 (2006): 529-538.
13. Fazel, Reza, et al. "Exposure to low-dose ionizing radiation from medical imaging procedures." New England Journal of Medicine 361.9 (2009): 849-857.
14. Presotto, L., et al. "Evaluation of image reconstruction algorithms encompassing time-of-flight and point spread function modelling for quantitative cardiac PET: phantom studies." Journal of Nuclear Cardiology 22.2 (2015): 351-363.
15. Mehranian, Abolfazl, and Habib Zaidi. "Accelerated time-of-flight (TOF) PET image reconstruction using TOF bin subsetization." Journal of Nuclear Medicine 56.supplement 3 (2015): 559.
16. Lecoq, Paul, et al. "Scintillation and inorganic scintillators." Inorganic Scintillators for Detector Systems: Physical Principles and Crystal Engineering (2006): 1-34.
17. Lecoq, Paul, Mikhael Korzhik, and Andrey Vasiliev. "Can transient phenomena help improving time resolution in scintillators?." IEEE Transactions on Nuclear Science 61.1 (2013): 229-234.
18. Spanoudaki, V. Ch, and C. S. Levin. "Investigating the temporal resolution limits of scintillation detection from pixellated elements: comparison between experiment and simulation." Physics in Medicine & Biology 56.3 (2011): 735.
19. Derenzo, Stephen E., Woon-Seng Choong, and William W. Moses. "Fundamental limits of scintillation detector timing precision." Physics in Medicine & Biology 59.13 (2014): 3261.
20. Cates, Joshua W., Ruud Vinke, and Craig S. Levin. "Analytical calculation of the lower bound on timing resolution for PET scintillation detectors comprising high-aspect-ratio crystal elements." Physics in Medicine & Biology 60.13 (2015): 5141.
21. Gundacker, S., et al. "On the comparison of analog and digital SiPM readout in terms of expected timing performance." Nuclear Instruments and Methods in Physics Research Section A: Accelerators, Spectrometers, Detectors and Associated Equipment 787 (2015): 6-11.
22. Nemallapudi, Mythra Varun, et al. "Sub-100 ps coincidence time resolution for positron emission tomography with LSO: Ce codoped with Ca." Physics in Medicine & Biology 60.12 (2015): 4635.
23. Slavík, Radan, et al. "All-optical phase and amplitude regenerator for next-generation telecommunications systems." Nature Photonics 4.10 (2010): 690-695.
24. Hecht, Jeff. Understanding fiber optics. Jeff Hecht, 2015.

25. Tomimoto, Shinichi, et al. "Ultrafast dynamics of lattice relaxation of excitons in quasi-one-dimensional halogen-bridged platinum complexes." Physical review B 66.15 (2002): 155112.
26. Weiner, Andrew. Ultrafast optics. Vol. 72. John Wiley & Sons, 2011.
27. Armani, Andrea M., et al. "Label-free, single-molecule detection with optical microcavities." Science 317.5839 (2007): 783-787.
28. Vollmer, Frank, and Lan Yang. "Review Label-free detection with high-Q microcavities: a review of biosensing mechanisms for integrated devices." Nanophotonics 1.3-4 (2012): 267-291.
29. Ingham, Joshua. Surface Temperature Equalisation Through Automated Laser Vaporisation of Thick Film Electrical Heating Elements. Diss. Lancaster University (United Kingdom), 2019.
30. Lee, Young-Jin, Dae-Hong Kim, and Hee-Joung Kim. "The effect of high-resolution parallel-hole collimator materials with a pixelated semiconductor SPECT system at equivalent sensitivities: Monte Carlo simulation studies." Journal of the Korean Physical Society 64.7 (2014): 1055-1062.
31. Wang, Yuli, Zehao Li, and Jianfeng Xu "Investigation of Pockels effect in optical property modulation-based radiation detection method for positron emission tomography." Medical Imaging 2019: Biomedical Applications in Molecular, Structural, and Functional Imaging. Vol. 10953. International Society for Optics and Photonics, 2019.
32. Tao, Li, et al. "Simulation studies to understand sensitivity and timing characteristics of an optical property modulation-based radiation detection concept for PET." Physics in Medicine & Biology 65.21 (2020): 215021.
33. Wang, Yuli, et al. "Detection sensitivity of optical property-based radiation detection for PET: refraction index modulation." 2020 IEEE Nuclear Science Symposium and Medical Imaging Conference (NSS/MIC). IEEE.
34. Vernon, S. P., et al. "X-ray bang-time and fusion reaction history at picosecond resolution using RadOptic detection." Review of Scientific Instruments 83.10 (2012): 10D307.
35. Wang, Yuli, et al. "Two-crossed-polarizers based optical property modulation method for ionizing radiation detection for positron emission tomography." Physics in Medicine & Biology 64.13 (2019): 135017.
36. Wang, Yuli, et al. "Investigation of optical property modulation based ionizing radiation detection method for PET: two-crossed-polarizers based method." 2019 IEEE Nuclear Science Symposium and Medical Imaging Conference (NSS/MIC). IEEE.
37. Ironside, Charlie. "Linear electro-optic effect, electroabsorption and electrorefraction." (2017).
38. Collett, Edward. "Field guide to polarization." Bellingham, WA: Spie, 2005.
39. Tao, Li, Henry M. Daghighian, and Craig S. Levin. "A promising new mechanism of ionizing radiation detection for positron emission tomography: modulation of optical properties." Physics in Medicine & Biology 61.21 (2016): 7600.
40. La Rocca, P., and F. Riggi. "Energetic photons through water: an undergraduate experiment with correlated γ-rays from 22Na." European Journal of Physics 37.4 (2016): 045804.
41. Wang, Yuli, et al. "Further investigations of a radiation detector based on ionization-induced modulation of optical polarization." Physics in Medicine & Biology 66.5 (2021): 055013.
42. Wang, Yuli, et al. "Approaches to improving the detection sensitivity of optical modulation based radiation detection method for positron emission tomography." 2019 IEEE Nuclear Science Symposium and Medical Imaging Conference (NSS/MIC). IEEE.

Multi-material Decomposition (m-MD) Based Spectral Imaging in Photon-Counting CT

Xiangyang Tang, Yan Ren, and Arthur E. Stillman

1 Introduction

The advent of dual-source dual-detector (DSDD) CT in mid-2000s not only made state-of-the-art multi-detector CTs (MDCT) more potent in cardiovascular imaging [1–3] but also provided the opportunity for spectral CT imaging. Up-to-date, the spectral (dual energy) CT implemented via energy-integration has been adding significant value in the clinic, but its capability is still limited by technological constraints such as mismatching in energy weighting [4–6], overlapping in source spectra [7–9], and limitation in radiation dose and electronic noise [10, 11]. Recently, there have been quite a few exciting technological advancements across the modalities in the clinic to make diagnostic imaging more potent over cardiovascular, oncologic, and neurovascular applications. Photon-counting spectral CT is one of them, mainly owing to its potential of improving soft tissue differentiation and multi-material decomposition (and thus multi-contrast agents) based spectral imaging for advanced clinical applications at high spatial resolution and low noise. In recognition of the value added by the energy-integration based spectral CT implemented via data acquisition at two distinct peak voltages, i.e., dual energy CT (DECT), the anticipation on photon-counting spectral CT's utility for advanced clinical applications is high.

Initially, spectral CT and X-ray imaging are proposed to be implemented in the projection domain via either photon-counting or energy-integration (e.g., fast/slow kV-switching or layered detector). The research and development (R&D) in spectral CT implemented via photon-counting detection (photon-counting spectral CT henceforth) is gaining the momentum [12–15], mainly owing to the immediate

X. Tang (✉) · Y. Ren · A. E. Stillman
Department of Radiology and Imaging Sciences, Emory University School of Medicine, Atlanta, GA, USA
e-mail: xiangyang.tang@emory.edu

benefit offered by the facilitation of spectral channelization via energy thresholding of the incident X-ray photons. Such a facilitation may improve the imaging performance of spectral CT substantially as the overlapping in source spectra existing in energy-integration detection can be, in principle, eradiated. Moreover, the facilitation in spectral channelization provides the opportunity to acquire projection data in more than two spectral channels. With the availability of multiple spectral channels, one may wonder how many channels are optimal to accomplish the tasks in clinical and preclinical applications. Obviously, such a fundamental question can only be answered based on an in-depth understanding of the fundamentals in physics for spectral imaging in photon-counting CT. Focusing on its implementation in the projection domain, we discuss multi-material decomposition (m-MD) based spectral imaging in photon-counting CT in this chapter, with the goal of providing necessary information to fully understand the fundamentals of this novel CT imaging method and the principles to guide its design, implementation, and optimization in practice.

2 Fundamentals in Physics for Spectral Imaging in Photon-Counting CT

In the energy range up to 1.02 MeV at which pair production may occur, an X-ray photon may undergo three independent interactions while it propagates in media: (1) photoelectric absorption, (2) free-electron Compton (incoherent) scattering, and (3) Rayleigh (coherent) scattering along with correction of electron-binding incoherent scattering [16–18]. The rule in physics states that the total cross section of interactions is a summation of the cross sections corresponding to each of the three interactions, which in turn leads to the fact that the mass attenuation of a material is a summation of the components corresponding to each of the three interactions. In reality, the tissues in human body are mixture of materials [19, 20]. At a location x in the three-dimensional (3-D) space, if the effective atomic number $Z(x)$ and effective electron density $N_e(x)$ of a material are known, the summation rule analytically states [16–18, 21–23]

$$\mu(x; E) = \frac{c \cdot N_e(x) \cdot Z^a(x)}{E^b} + N_e(x) \cdot \sigma_{KN}(E) + \frac{e \cdot N_e(x) \cdot Z^d(x)}{E^f}, \quad (1)$$

where E denotes the energy of X-ray photon in the beam and $\sigma_{KN}(E)$ the Klein–Nishina function. Parameters a–f are constants that can be determined via data fitting, but maintenance in their accuracy over a large energy range is extremely difficult [24]. Notably, Eq. (1) consists of three terms corresponding to the three interactions, but it is in fact tantamount to a two-parametric ($Z(x)$ and $N_e(x)$) model that characterizes the three interactions.

2.1 Interaction Decomposition

Based on the publication on human tissues' mass attenuation coefficient as function of energy [25] and their own experiments, Alvarez and Macovski made the assumption that the attenuation associated with Rayleigh (coherent) scattering and other interactions [19] in the energy range for diagnostic imaging can be omitted. In their pioneering work [26], Alvarez and Macovski proposed that within the energy range for diagnostic imaging and above the K-edge of materials in X-ray beam, the photoelectric absorption and Compton scattering can approximately span a two-dimensional (2-D) linear space, i.e.,

$$\mu\left(\boldsymbol{x};E\right) = K_1 \frac{\rho\left(\boldsymbol{x}\right)}{A\left(\boldsymbol{x}\right)} Z^n\left(\boldsymbol{x}\right) \cdot \frac{1}{E^3} + K_2 \frac{\rho\left(\boldsymbol{x}\right)}{A\left(\boldsymbol{x}\right)} Z\left(\boldsymbol{x}\right) \cdot \sigma_{KN}(E) \triangleq a_1\left(\boldsymbol{x}\right) \cdot f_{pe}(E)$$

$$+ a_2\left(\boldsymbol{x}\right) \cdot f_{KN}(E), \tag{2}$$

where K_1 and K_2 are constants, $n \approx 4$, and $\rho(\boldsymbol{x})$ denotes the material's mass density. $f_{pe}(\boldsymbol{x})$ and $f_{KN}(\boldsymbol{x})$ are the basis functions corresponding to the photoelectric absorption and Compton scattering, respectively. In mathematics, Eq. (2) states that the attenuation coefficient of a material can be defined as a vector, i.e., a linear combination of the attenuation coefficients of the two interactions [26], in the linear (interaction) space. Note that Eq. (2) is not just a reformation of Eq. (1) but a conceptual breakthrough from interaction summation to linear space spanning. Such a breakthrough is of theoretical prominence and has been the foundation of spectral CT and X-ray imaging for almost half a century, ranging from material mapping/cancellation [27–29] to virtual monochromatic imaging/analysis [30–32] in extensive clinical and preclinical applications. Alvarez and Macovski also showed that, via scanning under dual peak voltage (kVp), CT images corresponding to the photoelectric absorption and Compton scattering can be obtained separately, which may provide more information for diagnostic imaging than the conventional CT image that is a mixture of the two interactions [26].

2.2 Material Decomposition and Virtual Monochromatic Imaging

Alvarez et al. further showed via linear mapping that the interaction decomposition can be converted into material decomposition [30, 31]

$$\mu\left(\boldsymbol{x};E\right) = a_{m1}\left(\boldsymbol{x}\right) \cdot \mu_{m1}(E) + a_{m2}\left(\boldsymbol{x}\right) \cdot \mu_{m2}(E), \tag{3}$$

where the subscripts m_1 and m_2 denote two different materials. The importance of Eq. (3) cannot be overstated as it provides the freedom in selection of adequate basis functions for material decomposition in application-driven tasks. Also, such a linear mapping converts the two-parametric model (Eq. 1) into a two-material model that actually encloses the three interactions (photoelectric absorption and Compton and Rayleigh scatterings), which may be the underlying reason why the latter outperforms the former in accuracy as quantitatively evaluated and verified by Williamson et al. [21–23], though, as reported [26, 33], the interaction decomposition can outperform material decomposition in characterizing certain tissues, e.g., gray and white matters.

The mass densities $a_{m1}(x)$ and $a_{m2}(x)$ are not dependent on energy E. Given two basis materials, their mass attenuation coefficients $\mu_{m1}(E)$ and $\mu_{m2}(E)$ as the function of energy E are known in practice at sufficient accuracy. Thus, in principle, monochromatic image can be synthesized by setting the energy E in Eq. (3) (namely virtual monochromatic imaging henceforth), while virtual monochromatic analysis can be carried out by varying the energy level E in an interested energy range. Note that, as revealed in Eq. (3), a virtual monochromatic image no longer suffers from the so-called beam hardening artifact that has been one of the major image quality issues in the conventional CT.

2.3 Multi-material Decomposition Based Spectral Imaging

Iodine, with its K-edge (33.2 keV) falling in the energy range of diagnostic imaging, is utilized as the contrast agent in the vast majority (85%) [34] of CT scans for diagnostic imaging in clinics. With resort to singular value analysis (SVD), Lehmann and Alvarez studied how the dimensionality of material space changes if iodine is present in the materials to be imaged [19, 34]. They found that in presence of iodine, the dimensionality of the linear space for material decomposition becomes three and thus three basis materials, of which iodine is the necessary, have to be used to span the space for material decomposition (three-material decomposition henceforth) [35, 36]. By extending Eq. (3) into

$$\mu(x; E) = a_{m1}(x) \cdot \mu_{m1}(E) + a_{m2}(x) \cdot \mu_{m2}(E) + a_I(x) \cdot \mu_I(x), \qquad (4)$$

the three-material decomposition (3-MD) scheme is adopted by Roessl et al. for spectral imaging in photon-counting spectral CT that make use of iodine's K-edge effect [37–41]. Later on, 3-MD is further extended for four-material decomposition (4-MD) based spectral imaging of multi-contrast agents (iodine- and gadolinium-based) in photon-counting CT by extending Eq. (4) into

$$\mu(x; E) = a_{m1}(x) \cdot \mu_{m1}(E) + a_{m2}(x) \cdot \mu_{m2}(E) + a_I(x) \cdot \mu_I(E) + a_{Gd}(x) \cdot \mu_{Gd}(E), \qquad (5)$$

which means that the dimensionality of the material space enclosing biological tissues and materials with two distinct K-edges equals to four (see Sect. 2.2), and the K-edge materials have to be in the basis materials [9].

Up to now, the first two terms on the right side of Eqs. (4) and (5) have been assumed to be associated with two basis materials. However, in the work by Roessl and Schlomka, these two terms actually correspond to the photoelectric absorption and Compton scattering [38, 39], which may serve as an experimental verification that the basis function can be either a function of energy that characterizes X-ray's interaction with material or the material's attenuation coefficient. In other words, in presence of the material with its K-edge falling into the diagnostic energy range (K-edge material henceforth), one may choose interaction function and material's attenuation function in a mixed manner as the basis function, as long as the dimensionality of the basis function matches (equal to or larger than) that of the material space enclosing the materials to be imaged. Hence, we no longer distinguish between interaction and attenuation in basis functions despite the fact that such a hybrid decomposition may make the calibration process, in practice, complicated. Moreover, it is a prerequisite that the K-edge material must be included in the basis materials, otherwise the attenuation property of K-edge material, especially its behavior surrounding the K-edge, cannot be adequately characterized.

2.4 Dimensionality of Material Space and Its Determination

In practice, depending on the specific imaging task in practice, many material pairs, e.g., PMMA/Teflon [42], PMMA/Aluminum [16], and soft tissue/cortical bone [5], have been chosen as the basis materials for two-material decomposition (2-MD) of biological tissues ($1 \leq Z \leq 20$). Via principal component analysis, it has been found that the material decomposition carried out with the first and second principal components as the (virtual) basis materials perform the best in terms of differentiation of soft tissues in the human body. Williamson et al. reported in a series of papers [21–23] that a dual basis vector model (polystyrene/H_2O for low-Z materials or H_2O/$CaCl_2$ (23%) for high-Z materials) outperforms the two-parametric model (Eq. 2) in characterization of biological tissues. Various numerical approaches, such as polynomial fitting, iterative Newton–Raphson solution for non-linear integral equations, and maximum likelihood estimation, have been proposed for 2-MD. Up-to-date, all of these basis materials and numerical solutions behave very well in 2-MD, which, from our point of view, is mainly due to the fact that the dimensionality of the linear space for material decomposition of biological tissues equals to two, as investigated and verified by Lehmann and Alvarez through singular value analysis [19, 34, 35]. On the other hand, it is interesting to note that as many as 12 interactions may occur while X-ray photons are propagating in biological tissues, though the occurrence likelihood of vast majority of them is very low and thus can be omitted [19]. The dimensionality of the material space associated with the XCOM library has been further studied recently using a hypothesis-based approach and found to likely be up to four. Despite the likelihood associated with the interactions in addition to the photoelectric absorption and Compton scattering,

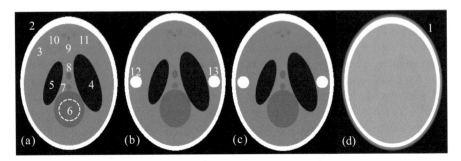

Fig. 1 Ground truth of Phan-0 (**a**), Phan-1 (**b**), and Phan-2 (**c**). The adipose layer enclosing the skull is visible in (**d**), while the ROI gauging the contrast, noise and CNR against the background (brain parenchyma) is indicated by the dashed circle in (**a**)

it has thus far been a common practice to assume that the dimensionality of linear space for material decomposition of biological tissues is two.

In analogue to what has been done by Lehmann and Alvarez [42], the SVD approach can be used to determine the dimensionality of material space for spectral imaging in photon-counting CT. Suppose a total of M materials are to be imaged in an application in the energy range [18 150] keV. Given the i-th material, let $\{\mu_{ij}\}$ denotes the sampling of its mass attenuation coefficient at 1 keV interval where $1 \leq j \leq N$ and N is the number of samplings in the energy range. A data matrix can be formed by sorting all of the sampled data row-wise into the matrix $A_{M \times N} = \{\mu_{ij}\}$. In principle, the singular value decomposition states that the matrix $A_{M \times N}$ can be decomposed into the product of three matrices [43–45]

$$A_{M \times N} = U_{M \times M} \Sigma_{M \times N} V_{N \times N}^{T}, \tag{6}$$

where $U_{M \times M}$ and $V_{N \times N}$ are orthogonal unity matrices and $\Sigma_{M \times N}$ is a rectangular diagonal matrix with the eigenvalues $\{\sigma_i : 1 \leq i \leq M\}$ of matrix $A_{M \times N}$ in descending order as its diagonal elements, i.e., $\{\Sigma_{ii} = \sigma_i\}$. By definition, the dimensionality of matrix $A_{M \times N}$ (i.e., the dimensionality of material space) is the number of the eigenvalues that is larger than zero.

Suppose there is a humanoid head phantom modified from the Shepp–Logan phantom, in which bony structures coexist with soft tissue (phan-0 in Fig. 1a). Two rods made of gadolinium at concentration 10 mg/ml are inserted in the brain parenchyma (phan-1 in Fig. 1b). Then, the rod on the left is replaced with iodine at 10 mg/ml, while that on the right remains unchanged (phan-2 in Fig. 2c). The mass attenuation coefficients of the soft tissues, cortical bone, and contrast materials in the phantom and their variation over energy are determined by consulting the authoritative publications [15, 42], e.g., the EPDL library [35]. Notably, a layer of adipose is added on top of the skull made of cortical bone to mimic the skin and subcutaneous fat (Fig. 1d). The geometry and material configuration of the phantoms in Fig. 1 are detailed in Table 1.

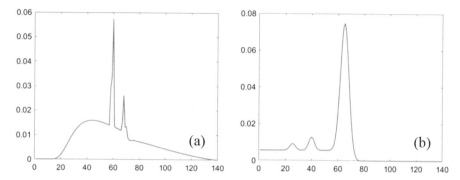

Fig. 2 Schematics showing the spectrum of an X-ray source (**a**) and the spectral response of a realistic detector (**b**)

Table 1 Parameters defining the geometry and material configuration of the humanoid head phantom modified from the Shepp–logan phantom [34] with the mass attenuation coefficients being set by consulting the EDPL library [35]

	Material	Center	Axis (a, b, c)	Angle (phi, theta, psi)
1	Adipose	$(0, 0, 0)$	$(0.73, 0.97, 0.9)$	$(0, 0, 0)$
2	Cortical bone	$(0, 0, 0)$	$(0.69, 0.92, 0.90)$	$(0, 0, 0)$
3	Brain gray	$(0, 0, 0.0184)$	$(0.6624, 0.874, 0.88)$	$(0, 0, 0)$
4	Brain white	$(-0.26, -0.25, 0)$	$(0.41, 0.21, 0.16)$	$(108, 0, 0)$
5	Brain white	$(0.22, -0.25, 0)$	$(0.31, 0.22, 0.11)$	$(-72, 0, 0)$
6	75% gray + 25% H_2O	$(0, -0.25, 0.35)$	$(0.21, 0.50, 0.25)$	$(0, 0, 0)$
7	75% gray + 25% H_2O	$(0, -0.25, 0.052)$	$(0.046, 0.046, 0.046)$	$(0, 0, 0)$
8	75% gray + 25% H_2O	$(-0.08, -0.25, -0.65)$	$(0.023, 0.02, 0.046)$	$(0, 0, 0)$
9	75% gray + 25% H_2O	$(0.06, -0.25, -0.65)$	$(0.046, 0.02, 0.023)$	$(0, 0, 0)$
10	75% gray + 25% H_2O	$(0, -0.25, -0.1)$	$(0.056, 0.10, 0.04)$	$(90, 0, 0)$
11	75% gray + 25% H_2O	$(0, -0.25, -0.605)$	$(0.02, 0.10, 0.02)$	$(90, 0, 0)$
12	Iodine rod	$(0.52, 0, 0)$	$(0.075, 0.075, 0.8)$	$(0, 0, 0)$
13	Gadolinium rod	$(-0.52, 0, 0)$	$(0.075, 0.075, 0.8)$	$(0, 0, 0)$

Itemized in the first row of Table 2 are singular values of the typical biological tissues in the human body. Via SVD, the dimensionality of the material space spanned by the typical biological tissues is determined as two which is consistent to what has been published in the literature [26, 35]. Presented in the second row are the singular values corresponding to the biological tissues in configuring Phan-0. Apparently, the dimensionality of Phan-0 is still two as all the tissues in it are just a subset of the tissues in human body (row 1). Listed in the third row are the singular values corresponding to Phan-1, showing that, with inclusion of Gadolinium (5 mg/ml), the dimensionality of material space increases from two to three, which is again consistent to what has been published in literature [26]. The singular values corresponding to Phan-2 are displayed in the fifth row, which

Table 2 Dimensionality of the typical biological tissues in human body [25, 46–48] and the tissues and materials that configure the phantoms used in the study (energy range: 18–150 keV; SV: singular value)

Materials to be imaged	SV_1	SV_2	SV_3	SV_4	SV_5	Dim.
Common biological tissues in human body [24, 25]	8.8583e+00 (1.0)	2.4994e−01 (2.8215e−02)	2.6310e−02 (2.9701e−03)	1.7657e−02 (1.9932e−03)	1.5322e−02 (1.7297e−03)	2
Adipose, white matter, gray matter, cortical bone	4.8678e+01 (1.0)	5.3443e+00 (1.0979e−01)	2.0031e−02 (4.1149e−04)	2.2550e−05 (4.6324e−07)	–	2
Adipose, white matter, gray matter, cortical bone, I (10 mg/ml)	4.9393e+01 (1.0)	6.5469e+00 (1.3255e−01)	1.1973e+00 (2.4240e−02)	1.8286e−02 (3.7022e−04)	2.2551e−05 (4.5656e−07)	3
Adipose, white matter, gray matter, cortical bone, Gd (10 mg/ml)	4.9526e+01 (1.0)	6.3207e+00 (1.2762e−01)	6.7173e−01 (1.3563e−02)	1.9990e−02 (4.0363e−04)	2.2543e−05 (4.5517e−07)	3
Adipose, white matter, gray matter, cortical bone, I (10 mg/ml), Gd (10 mg/ml)	5.0237e+01 (1.0)	7.2997e+00 (1.4531e−01)	1.2820e+00 (2.5519e−02)	6.6460e−01 (1.3229e−02)	1.8204e−02 (3.6236e−04)	4

shows that, with both iodine (5 mg/ml) and gadolinium (5 mg/ml) included, the dimensionality of materials becomes four.

2.5   Conditioning of Basis Materials and Its Assessment

Once the dimensionalities of material space and projection space are determined, the next task in implementing spectral CTs with photon-counting detection is the selection of basis materials. As reported in literature [8, 19, 20, 26, 29–31, 49, 50], the performance of material decomposition, especially the noise property, is quite stable over the selection of basis materials in 2-MD supported by two spectral channels. However, in our investigation of 3-MD and 4-MD thus far, it has been observed that its noise property is strongly dependent on the selection of basis materials [49, 50]. Moreover, we have noted that the material decomposition and virtual monochromatic imaging/analysis behave quite differently over basis material selection. Thus, aiming to further understand the fundamentals in photon-counting spectral CTs and providing guidelines on its analysis, design, and implementation, a systematic study of the effective dimensionality of material space and the conditioning of basis materials and their impact on spectral imaging (material decomposition and virtual monochromatic imaging/analysis) performance in photon-counting CT via analysis and simulation studies is necessary. Notably, by "conditioning," we refer to (1) sufficiency in dimensionality, (2) well-pose in the condition number of basis functions, and (3) matching of K-edge materials.

Given a set of basis materials, its dimensionality can be determined via SVD in exactly the same way as determining the dimensionality of the material space except for the only difference in forming the data matrix $A'_{M' \times N}$. By definition, the dimensionality of matrix $A'_{M \times N}$ (i.e., the dimensionality of basis materials) is the number of the eigenvalues of matrix $A'_{M \times N}$ that is larger than zero. In addition, the condition number (well-pose) of matrix $A'_{M \times N}$ (i.e., the condition number of basis materials) is defined as the square root of the ratio of the largest eigenvalue over the smallest nonzero eigenvalue, i.e., $\sigma'_{max}/\sigma'_{min}$. In linear algebra, the condition number of a system of linear equations is an indicator of the system's robustness over measurement error and/or noise [44]. It can also be considered as a quantity to assess the uniformity in the norm of the basis vectors that span a linear space [45]. Hence, intuitively, the condition number may be perceived as a quantity to characterize the extent at which the vectors that span the linear space are uncorrelated. In general, the larger the condition number, the worse the situation [44, 45]. Potentially, the minimum condition number can go down to unity, in which the involved vectors are not only uncorrelated but also orthogonal.

A number of sets of basis materials have been investigated by us as the basis materials to study the dimensionality of material space, conditioning of basis materials, and their impacts on imaging performance [51, 52]. In 3-MD based spectral CT, six sets of basis materials have been specifically designated: BM-0 (adipose, PMMA and Teflon), BM-1 (soft tissue, Gadolinium (10 mg/ml) and cortical bone), BM-2 (soft tissue, Gadolinium (20 mg/ml) and cortical bone), BM-3 (adipose, Gadolinium (20 mg/ml) and PMMA), BM-4 (Soft tissue, I (20 mg/ml), Gd (20 mg/ml) and cortical bone), and BM-5 (Adipose, I (20 mg/ml), Gd (20 mg/ml) and Teflon).

In the first case of 3-MD (first row of Table 3), the condition number corresponding to basis materials adipose, PMMA, and Teflon (BM-0) is larger than 3×10^4. Such a large condition number implies that the material decomposition carried out under this set of basis materials may be ill-posed. As demonstrated in more cases of 3-MD (rows 2–4 of Table 3), the condition number of basis materials becomes approximately 3-order smaller if Teflon, PMMA, and adipose are in turn replaced by cortical bone, gadolinium (10–30 mg/ml), and soft tissue. The fifth row in Table 3 designates a special case in which adipose and PMMA are utilized as the basis materials to boost the contrast between soft tissues (see Fig. 6 and Discussion for more detail). Moreover, as shown in rows 6–8 of Table 3, similar phenomenon occurs while iodine (10–30 mg/ml) is in turn replacing gadolinium (10–30 mg/ml). With increasing concentration in gadolinium (rows 2–4 in Table 3), the condition number of basis materials declines and does so in the cases wherein iodine is one of the basis materials (rows 6–8 in Table 3). In the case of 4-MD using cortical bone, gadolinium (20 mg/ml), iodine (20 mg/ml), and soft tissue as the basis materials (row 9 in Table 3), it is observed that, given gadolinium and iodine in the basis materials along with soft tissue and cortical bone, the condition number of basis materials is reasonably small even though soft tissue and cortical bone are replaced by adipose and Teflon (row 10 in Table 3), respectively.

2.6 Conditioning of Spectral Channelization and Its Assessment

In principle, the number of spectral channels can be conceived as the dimensionality of projection space. As material decomposition is a linear mapping from the projection space to A-space [20, 49, 50], the number of spectral channels in photon-counting spectral CT should be larger than or at least equal to that of the material space, depending on the specific task in which the interested tissues are to be decomposed into a linear combination of how many basis materials. More research has recently been carried out to address this fundamental question via various approaches. For example, based on signal detection theory, Alvarez studied the dimensionality of material space and the influence of the number of spectral channels on image quality (mainly signal-to-noise ratio (SNR)) [19, 28]. In practice, given a radiation dose, the number of X-ray photons falling into each spectral channel declines with the increasing number of channels, which may lead to degraded SNR in the basis image. Especially in the channel of lowest energy, reduction in the number of X-ray photons may induce photon-starvation that may offset the benefit offered by photon-counting spectral CT over the conventional

Table 3 Condition numbers of the basis materials employed in the study (energy range: 18–150 keV; SV: singular value; rows 1–8: 3-MD; rows 9–10: 4-MD; normalized singular value is in bracket)

	Basis materials	SV_1	SV_2	SV_3	SV_4	Condition
1	Adipose, PMMA, and Teflon (BM-0)	1.8515e+01 (1.0)	1.9881e+00 (1.0738e−01)	5.8465e−04 (3.1577e−05)	–	3.1668e+04
2	Soft tissue, Gd (10 mg/ml), and cortical bone (BM-1)	4.8878e+01 (1.0)	4.7498e+00 (9.7178e−02)	5.3173e−01 (1.0879e−02)	–	9.1922e+01
3	Soft tissue, Gd (20 mg/ml), and cortical bone (BM-2)	4.9518e+01 (1.0)	5.1824e+00 (1.0466e−01)	9.7901e−01 (1.9771e−02)	–	5.0579e+01
4	Soft tissue, Gd (30 mg/ml), and cortical bone	5.0337e+01 (1.0)	5.7120e+00 (1.1348e−01)	1.3222e+00 (2.6267e−02)	–	3.8070e+01
5	Adipose, PMMA, and Gd (20 mg/ml) (BM-3)	1.5342e+01 (1.0)	2.1152e+00 (1.3788e−01)	2.6951e−02 (1.7567e−03)	–	5.6925e+02
6	Soft tissue, I (10 mg/ml), and cortical bone	4.8743e+01 (1.0)	5.0891e+00 (1.0441e−01)	9.1948e−01 (1.8864e−02)	–	5.3011e+01
7	Soft tissue, I (20 mg/ml), and cortical bone	4.9164e+01 (1.0)	6.1021e+00 (1.2412e−01)	1.5460e+00 (3.1446e−02)	–	3.1800e+01
8	Soft tissue, I (30 mg/ml), and cortical bone	4.9691e+01 (1.0)	7.3744e+00 (1.4840e−01)	1.9146e+00 (3.8529e−02)	–	2.5954e+01
9	Soft tissue, I (20 mg/ml), Gd (20 mg/ml), and cortical bone (BM-4)	5.0639e+01 (1.0)	7.0196e+00 (1.3862e−01)	2.3250e+00 (4.5913e−02)	9.2075e−01 (1.8183e−02)	5.4997e+01
10	Adipose, I (20 mg/ml), Gd (20 mg/ml), and Teflon (BM-5)	2.4833e+01 (1.0)	2.8444e+00 (1.1454e−01)	1.3544e−01 (5.4539e−02)	7.1303e−01 (2.8713e−02)	5.9015e+01

CT. Hence, in a specific task, the number of spectral channels (dimensionality of projection space) in photon-counting spectral CT should be as few as possible as long as it is sufficient (equal to or larger than) to accommodate the dimensionality of basis materials.

In addition to being used to determine the dimensionality of material space and the conditioning of basis materials as presented above, the SVD approach can also be employed to study the conditioning of spectral channelization for spectral imaging in photon-counting CT. Suppose a total of M spectral channels in the energy range [18 150] keV are to be utilized for spectral imaging in photon-counting CT. Given the k-th spectral channel, let $\{q_{kj}\}$ denotes the sampling of its intensity profile ($D_k(E) \cdot N_0(E)$ in Eqs. (1) and (7)) over the entire energy range [18 150] keV at 1 keV interval, where $1 \leq j \leq N$ and N is the number of samplings in the energy range. A data matrix can be formed by sorting all of the sampled data row-wise into the matrix $Q_{M \times N} = \{q_{kj}\}$. In principle, the SVD states that the matrix $Q_{M \times N}$ can be decomposed into the product of three matrices [44]

$$Q_{M \times N} = U_{M \times M} \Sigma_{M \times N} V_{N \times N}^{T} \tag{7}$$

where $U_{M \times M}$ and $V_{N \times N}$ are orthogonal matrices and $\Sigma_{M \times N}$ is a rectangular diagonal matrix with the singular values $\{\lambda_i : 1 \leq i \leq M\}$ of matrix $Q_{M \times N}$ in descending order as its diagonal elements, i.e., $\{\Sigma_{ii} = \lambda_i\}$. By definition, the condition number (well-posedness) of matrix $Q_{M \times N}$, i.e., the conditioning of spectral channelization, is defined as square root of the ratio of the largest singular value over the smallest nonzero singular value, i.e., $\lambda_{max}/\lambda_{min}$ [44].

3   Data Acquisition and Image Formation for Spectral Imaging in Photon-Counting CT

3.1   Detector Spectral Response and Its Impact on Spectral Channelization

Suppose there is a photon-counting CT with its X-ray source spectrum plotted in Fig. 2a. The spectral channels are implemented via energy thresholding as [1 ~ 51, 52 ~ 68, 69 ~ 140] keV in 3-MD and [1 ~ 43, 44 ~ 58, 59 ~ 72, 73 ~ 140] in 4-MD. The spectral channels under ideal detector spectral response are plotted in Fig. 3a, b, while those under realistic response modulated by the spectral distortion (Fig. 2d) induced by charge-sharing, Compton scattering, and fluorescent escaping [40, 42] are in Fig. 3a′, b′. Notably, the distortion in detector's spectral response results in inter-channel spectral overlapping. Additionally, intentional overlap in adjacent spectral channels can be induced, and one such example at 25% overlapping is illustrated in Fig. 3c, c′. Since the distortion in detector's spectral response is inevitable in practice, all the results henceforth presented in this chapter are acquired

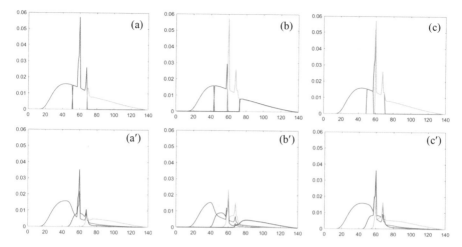

Fig. 3 Schematics showing the spectral channelization in 3-MD (**a, a'**) and 4-MD (**b, b'**) at 0% overlapping and 3-MD with 25% overlapping, under ideal (top row) and realistic (bottom row) detector spectral response

under realistic detector spectral response, unless the consideration of ideal detector spectral response is necessary.

3.2   Signal Detection in Photon-Counting CT

The detection of signal in a photon-counting detector can be modeled as [16, 52, 53]

$$I_k(L) = \int_{E_{\min}}^{E_{\max}} D_k(E) \, N_0(E) \exp\left(-\int_L \mu\,(\boldsymbol{x};\,E)\,dl\right) dE, \qquad (8)$$

where L denotes the path of X-ray beam, subscript k indexes the spectral channel for data acquisition, and $N_0(E)$ denotes the normalized spectrum in photon counts. $D_k(E)$ is the integrated effects of detector's efficiency $\eta(E)$ and spectral response $S_k(E)$ in the k-th spectral channel, i.e.,

$$D_k(E) = S_k(E)\eta(E). \qquad (9)$$

Analytically, the spectral response of the k-th channel is defined as

$$S_k(E) = \int_{E_{\min}^k}^{E_{\max}^k} R\left(E, E'\right) dE', \qquad (10)$$

where E denotes the incident energy. Under ideal detector spectral response, $R(E, E')$ is defined as a pulse function, but is assumed as a summation of Gaussian functions under realistic spectral response [40, 42]

$$
R\left(E, E'\right) = c_1(E) \left(\exp\left(-\left(E - E'\right)^2/2\sigma_1^2\right) + c_f(E) \exp\left(-(E - E_e)^2/2\sigma_2^2\right)\right.
$$
$$
\left. +c_2(E) \exp\left(-\left(E - E' + E_e\right)^2/2\sigma_2^2\right) + B\left(E, E'\right)\right). \tag{11}
$$

Taking the noise in data acquisition into consideration, the signal detection is modeled by turning Eq. (8) into

$$
I_k'(L) = \text{Poisson}\left(\int_{E_{\min}^k}^{E_{\max}^k} D_k(E)\, N_0(E) \exp\left(-\int_L \mu\left(x; E\right) dl\right) dE\right), \tag{12}
$$

where Poisson$(\cdot)$ denotes the output of a Poisson random generator [49].

3.3   Spectral Imaging in Photon-Counting CT

Reformatting Eq. (3) into a compact form

$$
\mu\left(x; E\right) = \sum_{i=1}^{K} a_i\left(x\right) \mu_i(E), \tag{13}
$$

Equation (4) can be rewritten as

$$
I_k(L) = \int_{E_{\min}^k}^{E_{\max}^k} D_k(E)\, N_0(E) \exp\left(-\sum_{p=1}^{P} A_p(L)\mu_p(E)\right) dE, \tag{14}
$$

where $A_p(L) = \int_L a_p(x) dl$ is the line integral of $a_p(x)$ ($p = 1, 2, \ldots, P$), the mass distribution corresponding to material p. Given the normalized source spectrum in Fig. 1 and detector efficiency and spectral response $D_k(E)$ in each channel, $A_p(L)$ can be obtained from $I_k(L)$ ($k = 1, 2, \ldots, K$) by solving the system of integral equations specified in Eq. (14) with the iterative Newton–Raphson algorithm [16], in which the initial condition is obtained by polynomial data fitting with its parameters determined by system calibration [1, 16, 32, 36, 37]. To assure convergence, the key parameters are set as: function tolerance $= 1e{-}6$; gradient tolerance $= 1e{-}6$; maximum number of iteration: 10,000; stopping criteria: both function and gradient tolerances are reached, the number of iteration reaches the maximum, or rerun the process with updated initialization. Once material decomposition is carried out, the basis images $a_p(x)$ can in turn be reconstructed from $A_p(L)$ ($p = 1, 2, \ldots, P$) using a filtered back-projection algorithm [40]. Subsequently, a virtual monochromatic image can be formed by setting the energy E in Eq. (3), while virtual monochromatic analysis can be carried out by varying E over a range that is specific to the imaging task.

4   3-MD Based Spectral Imaging in Photon-Counting CT

Let us assume a photon-counting CT works at 140 kVp, 1000 mA, and 1 rot/s gantry rotation speed and our focus is on soft tissue differentiation while bony tissue coexists in the head (relatively small dimension compared to thorax or abdomen). Hence, the criterion for spectral channelization is to assure that the photon counts in each spectral channel are roughly equal prior to their entering into the object to be imaged, though other criteria has been reported in the literature. In data acquisition, the photon-counting detector is a curved array at dimension 864×16 and pitch 1.024×1.092 mm^2. The source-to-iso and source-to-detector distances are 541.0 mm and 949.0 mm, respectively, leading to a nominal voxel size $0.5816 \times 0.5816 \times 0.625$ mm^3 in reconstructed image at 512×512 matrix.

4.1   Conditioning of Spectral Channelization and Its Impact on Image Quality

Given soft tissue, gadolinium (20 mg/ml) and cortical bone as the basis materials, the condition number and associated noise under both ideal and realistic detector spectral response and their variation over the cases of spectral overlap are quantitatively listed in Table 4 and graphically charted in Fig. 4. Regarding the condition number, it is noted that: (1) under ideal detector spectra response, the condition number corresponding to the case of no overlap reaches the minimum, while that corresponding to other cases grows with the increments in spectral overlapping (Fig. 4b); (2) under realistic detector spectral response, the condition number over all cases increases considerably, due to the distortion in detector's spectral response; (3) the distortion in detector's spectral response degrades the conditioning of spectral channelization more in the cases wherein no or small spectral overlapping occurs than the cases wherein relatively large spectral overlapping exists. In regard of noise, the distortion in detector's spectral response degrades the noise property of material specific imaging considerably, while the variation in the property of noise is consistent with that in the conditioning of spectral channelization.

Given soft tissue, gadolinium (20 mg/ml) and cortical bone as the basis materials, the material specific images corresponding to 0% (a_0–c_0 and a'_0–c'_0) and 25% (a_1–c_1 and a'_1–c'_1) of overlap in spectral response are shown in Fig. 5, providing a visual checking of the data in Table 3 and the charts in Fig. 4. The virtual monochromatic imaging at 18, 35, 50, and 65 keV obtained from the basis images displayed in Fig. 5 are presented in Fig. 6. As examples, the images correspond to 0% (a_0–f_0) and 25% (a_1–f_1) overlapping in detector's spectral response. It is observed that (1) the overlapping in detector's spectral channels degrades the noise in virtual monochromatic images, (2) the CNR reaches the highest around 50 keV, and (3) the distortion in detector's spectral response worsens the noise in virtual monochromatic images and leads to diminished CNR.

Table 4 Spectral channelization ([1 51] [52 68] [69 140] keV) in three-material decomposition based spectral imaging and resultant property in spectral conditioning and noise ($\sigma_t = \sqrt{\sigma_1^2 + \sigma_2^2 + \sigma_3^2}$) over spectral overlapping

	Ideal detector spectral response		Realistic detector spectral response	
Overlap (%)	Condition number	Noise	Condition number	Noise
0	2.1497	6.1555e−03	3.1186	9.3798e−03
5	1.8173	6.6851e−03	2.9356	9.9035e−03
10	1.8596	6.7127e−03	2.9757	1.0546e−02
20	1.9327	7.3553e−03	3.4716	1.1302e−02
25	2.5265	8.3463e−03	4.1277	1.4722e−02
30	3.6576	9.3408e−03	5.2150	1.2597e−02
40	4.5538	1.1019e−02	6.6331	1.4522e−02
50	5.4589	1.3193e−02	8.1986	1.6873e−02

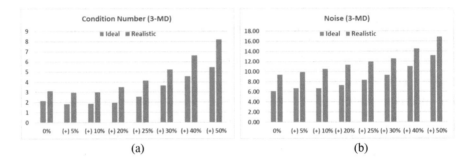

Fig. 4 The condition number (**a**) and the corresponding noise (**b**) in 3-MD under both ideal and realistic detector spectral response ("+" indicates overlap)

Under both ideal and realistic detector spectral response, the noise gauged at the targeted ROI and contrast-to-noise ratio against the background are plotted in Fig. 7 (a–a′) and (b–b′), respectively, in which the cases corresponding to up to 50% overlapping in detector spectral channels at 5% steps are enclosed. Again, the profiles plotted in Fig. 7 quantitatively confirm that the overlapping in spectral channels indeed degrades the noise in virtual monochromatic images cross the entire energy range, though almost no effect at the sweet spot (~70 keV) wherein the highest CNR is reached. Nevertheless, it is interesting and important to note that there exists a sweet spot in CNR around 50 keV, i.e., the K-edge of gadolinium. Notably, the overlapping in spectral channels improves the CNR at the sweet spots, but degrades the contrast-to-noise once leave the two sweet spots.

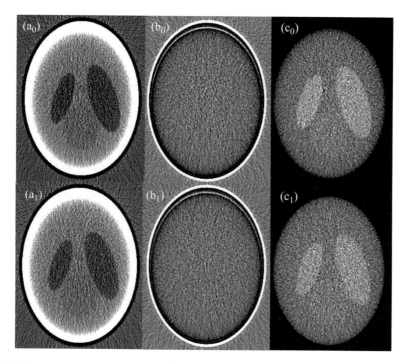

Fig. 5 Material specific (cortical bone: a_0, a_1; Gd (20 mg/ml): b_0, b_1; soft tissue: c_0, c_1) images in 3-MD over 0% (a_0–c_0) and 25% (a_1–c_1) overlap in spectral channels

4.2   Matching of Basis Materials

The material specific images corresponding to Phan-1and Phan-2 obtained under basis material set BM-1 (soft tissue, Gd at 10 mg/ml, and cortical bone) for 3-MD are shown in the top and bottom rows of Fig. 8 (WL: the level of brain parenchyma; WW: 8 times the noise of parenchyma), respectively. With gadolinium being included as the basis material, the material decomposition can be successfully carried out in Phan-1 that consists of two gadolinium rods, which is consistent with the data reported in the literature [4, 5]. However, artifacts (inaccuracy) appear in the area surrounding the iodine rod on the left side of Phan-2 (indicated by the white arrow in Fig. 8a_1), since iodine is not included as one of the basis materials. The virtual monochromatic images are formed from the basis images at 35, 50, 65, and 80 keV using Eq. (13) and the results are presented in Fig. 9. With no surprising, the inaccuracy (artifacts) in material decomposition due to the mismatching in basis materials pass on from the material specific images to the virtual monochromatic images.

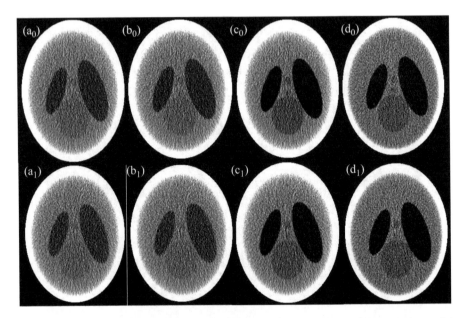

Fig. 6 Virtual monochromatic images at 18, 35, 50, and 65 keV from 3-MD (soft tissue, Gd (20 mg/ml) and cortical bone) at 0% (row 1) and 25% (row 2) spectral overlap

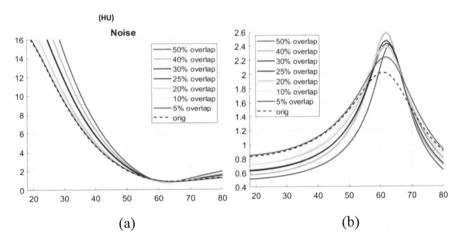

Fig. 7 Variation of noise (**a**) and CNR (**b**) gauged in virtual monochromatic analysis over energy, corresponding to the cases in Fig. 6, under realistic detector response (unit of all abscissa: keV)

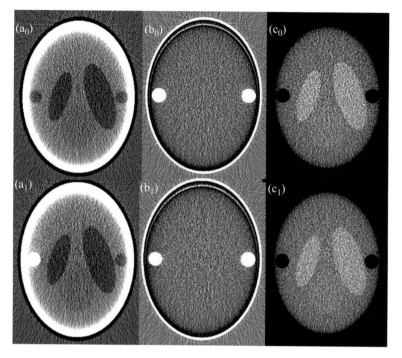

Fig. 8 Basis images of Phan-1 (top) and Phan-2 (bottom) from basis materials (cortical bone, Gd (20 mg/ml), and soft tissue)

4.3   Conditioning of Basis Materials and Its Impact on Imaging Performance

Given Phan-0, the basis images corresponding to various basis materials under realistic detector spectral response are placed in Fig. 10 (WL: the level of brain parenchyma; WW: 8 times the noise of parenchyma). The images in the top row $(a_0–c_0)$ correspond to the basis materials Teflon, PMMA, and adipose, in which, as predicted by the condition number in row 1 of Table 3, the noise is so severe that no internal structure of the head phantom is discernible. Using cortical bone, gadolinium (20 mg/ml), and soft tissue as the basis materials, the generated material specific images are in the second row $(a_1–c_1)$, in which the phantom's internal structure becomes discernible as the noise is reduced substantially. Moreover, if gadolinium (20 mg/ml) is kept but the other two basis materials are substituted with adipose and PMMA, the internal structure of Phan-0 becomes even more discernible $(a_2–c_2)$, particularly in the basis image corresponding to adipose (Fig. $10c_2$). However, since PMMA and adipose behave very similarly in their attenuation as function of energy, artifacts due to inaccurate material decomposition may appear, e.g., the bright peripheral areas indicated by the white arrow heads in Fig. $10c_2$. The variation in noise and resultant discernibility of internal structures in the head

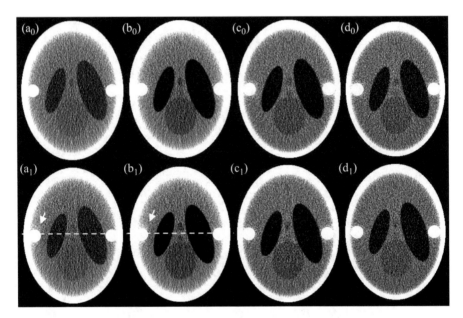

Fig. 9 VMI at 35, 50, 65, and 80 keV (left–right) of Phan-1 (top) and Phan-2 (bottom) obtained from 3-MD with BM-2 (cortical bone, 20 mg/ml Gd and soft tissue) under realistic spectral response

phantom align well with variation in the condition number of the basis materials itemized in Table 3. Quantitatively, the noise gauged in Fig. 10 is listed in Table 5.

The virtual monochromatic images of Phan-0 at 18, 25, 35, 50, 65, and 80 keV obtained from the basis images in Fig. 10 are shown in Fig. 11, (WL: the level of brain parenchyma; WW: 8 times the noise of parenchyma). It is observed that the best scenario case in CNR occurs around 50 keV where gadolinium's K-edge is located at. It is interesting and important to note that despite the noise in the material specific images varies dramatically due to the difference in the basis materials' conditioning, the noise in the virtual monochromatic images does not, which, may be owing to the intrinsic property in the correlation of noise in the material specific images [54].

5   4-MD Based Spectral Imaging in Photon-Counting CT

The material specific images corresponding to Phan-2 obtained under basis material set BM-4 (cortical bone, gadolinium (20 mg/ml), iodine (20 mg/ml), and soft tissue) are presented in the first row of Fig. 12 (WL: the level of brain parenchyma; WW: 8 times the noise of parenchyma), while their counterparts under basis material set BM-5 (Teflon, 20 mg/ml gadolinium, 20 mg/ml iodine, and adipose) are in the

Table 5 Noise (σ_1, σ_2, σ_3, σ_4) in the image corresponding to each basis material and the resultant total noise ($\sigma_t = \sqrt{\sigma_1^2 + \sigma_2^2 + \sigma_3^2}$ in 3-MD; $\sigma_t = \sqrt{\sigma_1^2 + \sigma_2^2}$ in 2-MD; $\sigma_t = \sqrt{\sigma_1^2 + \sigma_2^2 + \sigma_3^2 + \sigma_4^2}$ in 4-MD; the measurement of noise corresponding to ideal spectral response is presented in bracket)

	Basis materials	σ_1	σ_2	σ_3	σ_4	σ_t
1	Adipose, PMMA, and Teflon (BM-0)	(1.8135e+01) 1.2416e+01	(1.5508e+01) 1.2416e+01	(3.5406e−01) 2.8747e−01	–	(2.3864e+1) 1.7561e+01
2	Soft tissue, Gd (10 mg/ml), and cortical bone (BM-1)	(7.6896e−03) 1.1154e−02	(5.8727e−03) 1.1821e−02	(1.3848e−03) 1.1821e−02	–	(9.7743e−03) 2.0322e−02
3	Soft tissue, Gd (20 mg/ml), and cortical bone (BM-2)	(5.3147e−03) 7.1473e−03	(2.8853e−03) 5.8074e−03	(1.3898e−03) 3.0997e−03	–	(6.2050e−03) 9.7169e−03
4	Soft tissue, Gd (30 mg/ml), and cortical bone	(4.6839e−03) 6.2662e−03	(1.9066e−03) 3.8383e−03	(1.3908e−03) 3.1057e−03	–	(5.2448e−03) 7.9777e−03
5	Adipose, PMMA, and Gd (20 mg/ml) (BM-3)	(4.6309e−01) 1.0835e+00	(3.8133e−01) 8.9526e−01	(2.9078e−03) 5.9209e−03	–	(5.9989e−01) 1.4055e+00
6	Soft tissue, I (10 mg/ml), and cortical bone	(3.8834e−02) 7.0074e−02	(6.8703e−02) 1.3031e−01	(1.7887e−02) 3.5281e−02	–	(8.0920e−02) 1.5210e−01
7	Soft tissue, I (20 mg/ml), and cortical bone	(5.9368e−03) 7.9022e−03	(3.3968e−02) 6.3897e−02	(1.8196e−02) 3.5593e−02	–	(3.8989e−02) 7.3567e−02
8	Soft tissue, I (30 mg/ml), and cortical bone	(7.5598e−03) 1.6910e−02	(2.2457e−02) 4.2251e−02	(1.8237e−02) 3.5677e−02	–	(2.9901e−02) 5.7827e−02
9	Soft tissue, cortical bone, I (20 mg/ml), Gd (20 mg/ml) (BM-4)	(5.8311e−03) 7.5705e−03	(3.2781e−02) 2.3918e−01	(6.6436e−02) 4.4882e−01	(1.1289e−02) 4.1242e−02	(7.5165e−02) 5.1030e−01
10	Adipose, Teflon, I (20 mg/ml), and Gd (20 mg/ml) (BM-5)	(3.1791e−01) 1.6238e+00	(1.7872e−01) 9.1713e−01	(5.5780e−02) 2.6367e−01	(1.0411e−02) 2.7597e−02	(3.6907e−01) 1.8837e+00

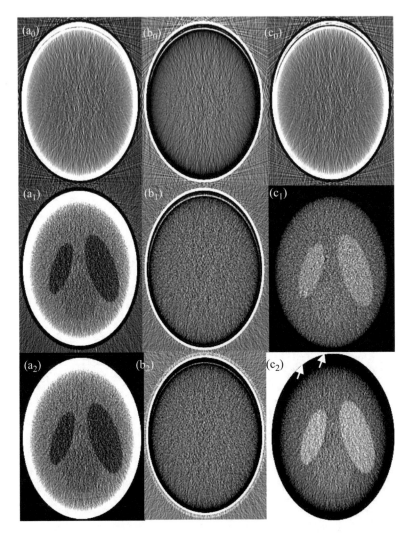

Fig. 10 Material specific images of Phan-0 corresponding to Teflon, PMMA, and adipose (a_0–c_0); bone, Gd (20 mg/ml), and tissue (a_1–c_1); Gd (20 mg/ml), PMMA, and adipose (a_2–c_2)

second row. Obviously, no artifacts (inaccuracy) exists around the iodine rod in the images, since, in either case, the dimensionality of basis materials is sufficient, and the necessary K-edge materials (Gd and I) are in place. However, streak artifacts (inaccuracy) exist at the skull and air interface in the images obtained from BM-4 (soft tissue, 20 mg/ml I, 20 mg/ml Gd, and cortical bone; row 1 in Fig. 12), which may be caused by the adipose at the skull–air interface. By replacing the cortical bone and soft tissue in basis material set BM-4 with Teflon and adipose to get basis material set BM-5, the streak artifacts almost disappear (row 2 in Fig. 12). The virtual monochromatic images of Phan-2 at 35, 50, 65, and 80 keV obtained from

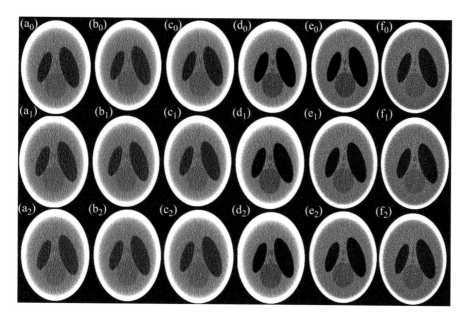

Fig. 11 Virtual monochromatic images at 18, 25, 35, 50, 65, and 80 keV of phan-0 obtained from 3-MD with basis materials: Teflon, PMMA, and adipose; bone, Gd (20 mg/ml), bone, and tissue; Gd (20 mg/ml), PMMA, and adipose

the basis images in Fig. 12 are shown in Fig. 13 (WL: the level of brain parenchyma; WW: 8 times the noise of parenchyma), in which no artifacts exists in the area surrounding the two rods made of K-edge materials.

6   Closing Remarks

With an extended coverage, we revisited the fundamentals in physics for material decomposition based spectral imaging in photon-counting CT. The facilitation of spectral channelization in photon-counting CT provides the unprecedented opportunity for m-MD based spectral imaging in which the conditioning of spectral channelization and basis materials plays a critical role in determining the imaging performance. We introduced the SVD approach that is utilized to assess the dimensionality of material space and the conditioning of basis materials and spectral channelization as well. Via phantom study by computer simulation, we presented the cases of 3-MD and 4-MD to illustrate m-MD based spectral imaging in photon-counting CT and variation of imaging performance over matching of basis materials, conditioning of spectral channelization and basis materials.

As illustrated in Fig. 4 and Table 4, the distortion in detector's spectral response that is inevitably induced by charge-sharing, Compton scattering, and fluorescent

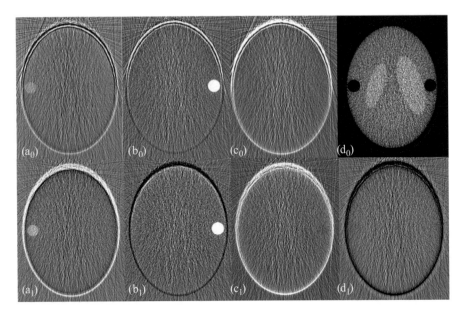

Fig. 12 Basis images of Phan-2 obtained from 4-MD under BM-4 (cortical bone, 20 mg/ml Gd, 20 mg/ml I and soft tissue; row 1) and BM-5 (Teflon, 20 mg/ml Gd, 20 mg/ml I, and soft tissue; row 2)

escaping leads to spectral overlapping in the spectral channels in data acquisition. Accordingly, the data acquired in adjacent channels become correlated, which in turn degrades the noise property of material specific imaging. Moreover, it is noted that a deliberate overlapping in spectral channels degrades the noise property in material specific images. This observation is consistent with the earliest prediction made by Alvarez and Macovski on the relationship between spectral separation and the noise in spectral images [19, 55], but seems contrary to that presented in reference [38]. Notably, in virtual monochromatic imaging, as shown in Fig. 7, the overlapping in spectral channels improves the property of noise and CNR at the sweet spot, whereas it degrades them in the energy range outside the sweet spot.

It has been shown in Figs. 8 and 9 that if a K-edge material (e.g., iodine or gadolinium, the most common materials to fabricate contrast agent for diagnostic imaging in the clinic) is present in the materials to be imaged, the identical material must also be one of the basis materials for material decomposition. This prerequisite of K-edge material matching is consistent to what has been reported in literature [37–40]. As long as the K-edge material is matched in the basis materials, its concentration almost does not matter in the accuracy of K-edge material decomposition, though the condition number of basis functions and resultant noise change over the concentration of K-edge materials.

We should have noted that the basis materials may be not only the biological tissues with their K-edges falling outside the energy range for diagnostic imaging

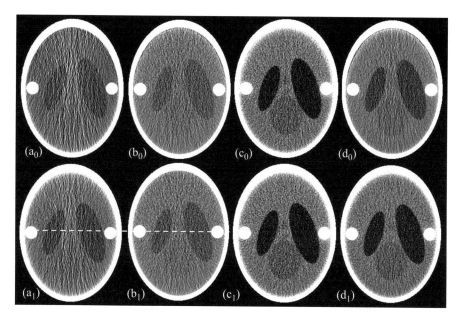

Fig. 13 VMI at 35, 50, 65, and 80 keV of Phan-2 obtained from 4-MD under basis material set BM-4 (row 1) and BM-5 (row 2)

(18–150 keV) but also the materials with their K-edge (iodine and gadolinium) within the energy range. As illustrated in Fig. 14, variation in the concentration of gadolinium from 10 to 30 mg/ml can effectively tune the correlation between each of the basis materials' mass attenuation coefficients and approaches the midway between soft tissue and cortical bone at the low energy end. Hence, it alters the condition number of basis materials and in turn the noise and CNR in material decomposition based spectral imaging. Notably, as the K-edges of iodine and gadolinium are at 33.2 keV and 50.2 keV, respectively, gadolinium may impose less impact on the signal detection of photon-counting detectors than iodine at the lowest energy channel, which starts at approximately 30 keV (Figs. 1 and 14).

It is to be noted that (as shown in Figs. 10 and 11) the conditioning of the basis material in 3-MD substantially affects the noise in the basis images but has only a moderate influence on virtual monochromatic imaging/analysis. For example, as illustrated in rows 1–4 of Table 3, the condition number of the basis material set BM-0 (Teflon, PMMA, and adipose) is roughly three-orders larger than that of the three basis material sets that consist of gadolinium. As shown in Fig. 10, the noise in the basis images behaves consistently with the condition numbers in Table 3 (rows 1, 2, 5). The visual inspection of noise in Fig. 10 is reinforced by the quantitative measurement of noise listed in Table 5. However, as demonstrated by the profiles plotted in Fig. 10, the behavior of noise and CNR in virtual monochromatic imaging/analysis is actually quite stable over the selection of basis materials in

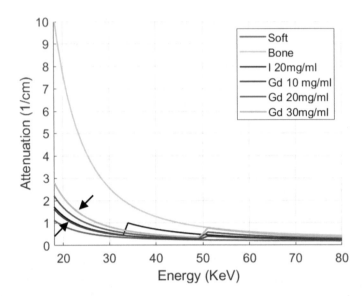

Fig. 14 Mass attenuation coefficients of soft tissue, cortical bone, iodine (10 mg/ml), and gadolinium (10–30 mg/ml)

3-MD. We speculate that the correlation in noise between each other of the basis images is behind this phenomenon.

A comparison of the basis images in the second and third rows of Fig. 10 deserves close attention. Given Gd (20 mg/ml) being one of the basis materials, the contrast of soft tissues against each other increases markedly if the two basis materials of soft tissue and cortical bone are substituted with adipose and PMMA. Referring to the third and fifth rows in Table 3, it is noted that such a substitution increases the condition number of the basis functions by roughly ten times, resulting in increased noise in those basis images (see row 5 in Table 5). On the other hand, such a selection of basis materials oriented for soft tissue differentiation leads to the fact that the increase in contrast outpaces that in noise. As a result, the internal structure of the head phantom, which is mainly configured of soft tissues, becomes more discernible than the case associated with basis material set BM-2 (soft tissue, Gd (20 mg/ml), and cortical bone). Meanwhile, however, it should be noted that a worsened conditioning in the basis materials may lead to inaccuracy in material decomposition, which, in turn, manifests itself as artifacts in basis images (Fig. $10c_1$). Hence, we point out that the optimization of basis materials and its effect on the performance of material decomposition based spectral imaging of soft tissues (e.g., protein and adipose) while bony tissue coexists, such as accuracy of material decomposition, contrast, and noise, is task-dependent and deserves further study.

Prior to ending the chapter, we want to make the following two remarks. First, by "conditioning," we underscore three factors: (1) sufficiency in dimensionality, (2) well-posedness in the condition number of basis functions, and (3) matching of K-edge materials. Second, given identical radiation dose, the noise in the basis images via 4-MD is significantly larger than that in its 3-MD counterparts (see Table 5). Hence, as long as it is not smaller than the dimensionality of material space, the dimensionality of the projection space (number of spectral channels) in photon-counting CT for spectral imaging of soft tissues while bony tissue coexists should be equal to the dimensionality of material space and that of the projection space. Any additional spectral channels may not provide meaningful information but instead increase the system's complexity and degrade its property in noise.

References

1. Krauss B, Schmidt B, Flohr TG. Dual Source CT. In: Johnson T, Fink C, Schönberg SO, Reiser MF, eds. *Dual Energy CT in Clinical Practice.* doi: https://doi.org/10.1007/174_2010_44 Berlin, Heidelberg: Springer Berlin Heidelberg; 2011:11-20.
2. Holmes DR, Fletcher JG, Apel A, et al. Evaluation of non-linear blending in dual-energy computed tomography. *European Journal of Radiology.* 2008;68(3):409-413.
3. Flohr TG, Leng S, Yu L, et al. Dual-source spiral CT with pitch up to 3.2 and 75 ms temporal resolution: Image reconstruction and assessment of image quality. *Medical Physics.* 2009;36(12):5641-5653.
4. Tapiovaara MJ and Wagner RF 1985. SNR and DQE analysis of broad spectrum x-ray imaging. *Phys. Med. Biol.* 30(6): 519-529.
5. Cahn RN, Cederstrom B, Danielsson M., Hall A., Lundqvist and Nygren D., Detective quantum efficiency dependence on x-ray energy weighting in mammography, *Medical Physics.* 1999;26(12):2680-2683.
6. Butler APH, Anderson NG, Tipples R, et al. Bio-medical X-ray imaging with spectroscopic pixel detectors. *Nuclear Instruments and Methods in Physics Research Section A: Accelerators, Spectrometers, Detectors and Associated Equipment.* 2008;591(1):141-146.
7. Alvarez RE, Seibert JA and Thompson SK, Comparison of dual energy detector system performance, *Medical Physics.* 2004;31(3):556-565.
8. Kelcz F, Joseph PM, Hilal SK. Noise considerations in dual energy CT scanning. *Medical Physics.* 1979;6(5):418-425.
9. Primak AN, Ramirez Giraldo JC, Liu X, Yu L, McCollough CH. Improved dual-energy material discrimination for dual-source CT by means of additional spectral filtration. *Medical Physics.* 2009;36(4):1359-1369.
10. Whiting BR, Massoumzadeh P., Earl OA, O'Sullivan JA, Snyder DL and Williamson JF, Property of preprocessed sinogram data in x-ray computed tomography. *Medical Physics.* 2006;33(9):3290-3303.
11. Duan X., Wang J. Leng S. *et al,* Electronic noise in CT detectors. *American Journal of Roentgenology* 2013;201(4):W626-632.
12. Taguchi K, Iwanczyk JS. Vision 20/20: Single photon counting x-ray detectors in medical imaging. *Medical Physics.* 2013;40(10):100901.
13. Willemink MJ, Persson M, Pourmorteza A, Pelc NJ, Fleischmann D. Photon-counting CT: Technical Principles and Clinical Prospects. *Radiology.* 2018;289(2):293-312.
14. Chandra N, Langan DA. Gemstone Detector: Dual Energy Imaging via Fast kVp Switching. In: Johnson T, Fink C, Schönberg SO, Reiser MF, eds. *Dual Energy CT in Clinical Practice.* doi: https://doi.org/10.1007/174_2010_35 Berlin, Heidelberg: Springer Berlin Heidelberg; 2011:35-41.
15. Krauss B, Schmidt B, Flohr TG. Dual Source CT. In: Johnson T, Fink C, Schönberg SO, Reiser MF, eds. *Dual Energy CT in Clinical Practice.* doi: https://doi.org/10.1007/174_2010_44 Berlin, Heidelberg: Springer Berlin Heidelberg; 2011:11-20.

16. Spiers FW, Effective atomic number and energy absorption in tissues. *Radiology.* 1946;19(218):56-63.
17. Rutherford RA, Pullan BR and Isherwood I., Measurement of effective atomic number and electron density using an EMI scanner. *Neuroradiology.* 1976;11(1):15-21.
18. Williamson JF, Deibel FC, and Morin RL, The significance of electron binding corrections in Monte Carlo photon transport calculations. *Physics in Medicine & Biology.* 1984;29(9):1063-1073.
19. Alvarez RE. (1976) *Extraction of energy-dependent information in radiography.* (Doctoral dissertation) Stanford University, Stanford, CA.
20. Alvarez RE, Dimensionality and noise in energy selective x-ray imaging. *Medical Physics.* 2013;40(11):111909.
21. Williamson JF, Li S., Devic S., Whiting BR and Lerma FA, On two-parameter models of photon cross sections: Application to dual-energy CT imaging. *Medical Physics.* 2006;33(11):4115-4129.
22. Han D., Siebers JV and Williamson JF, A linear, separable two-parameter model for dual energy CT imaging of proton stopping power computation. *Medical Physics.* 2016;43(1):600-612.
23. Han D., Porras-Chaverri, O'Sullivan, Politte DG and Williamson JF, Technical Note: On the accuracy of parametric two-parameter photon cross-section models in dual energy-CT applications. *Medical Physics.* 2017;44(6):2338-2346.
24. White DR, An analysis of the Z-dependence of photon and electron interactions. *Physics in Medicine & Biology.* 1977;22(2):219-228.
25. Phelps ME, Hoffman EJ and Ter-Pogossian, Attenuation coefficients of various body tissues, fluid, and lesions at photon energies of 18 to 136 keV, *Radiology,* 1975;117(3): 573-583.
26. Alvarez RE, Macovski A. Energy-selective reconstructions in X-ray computerised tomography. *Physics in Medicine & Biology.* 1976;21(5):733-744.
27. Brody WR, Butt G., Hall A. and Macovski A, A method for selective tissue and bone visualization using dual energy scanned projection radiography. *Medical Physics.* 1981;8(3):353-357.
28. Lehmann LA, Alvarez RE, Macovski A, Brody WR, Pelc NJ, Riederder SJ and Hall AL, Generalized image combinations in dual kVp digital radiography. *Medical Physics.* 1981;8(5):659-667.
29. Vetter JR, Perman WH, Kalender WA, Mazess RB and Holden JE, Evaluation of a prototype dual-energy computed tomographic apparatus. II. Determination of vertebral bone mineral content. *Medical Physics.* 1986;13(3):340-343.
30. Alvarez R, Seppi E. A Comparison of Noise and Dose in Conventional and Energy Selective Computed Tomography. *IEEE Transactions on Nuclear Science.* 1979;26(2):2853-2856.
31. Kalender WA, Perman WH, Vetter JR and Klotz E, Evaluation of a prototype dual-energy computed tomographic apparatus. I. Phantom studies. *Medical Physics.* 1986;13(3):334-339.
32. Yu L, Leng S, McCollough CH. Dual-Energy CT–Based Monochromatic Imaging. *American Journal of Roentgenology.* 2012;199(5_supplement):S9-S15.
33. Weaver JB and Huddleston AL, Attenuation coefficients of body tissues using principal-components analysis. *Medical Physics.* 1985;12(1):40-45.
34. Lehmann LA and Alvarez RE. (1986) *Energy-selective radiography: A review.* (Plenum Press, New York, 1986), 145-187.
35. Roth E. (1984) *Multiple-energy selective contrast material imaging.* (Doctoral dissertation) Stanford University, Stanford, CA.
36. Sukovle P and Clinthorne NH, Basis material decomposition using triple-energy X-ray computed tomography, IMTC/99. Proceedings of the 16th IEEE Instrumentation and Measurement Technology Conference (Cat. No.99CH36309), Venice, 1999, pp. 1615-1618 vol.3, doi: https://doi.org/10.1109/IMTC.1999.776097.
37. Roessl E. and Proksa R. K-edge imaging in x-ray computed tomography using multi-bin photon counting detectors. *Physics in Medicine & Biology.* 2007;52(15):4679-4696.
38. Roessl E, Ziegler A and Proksa R, On the influence of noise correlations in measurement data on basis image noise in dual-energy like x-ray imaging. *Medical Physics.* 2007;34(3):959-966.

39. Schlomka JP, Roessl E., Dorscheid R. *et al* Experimental feasibility of multi-energy photon-counting K-edge imaging in pre-clinical computed tomography. *Physics in Medicine & Biology*. 2008;53(15):4031-4047.
40. Roessl E, Brendel B, Engel K, Schlomka J, Thran A and Proksa R, Sensitivity of photon-counting based K-edge imaging in x-ray computed tomography. *IEEE Transactions on Medical Imaging*. 2011;30(9):1678-1690.
41. Zhao W, Vernekohl, Han F *et al*, A unified material decomposition framework for quantitative dual- and triple-energy CT imaging. *Medical Physics*. 2018;45(7):2964-2977.
42. Ehn, S. (2017). Photon-counting hybrid-pixel detectors for spectral X-ray imaging applications (Doctoral dissertation). Retrieved from https://mediatum.ub.tum.de/doc/1363593/ 1363593.pdf..
43. Leon SJ. *Linear Algebra with Applications*. doi: https://doi.org/10.1007/978-3-540-39408-2. Seventh ed: Pearson Prentice Hall; 2006.
44. Tang X. and Ren Y., "On the conditioning of basis materials (functions) and its impact on multi-material decomposition based spectral imaging in photon-counting CT," *Med. Phys.*, v.48(3), 1100-1116, 2021.
45. Ren Y., Xie H., Long W., Yang X. Tang X., "On the conditioning of spectral channelization (energy binning) and its impact on multi-material decomposition based spectral imaging in photon-counting CT," *IEEE Transactions on Biomedical Engi.*, accepted, in press, 2021.
46. Woodard HQ and White DR, "The composition of body tissues," British J. Radiology, 59(12):1209-1219, 1986.
47. White DR, Griffith RV, Wilson IJ. ICRU Report 46: Photon, electron, proton and neutron interaction data for body tissues. *Journal of the International Commission on Radiation Units and Measurements*. 1992;os24(1).
48. Cullen DE, Hubbell JH and Kissel L 1997 EPDL97: the evaluated photon data library Lawrence Livermore National Laboratory Report UCRL-50400 vol 6, rev 5.
49. Alvarez RE, Near optimal energy selective x-ray imaging system performance with simple detectors. *Medical Physics*. 2010;37(2):822-841.
50. Alvarez RE, Invertibility of the dual energy x-ray data transform. *Medical Physics*. 2019;46(1):93-103.
51. Ren Y., Xie H., Long W., Yang X. and Tang X. (2020) Optimization of basis material selection and energy binning in three material decomposition for spectral imaging without contrast agents in photon-counting CT, *SPIE Proc.* vol. 11312, 113124X (8 pages), doi: https://doi.org/ 10.1117/12.2549678.
52. Tang X, Ren Y, Xie H and Long W. (2020) Three material decomposition for spectral imaging without contrast agent in photon-counting CT - Modeling and feasibility study, *SPIE Proc.* vol. 11312, 113124W (9 pages), doi: https://doi.org/10.1117/12.2549660.
53. De Man B., Basu S., Chandra Naveen, Dunham B., Edic P., Iatrou M., McOlash S., Sainath P., Shaughnessy C., Tower Brenton and Williams E. (2007) CatSim: A new computer assisted tomography simulation environment, *SPIE Proc.* vol. 6510, 65102G (8 pages), doi: https:// doi.org/10.1117/12.710713.
54. Tang X. and Ren Y., "Noise correlation in multi-material decomposition-based spectral imaging in photon-counting CT," accepted by The 16th Fully 3-D Image Reconstruction in Radiology and Nuclear Medicine, *Leuven, Belgium*, July 19-23, 2021.
55. Buzug TM. *Computed Tomography: From Photon Statistics to Modern Cone-Beam CT*. doi: https://doi.org/10.1007/978-3-540-39408-2. First ed: Springer-Verlag Berlin Heidelberg; 2008.

X-Ray Multispectral CT Imaging by Projection Sequences Blind Separation Based on Basis-Effect or Basis-Material Decomposition

Ping Chen and Xiaojie Zhao

1 Introduction

X-ray computed tomography (CT) plays an irreplaceable role in medical diagnosis and nondestructive testing [1]. Compared with conventional CT, multispectral CT can suppress beam-hardening artifacts and identify the components [2, 3]. Multispectral CT obtains projections with different energies by a hardware, such as photon-counting detectors [4], dual-layer or multilayer detectors [5], voltage switching [6, 7], or many X-ray sources with different energies [8]. CT reconstruction is used to obtain a distribution of the physical parameters, which can realize tissue characterization, and material separation [9]. However, these methods have limitations at the system cost, imaging efficiency, and practical applications. Therefore, based on conventional X-ray imaging systems, multispectral CT is still research focus. In this area, there are two major methods: multi-spectrum statistical iterative and projection domain separation.

The iterative reconstruction method uses multi-spectrum statistical iteration algorithm, based on physical properties of multispectral X-ray imaging [10]. During the iterations, prior information can be used to improve the CT image quality [11]. Elbakri and Fessler [12] proposed a CT statistical reconstruction model based on a physical model. Xu et al. [13] proposed an iterative reconstruction method for dual-energy CT based on penalized-likelihood criteria. Long and Fessler [14] proposed a penalized-likelihood method with edge-preserving regularization, which not only reconstructed the image of a composite material but also significantly reduced the CT image noise. Chen et al. [15] proposed a heuristic adaptive steepest descent nonconvex projection onto convex sets (ASD-NC-POCS) algorithm to reconstruct the image of the basis material. In iterative reconstruction, there is also another

P. Chen (✉) · X. J. Zhao
North University of China, Taiyuan, Shanxi, China

important type, so-called iteration image domain decomposition method. That use the reconstructions of all scans (with different energy) to decompose the voxels into different materials based on the set of the different attenuation values [16, 17]. However, the application of these methods has limitations because of the prior of X-ray spectrum, complex model, complex calculation, etc.

The projection separation method separates decomposition and reconstruction processes. It first decomposes the line integrals of the base maps (material-specific line integrals) from measurements [18]. This method can process projection data from every projection angle in parallel with high computing efficiency. Then the CT reconstruction algorithm may be a conventional algorithm that can be easy to implement [19, 20]. The challenge of projection separation is that the multi-energy X-ray attenuation model is nonlinear, and knowledge of the X-ray energy spectrum is necessary. In actual CT imaging, however, the energy spectrum usually cannot be directly measured, and the simulated energy spectrum differs from the actual spectrum [21].

Note that each element response of the detector in a conventional CT system contains attenuation information of photons at all transmission energies of each continuous X-ray spectrum. Energy spectrum separation is equivalent to separating the mono-energetic projection or narrow-spectrum ones. This separation is similar to the "cocktail party" problem. The "cocktail party" problem is normally solved by the blind separation method. Blind source separation refers to the separation of a set of source signals from a set of mixed signals under a blind scenario where both the sources and the mixing methodology are unknown. Based on the concept of blind separation, an optimization model can be created to decompose multi-voltage X-ray images and obtain narrow-spectrum projections. Based on physical models, Wei et al. proposed a weighted least-square blind separation model with nonnegative constraints, and derived a solution method by the first-order Karush–Kuhn–Tucker (KKT) condition [22, 23]. The method can improve the differentiability of components with the similar attenuation characteristics. However, although the method has certain effects for reducing hardening artifacts and improving material contrast, many decomposed projections are not narrow-spectrum projections and have higher noise than the original projections. Moreover, the energy directionality of separation projections is uncertain because the update of attenuation coefficients is not limited by the photon energies during the solution of the separation model. Then based on the relationship between X-ray energy and material attenuation coefficient, the quantitative accuracy of material component will be lower.

For the separated narrow-spectrum projection, we can preset the energy interval, and the basis-effect decomposition or basis-material decomposition with component prior can use this presupposed energy to calculate the attenuation coefficient. The energy of every narrow-spectrum projection can be subsequently determined. So, this chapter focuses on improving the separation model based on basis-effect or basis-material. The remainder of the chapter is organized as follows. In the next section, an X-ray multispectral forward model and the existing blind separation model are introduced. In the third section, the improved model based on basis-effect is introduced and the real experiments are used to verify the proposed method. In the fourth section, blind separation model based on basis-material and component prior is developed. Similarly, several experiments are used to verify the method. The final section summarizes the content of this chapter.

2   Blind Separation of Projection Sequence

2.1   X-Ray Multispectral Forward Model

In a conventional X-ray CT system, when an X-ray beam traverses an object, according to the Beer–Lambert theorem, the ideal projection measurements at the pixel position $u \in \Phi$ of the detector can be expressed as

$$\bar{I}(u) = I^0(u) \int_E S(E) \exp\left(-\int_{L(u)} \mu(x, E) \, dl\right) dE, \tag{1}$$

where $L(u)$ denotes a ray path depending on the imaging geometry (parallel beam, fan beam, etc.), $I^0(u)$ denotes the incident X-ray intensity at the ray path $L(u)$, $\mu(x, E)$ denotes the linear attenuation coefficient (LAC) for the object at energy E and spatial position x and $S(E)$ denotes the normalized spectrum function. The notations Φ represent the detector plane.

Depending on the materials, the attenuation coefficient can be expressed as a linear combination of K basis materials [24, 25]:

$$\mu(x, E) = \sum_{k=1}^{K} \mu_k(E) b_k(x). \tag{2}$$

where $\mu_k(E)$ denotes the LAC function of material k independent of the object space and $b_k(x)$ is the spatial distribution of material k.

Each pixel position of the detector contains the attenuation information of photons at all transmitted energies of the continuous X-ray spectrum, which provides a basis for extracting narrow-spectrum projections from polychromatic measurements. The X-ray spectrum is divided into R narrow-energy-width spectra, each of which has a narrow energy range of $[E_{r-1}, E_r]$, where $r = 1, \ldots, R, E_0 = 0$, $E_R = E_{max}$ and E_{max} denotes the maximum energy of the X-ray photons. Generally, each narrow energy range can be limited to 10 keV. According to the generalized first integral mean value theorem, Eq. (1) can therefore be expressed as

$$\bar{I}(u) = I^0(u) \sum_{r=1}^{R} s_r \exp\left(-\sum_{k=1}^{K} \mu_k(\xi_r) d_k(u)\right), \tag{3}$$

where $d_k(\boldsymbol{u}) = \int_{L(\boldsymbol{u})} b_k(\boldsymbol{x}) dl$, $s_r = \int_{E_{r-1}}^{E_r} S(E) dE$ and $\xi_r \in [E_{r-1}, E_r]$. Since $S(E)$ is the normalized spectrum function, $\sum_{r=1}^{R} s_r = 1$ is obtained and called the sum-to-one constraint.

In short, $p_r(\boldsymbol{\Phi}) = \sum_{k=1}^{K} \mu_k(\xi_r) d_k(\boldsymbol{\Phi})$ is the r-th decomposition projection, and $p_r(\boldsymbol{\Phi})$'s reconstructed image is the r-th reconstructed image. Due to reflecting the attenuation information of the object on the narrow-energy-width spectrum, $p_r(\boldsymbol{\Phi})$ is the narrow-spectrum projection. An unknown X-ray spectrum is considered; that is, the spectral weight s_r is blind.

Assume that the number of voltage is N, and E_{max} is still the maximum energy of the X-ray photons under all spectra. The same narrow-energy-width spectrum division $[E_{r-1}, E_r]$ and energy value ξ_r are taken for each projection measurement of each spectrum. The index n is introduced to represent the variables $\overline{I}_n(\boldsymbol{u})$, $I_n^0(\boldsymbol{u})$ and s_{nr} of different spectra. In addition, the spatial distribution of the detector pixels is discretized. Suppose that the detector is composed of a pixel array P located at $\boldsymbol{u}(p)$, $p = 1, \ldots P$, the number of angles during CT scanning is Y, and $M = PY$ is the total number of rays. Thus, the following multispectral discretization model is obtained:

$$\overline{I}_{nm} = I_{nm}^0 \sum_{r=1}^{R} s_{nr} \exp\left(-\sum_{k=1}^{K} \mu_k(\xi_r) d_{km}\right). \tag{4}$$

Let $\boldsymbol{S} = (s_{nr})_{N \times R}$, $\boldsymbol{U} = (\mu_k(\xi_r))_{R \times K}$, $\boldsymbol{D} = (d_{km})_{K \times M}$, $\overline{\boldsymbol{F}} = \left(\overline{I}_{nm}/I_{nm}^0\right)_{N \times M}$. Therefore, the ideal projection measurement matrix is expressed as

$$\overline{\boldsymbol{F}} = \boldsymbol{S} \exp(-\boldsymbol{U}\boldsymbol{D}) \tag{5}$$

Equation (4) is called the X-ray multispectral forward model.

2.2 Existing Separation Model

In Eq. (5), NM equations exist. $\boldsymbol{S}$, $\boldsymbol{U}$, and $\boldsymbol{D}$ are unknown variables, and $NR + RK + KM$ unknown variables exist. Of course, the separated narrow-spectrum projection $\boldsymbol{U}\boldsymbol{D}$ can be considered as one unknown variable to directly solve. Then, the number of unknown variables is $NR + RM$. In general, considering the practicality and experimental efficiency, the number N with different voltage is lower, the number of narrow-energy spectra R is higher than N. Then, $NR + RM$ is substantially larger than NM. Then Eq. (5) is difficult to solve. Therefore, $\boldsymbol{U}$ and $\boldsymbol{D}$ are solved separately. So, for the solution of Eq. (5), the number of equations should be greater than the number of unknown variables:

$$NM > NR + RK + KM \tag{6}$$

To satisfy Eq. (6), projections with N ($N > 1$) voltages should be obtained and determined by the dimensions M, R, and K.

Additionally, S, U, and D are nonnegative matrices for every element. Equation (5) is a special nonnegative matrix factorization (NMF) model, which can covert an optimization model, and can be solved by KKT condition [26]. If I_{nm} is the measured projection measurement data and let $F = \left(I_{nm}/I_{nm}^0\right)_{N \times M}$, that can convert the optimization model whose objective function is the minimal sum of the squared error (BSM-MSE) [22]:

$$\min_{S,U,D} \|F - S \exp(-UD)\|_F^2 \quad \text{s.t.} \quad S \geq O_{N \times R}, U \geq O_{R \times K}, D \geq O_{K \times M}$$
(7)

where O is zero matrix, $\|\cdot\|_F$ is the Frobenius norm of matrix. For Eq. (7), in references [22], projections with the superior narrow-spectrum characteristics can be obtained. However, the separated narrow-spectrum projection is expressed by UD, and U is the ray energy-dependent variable. Namely, in the optimization, the variables U have no energy constraint, and the corresponding energy of the attenuation coefficient may deviate from the actual value.

3 Blind Separation Mode Based on Basis-Effect

In Eq. (7), the variable U is related to the material and X-ray energy, which can be decomposed by the basis-effect of the photoelectric effect and the Compton effect [27]. For the decomposed variable, the photoelectric effect and the Compton effect are known when the ray energy is known. And for the blind separation model, the narrow-spectrum projection is expected to obtain at the every ray energy. Therefore, if the separation energy is preset, the photoelectric effect and the Compton effect are known. Then, the variable U will have energy constraint, which can make the separated narrow-spectrum projections UD have energy directivity.

3.1 Projection Observation Model Based on Basis-Effect

X-ray attenuation includes photoelectric effects, Compton scattering, and Rayleigh scattering. For the range of 10 keV to 1 MeV, the photoelectric effect and the Compton effect are main [24]. So, the attenuation coefficient can be decomposed into a linear combination of the two basis effects:

$$\mu_k(\xi_r) = a_c(k)f_{KN}(\xi_r) + a_p(k)f_p(\xi_r)$$
(8)

where c and p represent the Compton effect and the photoelectric effect, respectively; functions $f_{KN}(E)$ and $f_p(E)$ are only related to a predefined narrow-spectrum energy for separation; and $a_c(k)$ and $a_p(k)$ are only related to the material. Let

$$
U' = \begin{bmatrix} f_{KN}(\xi_1) & f_p(\xi_1) \\ \vdots & \vdots \\ f_{KN}(\xi_R) & f_p(\xi_R) \end{bmatrix}, \quad D' = AD = \begin{bmatrix} a_{c1} & \cdots & a_{cK} \\ a_{p1} & \cdots & a_{pK} \end{bmatrix} D \tag{9}
$$

where U' is an $R \times 2$ matrix and D' is a $2 \times M$ matrix. If the separated energy interval is presupposed, U' is a known variable. Due to the narrow energy interval, the attenuation coefficient changes less in this interval. Therefore, let $\xi_r = (E_{r-1} + E_r)/2$. $a_c(k)$ and $a_p(k)$ are unknown variables, D is unknown, and D' is a new unknown variable.

$$
U'D' = \begin{bmatrix} f_{KN}(\xi_1) & f_p(\xi_1) \\ \vdots & \vdots \\ f_{KN}(\xi_R) & f_p(\xi_R) \end{bmatrix} \begin{bmatrix} a_{c1} & \cdots & a_{cK} \\ a_{p1} & \cdots & a_{pK} \end{bmatrix} D = \begin{bmatrix} \mu_1(\xi_1) & \cdots & \mu_K(\xi_1) \\ \vdots & \vdots & \vdots \\ \mu_1(\xi_R) & \cdots & \mu_K(\xi_R) \end{bmatrix} D = UD \tag{10}
$$

From Eq. (10), $U'D'$ is the narrow-spectrum projection, which is equivalent to UD. Thus, model (5) can be rearranged as follows:

$$
\overline{F} = S \exp\left(-U'D'\right). \tag{11}
$$

According to Eq. (11), because the attenuation is converted to the expression of the photoelectric effect and the Compton effect, the number of unknown variables decreases, whose dimension is $NR + 2M$. Also, if solve Eq. (10), the relevant dimension should meet with Eq. (12)

$$
NM \geq NR + 2M. \tag{12}
$$

Compare Eq. (6), the unknown variable has been reduced. It will be easier to solve.

3.2 Projection Separation Model

Based on BSM-MSE model (Eq. 7), the blind separation mode based on basis-effect (BSM-BE) model is built:

$$
\min_{S, D'} \left\| F - S \exp\left(-U'D'\right) \right\|_F^2 \quad \text{s.t.} \quad S \geq O_{N \times R}, \, D' \geq O_{2 \times M}. \tag{13}
$$

For Eq. (13), according to the set ray energy parameters, the separated energy interval can be preset to calculate $f_{KN}(E)$ and $f_p(E)$. Then, U' is known. So, Eq. (13) has two unknown variables S and D'. According to Eq. (6), Eq. (13) also can be solved by KKT condition. Due to the preset energy interval and the known variable U', the new narrow-spectrum projection will have energy constraint. The update formula is as follows:

$$S \leftarrow S \odot \left(\frac{S \exp\left(-U'D'\right)\left(\exp\left(-U'D'\right)\right)^{\mathrm{T}}}{F \exp\left(-U'D'\right)} \right), \tag{14}$$

$$D' \leftarrow D' \odot \left(\frac{\left(U'\right)^{\mathrm{T}}\left(\left(S^{\mathrm{T}}\left(S \exp\left(-U'D'\right)\right)\right) \odot \exp\left(-U'D'\right)\right)}{\left(U'\right)^{\mathrm{T}}\left(\left(S^{\mathrm{T}}F\right) \odot \exp\left(-U'D'\right)\right)} \right). \tag{15}$$

3.3 Experiments and Analysis

A cylinder sample consisting of silicon and aluminum materials is selected because of the similar attenuation coefficients of silicon and aluminum. The cross-sectional diameter is 40 mm, and the diameter of the aluminum core cylinder is 30 mm.

In the experiment, according to the material and dimension and the experience in ray imaging, the appropriate X-ray source tube voltage is 130 kV and the tube current is 1.5 mA. And to make the separation model solvable, according to Eq. (12), three X-ray source tube voltages are needed. In the process of solving Eq. (13), the separation energy interval needs to be preset. In order to get the more attenuation information of the narrow-spectrum, the separation energy interval is set to 10 keV, and three voltages (120, 130, and 140 kV) and one tube current (1.5 mA) are selected to obtain projection sequences. The experimental system is a CT system that consists of a 450 kV GE X-ray source (GE ISOVOLT-450) and a digital detector array (PerkinElmer XRD1621). The distance from the source to the detector (SDD) is 140 cm, and the distance from the source to the object (SOD) is 120 cm. For rotation, the angular sampling interval is 1°, and the total angular distance is 360°. For the captured projection, the classic algebraic reconstruction algorithm (ART) is used to reconstruct to get the CT image of the object.

For three voltages (120, 130, and 140 kV), the directly reconstructed CT images and grayscale curves of the labeled line are shown in Fig. 1a–c. From these three figures, the beam-hardening artifacts are severe, the component analysis is difficult. Then, the blind separation model is used to separate projection measurement. In the separation model, the maximum number of the iteration is 500, and the convergence factor is 10^{-2}. For the starting value setting, considering the nonconvex of the separation model, the value is taken from the reconstructed CT image of the directly

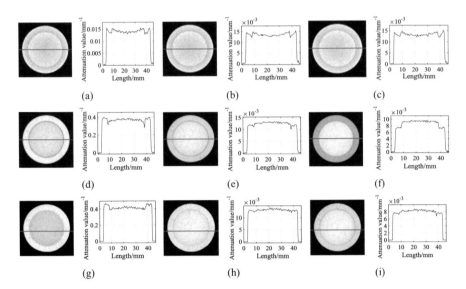

Fig. 1 The directly reconstructed images and grayscale curves of the labeled line (**a–c**) at 120, 130, and 140 kV. (**d–f**) are three reconstructed images of separated projection by BSM-MSE model. (**g–i**) are three reconstructed images of separated projection by BSM-BE model, corresponding to energy 30, 60, 100 keV respectively

acquired projection and the starting value of the X-ray energy spectrum is taken by the Spectrum GUI_1.03 software.

By BSM-MSE model, part of reconstructed images of separation projection and their grayscale curves of the labeled line are shown in Fig. 1d–f. The blind separation can suppress beam-hardening artifacts, and generate the narrow-spectrum imaging, which is reflected by gray change of silicon and aluminum in CT image (Fig. 1d–f). But, because of no energy constraint, the separated energy cannot be directly determined. The reconstruction images need to be rearranged based on the gray of aluminum and silicon. Also, the results of BSM-BE model are shown in Fig. 1g–i. The BSM-BE model can better suppress beam-hardening artifacts, determine the separated energy directly. But in some separation CT images, it is difficult to distinguish aluminum and silicon. This is because the attenuation coefficients between aluminum and silicon only have the smaller difference. So we further analysis the deviation of the attenuation coefficients between the NIST data and the calculated value. The deviation of BSM-MSE model and BSM-BE model is shown in Fig. 2.

The BSM-BE model produces a smaller deviation (Fig. 2a, b, d, e). We calculate the relative difference δ of the attenuation coefficients between aluminum and silicon by using the following formula:

$$\delta = \frac{\mu_{Al} - \mu_{Si}}{\mu_{Al}} \tag{16}$$

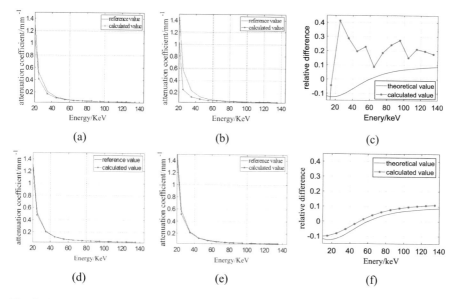

Fig. 2 Attenuation coefficient curves (**a**) and (**b**) obtained by BSM-MSE model. (**c**) is contrast of silicon and aluminum about the relative difference of the attenuation coefficients in the reconstructed image by BSM-MSE model. (**d–f**) are attenuation coefficients and contrast of silicon and aluminum obtained by BSM-BE model. The first and second columns are attenuation coefficients of silicon and aluminum, respectively

where μ_{Al} is the attenuation coefficient of aluminum and μ_{Si} is the attenuation coefficient of silicon. The results obtained by BSM-MSE and BSM-BE are shown in Fig. 2c, f.

In Fig. 2c, the variation trend for the contrast obtained by the BSM-MSE model is chaotic and does not match the trend of theoretical value. However, based on the BSM-BE model, the variation trend (Fig. 2f) can match this trend. So the attenuation coefficients for the BSM-BE model have higher precision and better energy feature than BSM-MSE model.

In order to prove that the BSM-BE model can be applied to the other objects, two other verification experiments are performed. The first experiment is the single material of aluminum foam. According to the size and material, the scanning voltages are 100, 110, and 120 kV. The experiment results of BSM-BE model are shown in Fig. 3a–c.

From Fig. 3a–c, the proposed algorithm also can suppress the beam-hardening artifact and the consistency of gray level has been improved obviously. In Fig. 3a, the relative gray deviation between the center and edge is 0.1506, which are labeled by the red arrows. But in Fig. 3b, c, they are 0.0145 and 0.0146 at the same labeled area. The consistency increases by 90%.

The second verification experiment is multi-material sample, which is the simulated energetic material. The ingredients are aluminum powder and flour, which

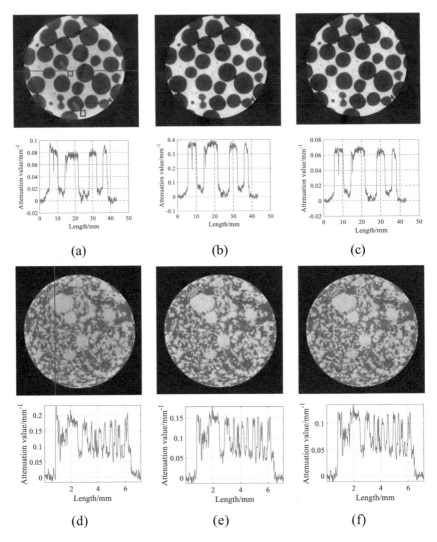

Fig. 3 The directly reconstructed image (**a**) at 110 kV and the 3th and 6th reconstructed images (**b**)-(**c**) decomposed by BSM-BE model in the aluminum foam experiment. The directly reconstructed image (**d**) at 70 kV and the 6th and 7th reconstructed images (**e, f**) decomposed by BSM-BE model in the energetic material experiment. The graph below each reconstructed image represents the gray value at the blue line

is die-cast. According to the size and materials, the scanning voltages are 70, 80, and 90 kV. The experiment results of BSM-BE model are shown in Fig. 3d–f.

For the energetic material, CT image of direct reconstruction has the lower contrast and beam-harden artifact (Fig. 3d). By separation of BSM-BE model, CT image quality has been improved (Fig. 3e, f).

4 Blind Separation Model Based on Basis-Material and Component Prior

The BSM-BE method introduces the energy prior through basis-effect decomposition, so that narrow-spectrum projections have clear energy directivity. Similarly, by introducing the composition prior, the blind separation model based on basis-material decomposition can also determine the energy orientation of narrow-spectrum projection. Therefore, this section describes a new blind separation method based on basis-material and component prior.

4.1 Regularized Maximum Likelihood Estimation with Constraints

According to component prior, when the energy value ξ_r is set, the attenuation coefficient $\mu_k(\xi_r)$ is known. The energy property of each narrow-spectrum projection is only related to $\mu_k(\xi_r)$. Therefore, the energy orientation of narrow-spectrum projection can be set in advance and remains unchanged during the solution process. Same as Sect. 3.1, set $\xi_r = (E_{r-1} + E_r)/2$. In order to simplify the formula derivation, rearrange the matrices S and D to obtain the following vector:

$$s = \left[s_{1,1} \cdots s_{N,1} \cdots\cdots s_{1,R} \cdots s_{NR}\right]^{\mathrm{T}}, d = \left[d_{1,1} \cdots d_{K,1} \cdots\cdots d_{1,M} \cdots d_{KM}\right]^{\mathrm{T}}. \tag{17}$$

Simultaneously, let

$$\overline{I} = \left[\overline{I}_{1,1} \cdots \overline{I}_{N,1} \cdots\cdots \overline{I}_{1,M} \cdots \overline{I}_{NM}\right]^{\mathrm{T}}, I = \left[I_{1,1} \cdots I_{N,1} \cdots\cdots I_{1,M} \cdots I_{NM}\right]^{\mathrm{T}}. \tag{18}$$

Adjust the model (5) to get the following model:

$$\overline{I} = \mathcal{F}(s, d), \tag{19}$$

where $\mathcal{F}$ denotes the nonlinear mapping induced by Eq. (4). Due to the introduction of component prior, $\mu_k(\xi_r)$ is known. Thus, the model (19) only contains unknown vectors s and d. The projection measurements are assumed to be subject to a Poisson distribution, i.e.,

$$I_{nm} \sim P\left(\lambda = \overline{I}_{nm}\right), \tag{20}$$

where $P(\lambda)$ denotes a Poisson distribution with a mean of λ.

To decompose narrow-spectrum projections from polychromatic attenuation measurements, it is necessary to solve the variables s and d. To this end, according to the maximum likelihood principle, the constraint optimization problem is established by

$$\min_{s,d} \frac{1}{2}\|I - \mathcal{F}(s,d)\|_W^2 + \alpha \mathcal{G}(d),$$

$$\text{s.t. } s \geq O_{NR\times1}, d \geq O_{NM\times1}, \sum_{r=1}^{R} s_{n,r} = 1, \forall n, \tag{21}$$

where $\|x\|_W^2 = x^T W^T W x$. The matrix W is a weight matrix, which is chosen as $W = 1/\sqrt{I}$ so that $\|I - \mathcal{F}(s,d)\|_W^2/2$ is the weighted least squares fidelity term applicable to the Poisson distribution. Due to the ill-posed nature of CT decomposition problems, a regularization function $\mathcal{G}$ is introduced, where α denotes the regularization parameter.

Given the differences in component priori, the following regularization function is used:

$$\mathcal{G}(d) = \sum_{k=1}^{K} \beta_k \mathcal{G}_k (Ld_k), \tag{22}$$

where $d_k = [d_{k,1} \cdots d_{kM}]^T$. β_k and $\mathcal{G}_k$ are the regularization parameter and regularization function of the kth component, respectively. The linear transformation L is chosen as a discrete Laplace.

The internal compositions of most objects in nature are generally continuous distributions (components with uniform mixing of several materials can be regarded as a material), so the difference in the thickness d_{km} and $d_{k(m+1)}$ of each material on the adjacent ray path is small. Consequently, given the prior of smoothness, a quadratic potential function is used as the regularization function; that is, $\mathcal{G}_k (Ld_k) = \|Ld_k\|_2^2$.

In addition, to ensure that Eq. (21) is not undetermined, N, R, K, and M must meet the following inequality:

$$NM > NR + KM. \tag{23}$$

4.2 Optimization Algorithm

In Eq. (21), although the feasible domain determined by the constraints is a convex set, the objective function is a nonconvex function. The optimization algorithm below aims to find an interesting (local) solution to the problem, and thus, an effective initialization approach is necessary.

Due to the differences in number scale and constraints, the solved variables are naturally divided into two blocks, s and d. Inspired by the block coordinate descent approach, s and d are alternately iterated using different algorithms to seek the solution of Eq. (21), where NMF and the Gauss–Newton (GN) algorithm [28] are used to update S and D, respectively.

NMF is a novel and effective blind separation method with inherent nonnegative constraints, and the sum-to-one constraint is introduced to NMF by employing a method in the literature [29]. The vector d is much larger than that of s, and the objective function for d is nonlinear in the framework of alternating optimization. Using a first-order solver together with s will slow the convergence speed of the algorithm as a whole. Given that the GN algorithm is a classic nonlinear optimization method with second-order convergence speed, it is adopted as the solver of d.

The proposed method minimizes Eq. (21) by alternately updating s and d using steps (1) and (2) below, where the ith iteration proceeds as follows:

1. (NMF) The vector d is fixed as d^{i-1}, and NMF is adopted to deal with the following subproblem of Eq. (21):

$$\min_{s} \frac{1}{2} \left\| I - \mathcal{F}\left(s, d^{i-1}\right) \right\|_{W}^{2}, \text{ s.t. } s \geq O_{NR \times 1}, \sum_{r=1}^{R} s_{n,r} = 1, \forall n. \tag{24}$$

Regardless of the sum-to-one constraint, the following formula can be obtained from the literature [30]:

$$s_{nr}^{i} = s_{nr}^{i-1} \left(\sum_{m=1}^{M} I_{nm}^{0} t_{rm}^{i-1} \right) / \left(\sum_{m=1}^{M} \frac{\left(I_{nm}^{0}\right)^2}{I_{nm}} t_{rm}^{i-1} \left(\sum_{r=1}^{R} s_{nr}^{i-1} t_{rm}^{i-1} \right) \right), \forall n, r, \tag{25}$$

where $t_{rm}^{i-1} = \exp\left(-\sum_{k=1}^{K} \mu_k \left(\xi_r\right) d_{km}^{i-1}\right)$. In addition, to satisfy the sum-to-one constraint, I is extended to

$$I_A = \left[I^{\mathrm{T}} \ I_{1,M+1} \cdots I_{N,M+1} \right]^{\mathrm{T}}, I_{n,M+1} = I_{n,M}^{0} \delta, \forall n, \tag{26}$$

and let $t_{r(M+1)}^{i-1} = \delta, \forall r, I_{n(M+1)}^{0} = I_{nM}^{0}, \forall n$. The iterative formula of Eq. (25) is as follows by expanding M in Eq. (27) to $M + 1$:

$$s_{n,r}^{i} = \frac{s_{n,r}^{i-1} \left(\sum_{m=1}^{M} I_{nm}^{0} t_{rm}^{i-1} + I_{nM}^{0} \delta \right)}{\sum_{m=1}^{M} \frac{\left(I_{nm}^{0}\right)^2}{I_{nm}} t_{rm}^{i-1} \left(\sum_{r=1}^{R} s_{nr}^{i-1} t_{rm}^{i-1} \right) + I_{nM}^{0} \delta \sum_{r=1}^{R} s_{nr}^{i-1}}, \forall n, r, \tag{27}$$

where δ controls the impact of the sum-to-one constraint. The larger that δ is, the closer that $\sum_{r=1}^{R} s_{nr}$ is to 1.

2. (GN) The vector s is fixed as s^{i}, and the GN algorithm is employed to solve the following subproblem of Eq. (21):

$$\min_{d} \frac{1}{2} \left\| I - \mathcal{F}\left(s^i, d\right) \right\|_W^2 + \alpha \mathcal{G}\left(d\right), \text{ s.t. } d \geq O_{NM \times 1}. \tag{28}$$

For nonnegative constraints, an interior point (IP) method [31] is used. The logarithmic barrier function is written as

$$P\left(d, u\right) = \frac{1}{2} \left\| I - \mathcal{F}\left(s^i, d\right) \right\|_W^2 + \alpha \mathcal{G}\left(d\right) - u \sum_{m=1}^{M} \sum_{k=1}^{K} \ln\left(d_{km}\right), \tag{29}$$

where u denotes the barrier parameter. For a given barrier parameter u^j, the iteration formula of Eq. (29) is as follows:

$$d^i = d^{i-1} + \gamma^i \delta^i, \tag{30}$$

where $\delta^i \in \mathbf{R}^{KM}$ denotes the GN direction and $\gamma^i \in \mathbf{R}$ denotes the search step. The GN direction is obtained by solving the following equation:

$$\left(\left(J^{i-1}\right)^{\mathrm{T}} W^{\mathrm{T}} W J^{i-1} + \alpha \nabla^2 \mathcal{G}\left(d^{i-1}\right) + u^j C^{i-1} C^{i-1} \right) \delta^i = -\nabla P\left(d^{i-1}, u^j\right), \tag{31}$$

where $C^{i-1} = \operatorname{diag}\left(1/d^{i-1}\right)$, J^{i-1} denotes the Jacobi matrix of $\mathcal{F}(s^i, d)$ at d^{i-1}, $\nabla^2 \mathcal{G}\left(d^{i-1}\right)$ denotes the Hessian matrix of $\mathcal{G}$ at d^{i-1} and $\nabla P(d^{i-1}, u^j)$ denotes the gradient of $P(d, u^j)$ at d^{i-1}. $\nabla P(d^{i-1}, u^j)$ is given by

$$\nabla P\left(d^{i-1}, u^j\right) = -\left(J^{i-1}\right)^{\mathrm{T}} W^{\mathrm{T}} W \left(I - \mathcal{F}\left(s^i, d^{i-1}\right)\right) + \alpha \nabla \mathcal{G}\left(d^{i-1}\right) - u^j \frac{1}{d^{i-1}}, \tag{32}$$

where $\nabla \mathcal{G}\left(d^{i-1}\right)$ denotes the gradient of $\mathcal{G}$ at d^{i-1}.

The Jacobi matrix $J \in \mathbf{R}^{NM \times KM}$ introduced in Eq. (31) is defined as

$$J = \frac{\partial \mathcal{F}\left(s^i, d\right)}{\partial d^{\mathrm{T}}}. \tag{33}$$

Let $\bar{I}^i = \mathcal{F}\left(s^i, d\right) = \left[\bar{I}_{1,1}^i \cdots \bar{I}_{N,1}^i \cdots \cdots \bar{I}_{1,M}^i \cdots \bar{I}_{NM}^i\right]^{\mathrm{T}}$; then,

$$\frac{\partial \bar{I}_{nm}^i}{\partial d_{km'}} = \begin{cases} 0, & m \neq m' \\ l_{nkm}^i, & m = m' \end{cases} \tag{34}$$

with

$$l_{nkm}^i = -I_{nm}^0 \sum_{r=1}^R s_{nr}^i \mu_k \left(\xi_r\right) \exp\left(-\sum_{k=1}^K \mu_k \left(\xi_r\right) d_{km}\right). \tag{35}$$

Thus, J can be represented as a block diagonal matrix

$$J = \mathrm{diag}\left(\left[J_1 \quad \cdots \quad J_M\right]\right), \tag{36}$$

where $J_m \in \mathbf{R}^{N \times K}$ and $(J_m)_{n,k} = l_{n,k,m}^i$. Obviously, J is a sparse matrix with only NKM nonzero elements.

The gradient of the regularization function $\mathcal{G}$ is calculated as follows:

$$\nabla \mathcal{G}\left(d\right) = \sum_{k=1}^K \beta_k \nabla_k \mathcal{G}_k \left(Ld_k\right) \otimes e_k, \tag{37}$$

where ∇_k denotes the gradient with respect to d_k and is defined as

$$\nabla_k = \left[\frac{\partial}{\partial d_{k,1}} \cdots \frac{\partial}{\partial d_{k,M}}\right]^{\mathrm{T}}, \tag{38}$$

$\nabla_k \mathcal{G}_k \left(Ld_k\right) = 2L^{\mathrm{T}} L d_k$, e_k denotes a K-dimensional natural basis vector, and $\otimes$ denotes the Kronecker product.

Similarly, the Hessian matrix of the regularization function $\mathcal{G}$ is given as

$$\nabla^2 \mathcal{G}\left(d\right) = \sum_{k=1}^K \beta_k \nabla_k^2 \mathcal{G}_k \left(Ld_k\right) \otimes e_k \otimes e_k^{\mathrm{T}}, \tag{39}$$

where $\nabla_k^2 \mathcal{G}_k \left(Ld_k\right) = 2L^{\mathrm{T}} L$.

The selection of the search step γ^i will determine whether d^i in Eq. (30) can be sustained as a feasible solution. Backtracking linear search [32] is used to determine the step size as follows:

1. Let $\overline{\gamma}^i = \max\left\{1 / \left(1 + \sqrt{-\nabla P\left(d^{i-1}, u^j\right)^{\mathrm{T}} \delta^i}\right), \gamma_0\right\}$, where γ_0 denotes a constant.

2. Select $\gamma^i = \tau^c \overline{\gamma}^i$, where $c \geq 0$ is the smallest integer that enables γ^i to meet conditions Eqs. (40) and (41), $\tau \in (0, 1)$ denotes the step size adjustment parameter and τ' denotes the backtracking parameter.

$$d^{i-1} + \gamma^i \delta^i \geq O_{KM \times 1}. \tag{40}$$

$$P\left(d^{i-1} + \gamma^i \delta^i, u^j\right) \leq P\left(d^{i-1}, u^j\right) + \gamma^i \tau'\left(\nabla P\left(d^{i-1}, u^j\right)\right)^{\mathrm{T}} \delta^i. \quad (41)$$

Iterative formulas of Eqs. (27) and (30) iterate alternately. When $\|d^i - d^{i-1}\|_2 < \varepsilon_1 \|d^i\|_2$ and $M \times u^j \geq \varepsilon_2$ meet, the barrier parameter is updated to $u^{j+1} = \theta u^j$; then, proceed to the $(i+1)$-th iteration, where $\varepsilon_1 > 0$ denotes the GN stop threshold, $\varepsilon_2 > 0$ denotes the IP stop threshold and θ denotes the IP adjustment factor. The iteration continues until $M \times u^j < \varepsilon_2$ or $i > T$ is satisfied, where T denotes the maximum number of iterations.

The proposed algorithm is referred to as the NMF-GN algorithm for simplicity.

4.3 Experiments and Analysis

To evaluate the proposed NMF-GN algorithm, real experiments are presented. The reconstruction algorithm adopts the filtered backprojection (FBP). Set $\gamma_0 = 1$, $\tau = 0.8$, $\tau' = 0.01$, $\varepsilon_1 = 10^{-4}$, $u^0 = 0.01$, $\theta = 0.1$, $\varepsilon_2 = 1$ and $T = 500$ in the GN step. The values of these parameters are relatively broad because they are not sensitive to the solved results. The regularization parameter α is selected by the L-curve method [33].

An effective initialization approach is provided for the nonconvexity of Eq. (21). The initial value d^0 is obtained by projecting forward the digitized model based on the imaging parameters, in which a slight disturbance is added to each projection value. The digital model is determined by segmenting an FBP reconstructed image of polychromatic projections by thresholds under a certain spectrum. In particular, when the gray values of two component materials in the image overlap (such as silicon and aluminum), these two materials are divided into "the same material," and d_k of both materials is one-half of the calculated result. The initial value s^0 is calculated in advance by Eq. (27). In the calculation of s^0, the vector d is fixed to d^0 and the initial value of s satisfies

$$\sum_{r=1}^{R} s_{n,r} = 1, \quad s_{n,r} = 0, \forall (n, r) \in \left\{(n, r) \mid E_{n,\max} \leq E_{r-1}\right\}, \quad (42)$$

where $E_{n,\max}$ is the maximum energy of the nth energy spectrum. The initial value is determined by using the spectra of SpectrumGUI_1.03 software in the experiments. Stop the iteration and output s^0 when the maximum number of iterations, 1000, is reached or $\|s^t - s^{t-1}\|_2 < \|s^t\|_2 \times 10^{-5}$ is satisfied. Since Eq. (27) is multiplicative, s_{nr} set to 0 will always be constant. The parameter $\delta = \sum_{m=1}^{M} t_{1,m}^0$ is employed in this method.

Three samples were scanned using a YXLON FF20CT device and labeled as samples 1 to 3. The flat panel detector used by the device has 1122×1122 pixels, and each pixel is 0.127 mm $\times$ 0.127 mm in size. The three samples were scanned at

360, 360, and 1080 angles from 0 to 360 at equal intervals. For the sake of simplicity, only the fan-beam data at the center of the source were processed.

Sample 1 is composed of four irregular small pieces (two silicon, two aluminum). The scan voltages are 40, 50, and 60 kV, and the current is 100 µA. To reduce the noise at low voltage, the exposure time was chosen as 2 s, 1 s, and 0.5 s, respectively. The SDD is 600 mm, and the SOD is 80 mm. The directly reconstructed image at 50 kV is shown in Fig. 4a. Polychromatic attenuation measurements obtained by CT scanning were decomposed by the proposed NMF-GN algorithm. The energy range [0, 60] was equally divided into 12 intervals; that is, the number of narrow-energy-width spectra is $R = 12$, and let $E_r = r \times 5$. The regularization parameters were chosen as $\alpha = 10$ and $\beta_k = 1$ during the decomposition process.

The 6th and 12th decomposition projections were selected from the decomposition results, and their reconstructed images are shown in Fig. 4b, c. Compared with direct reconstruction in Fig. 4a, the reconstructed images in Fig. 4b, c after decomposition not only reduce the cupping artifacts but also show that the LAC of silicon is higher than that of aluminum in one CT image and close to that of aluminum in another CT image (Fig. 4b, c). Thus, the proposed algorithm avoids overlap of the gray values of the two materials and enables the two materials to be distinguished.

Sample 2 is also composed of four irregular small pieces (two magnesium, two aluminum), and the attenuation characteristics of the two materials constituting the sample differ greatly. The scan voltages are 60, 80, and 100 kV, and the currents and exposure times are 100, 80, and 60 µA and 833, 500, and 333 ms, respectively. The SDD is 750 mm, and the SOD is 120 mm. The directly reconstructed image at 60 kV is shown in Fig. 4d. The energy range [0, 100] was equally divided into 20 intervals; that is, the number of narrow-energy-width spectra is $R = 20$, and let $E_r = r \times 5$. The regularization parameters were set to $\alpha = 10$ and $\beta_k = 1$ during the decomposition process. The 8th and 16th decomposition projections were selected from the decomposition results, and their reconstructed images are shown in Fig. 4e, f.

Similar to Sect. 3.3, an aluminum foam is used as sample 3. The scan voltages are 110 and 120 kV, and the currents and exposure times are 55, 50 µA and 454, 400 ms, respectively. The SDD is 780.58 mm, and the SOD is 260 mm. The energy range [0, 120] was equally divided into 24 intervals; that is, the number of narrow-energy-width spectra is $R = 24$, and let $E_r = r \times 5$. The regularization parameters were set to $\alpha = 1$ and $\beta_k = 1$ during the decomposition process. The directly reconstructed image at 110 kV and the 12th reconstructed image are shown in Fig. 5a, b.

Similar to the results for sample 1, the cupping artifacts in the reconstructed image of decomposition projection of samples 2 and 3 are weakened, and the gray values of the same material change relatively steadily (Figs. 4 and 5).

Bean hardening artifacts cause large fluctuations in the gray values of the same material area, and the standard deviation (STD) can measure the fluctuation and then quantify the degree of hardening artifacts. Because there are differences in the gray levels of different reconstructed images, the coefficient of variation (CV) was selected instead of the STD. In the case of similar noise, the smaller the CV, the

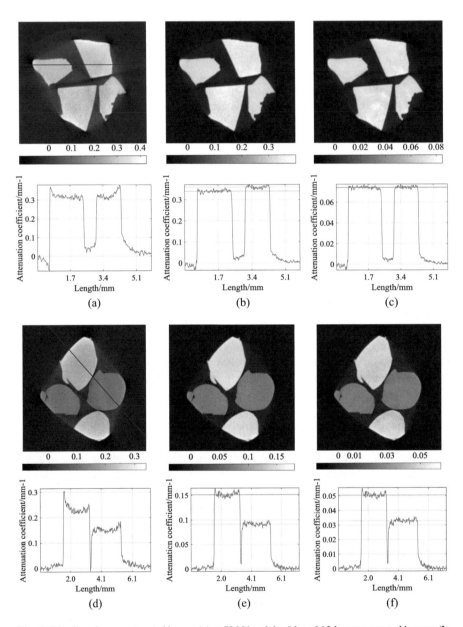

Fig. 4 The directly reconstructed image (**a**) at 50 kV and the 6th and 12th reconstructed images (**b**, **c**) decomposed by the NMF-GN algorithm in the sample 1 experiment. The directly reconstructed image (**d**) at 60 kV and the 8th and 16th reconstructed images (**e**, **f**) decomposed by the NMF-GN algorithm in the sample 2 experiment. The graph below each reconstructed image represents the gray value at the blue line. The green and red lines in (**b**) and (**c**) represent the grayscale mean levels of silicon and aluminum, respectively. The green and red lines in (**e**) and (**f**) represent the grayscale mean levels of aluminum and magnesium, respectively

Table 1 The means, STDs, and CVs of each material region in the reconstructed images of sample 1

CT image	50 kV		6		12	
Material	Si	Al	Si	Al	Si	Al
Mean	0.3339	0.3225	0.3621	0.3427	0.0754	0.0738
STD	0.0208	0.0147	0.0087	0.0083	0.0023	0.0019
CV	**0.0671**	**0.0509**	0.0241	0.0241	0.0304	0.0259

Table 2 The means, STDs, and CVs of each material region in the reconstructed images of samples 2 and 3. The last two columns show the numerical results of sample 3

CT image	60 kV		8		16		110 kV	12
Material	Mg	Al	Mg	Al	Mg	Al	Al Foam	Al Foam
Mean	0.1512	0.2439	0.0916	0.1551	0.0326	0.0512	0.0810	0.0616
STD	0.0106	0.0181	0.0039	0.0053	0.0015	0.0020	0.0093	0.0045
CV	**0.0702**	**0.0740**	0.0428	0.0339	0.0460	0.0395	**0.1146**	0.0726

lower the degree of hardening artifacts. Tables 1 and 2 show the mean values, STDs, and CVs of each material region in the partially reconstructed images of the three samples. The symbol 50 kV represents the directly reconstructed image at 50 kV. The last two columns in Table 2 show the numerical results of sample 3. The CVs after decomposition are significantly smaller than those in the directly reconstructed images (Tables 1 and 2).

To evaluate the relationship between the reconstructed image of decomposition projection and the narrow energy spectrum, Fig. 6 shows the curves of the LAC of each material as a function of energy. Figure 6a, b show the LAC curves of the materials Si and Al in sample 1. Figure 6c, d show the LAC curves of the materials Mg and Al in sample 2. The blue lines represent the true values of the LAC at the middle energy of narrow-energy-width spectra. The red lines represent the mean LACs of reconstruction images corresponding to narrow-energy-width spectra. There is a small error between the true value and the calculated value of the LAC, and the change trend of the two is basically the same (Fig. 6).

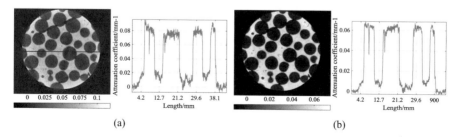

(a)   (b)

Fig. 5 The directly reconstructed image (**a**) at 110 kV and the 12th reconstructed image (**b**) for sample 3 decomposed by the NMF-GN algorithm. The grayscale curves of the two reconstructed images at the blue line position are placed on the right side of reconstructed image

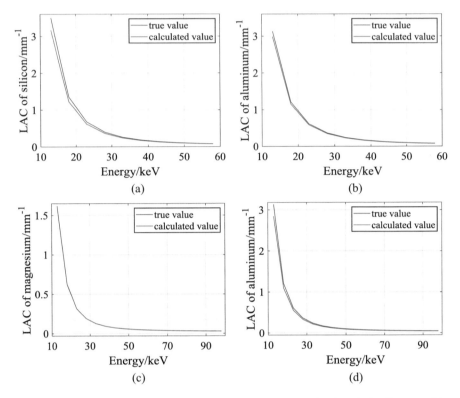

Fig. 6 The LAC curves of each material as a function of energy. (**a, b**) show the LAC curves of the materials Si and Al in sample 1. (**c, d**) show the LAC curves of the materials Mg and Al in sample 2. The blue lines represent the true values of the LAC at the middle energy of narrow-energy-width spectra. The red lines represent the mean LACs of reconstruction images corresponding to narrow-energy-width spectra

The scale of hardening artifacts in reconstructed images is an important feature for distinguishing polychromatic projections from narrow-spectrum projections. In the above experiments, the lower levels of hardening artifacts in reconstructed image of decomposition projection conform to the characteristics of the reconstructed image of the narrow-spectrum projection (Figs. 4 and 5). The contrast of materials with approximately LACs was improved due to the reduction in hardening artifacts (Fig. 4b). The error between the true value and the calculated value of the LAC in each narrow-energy-width spectrum is relatively small, and the trends of the true value and the calculated value are basically the same (Fig. 6). This fact verifies that each decomposition projection has a clear orientation of narrow-energy-width spectrum, which conforms to the characteristics of narrow-spectrum projection.

5 Conclusion

Blind separation of projection sequences with varying energy can obtain narrow-spectrum projections. In order to improve the energy directivity of separated projections, we developed two projection blind separation methods of multispectral CT for obtaining narrow-spectrum projections from measured polychromatic attenuation data. Since the spectrum is updated as an unknown with other variables, detailed information about the spectrum is not required.

The first method is a blind separation model based on basis-effect (BSM-BE). By basis-effect decomposition, the linear attenuation coefficient can be decomposed into a weighted sum of the photoelectric effect and the Compton effect. When the solved energy interval is preset, the blind separation model can convert to the model with the ray energy constraint. To verify the effectiveness of the proposed method, the sample consisting of silicon and aluminum with similar attenuation coefficients is employed as the test object. The experiment results show that the proposed model achieved results similar to the mono-energetic CT imaging, and the beam-hardening artifacts can be suppressed. Also the accuracy of the attenuation coefficient is higher. From the experiment of two other samples (single material sample and multi-material sample), the proposed BSM-BE model can get the better narrow-spectrum projections.

The second method is a blind separation model based on basis-material and component prior. By basis-material decomposition, component prior is introduced into the X-ray multispectral forward model and is critical to determining the energy orientation of the narrow-spectrum projection. Using the statistical properties of the measured data, a regularized weighted least squares optimization problem with constraints is established. An alternating algorithm called NMF-GN is combined with the proposed initialization approach to solve the proposed nonconvex optimization problem. Real experiments show that the reconstructed images obtained by decomposition have smaller hardening artifacts and that the decomposition projections are energy-dependent on the narrow-energy-width spectrum. This indicates that the decomposed projections have the characteristics of a true narrow-spectrum projection. The proposed algorithm determines their energy orientation and improves the accuracy of obtaining the narrow-spectrum projections.

It is worth noting that, for high density materials or large size object, the higher voltage is required. Then there will be scattering effect, and the attenuation coefficient difference is smaller. These will affect the separation effectiveness. So, for high density materials or large size object, the corresponding separation methods need to be further studied.

References

1. Li Y, Li K, Zhang C, et al. Learning to Reconstruct Computed Tomography Images Directly from Sinogram Data under A Variety of Data Acquisition Conditions. *IEEE Transactions on Medical Imaging*, 2019, 38(10): 2469–2481.

2. Fang L, Yin X, Wu L, et al. Classification of microcrystalline celluloses via structures of individual particles measured by synchrotron radiation X-ray micro-computed tomography. *International Journal of Pharmaceutics*, 2017, 531(2):658-667.

3. Ketcham R A, Hanna R D. Beam hardening correction for X-ray computed tomography of heterogeneous natural materials. *Computers & Geosciences*, 2014, 67(4):49-61.

4. Taguchi K, Iwanczyk J S. Vision 20/20: Single photon counting x-ray detectors in medical imaging. *Medical Physics*, 2013, 40(10):100901.

5. Dong W K, Kim H K, Youn H, et al. Signal and noise analysis of flat-panel sandwich detectors for single-shot dual-energy x-ray imaging. SPIE Medical Imaging. *International Society for Optics and Photonics*, 2015. p. 94124A.

6. Zou Y, Silver M D. Analysis of fast kV-switching in dual energy CT using a pre-reconstruction separation technique. *Proceedings of SPIE - The International Society for Optical Engineering*, 2008, 6913:691313-691313-12.

7. Szczykutowicz T P, Chen G H. Dual energy CT using slow kVp switching acquisition and prior image constrained compressed sensing. *Physics in Medicine & Biology*, 2010, 55(21):6411-6429.

8. Sebastian F, Stefan K, Stefan S, et al. Performance of today's dual energy CT and future multi energy CT in virtual non-contrast imaging and in iodine quantification: A simulation study. *Medical Physics*, 2015, 42(7):4349.

9. Kim K, Ye J C, Worstell W, et al. Sparse-view spectral CT reconstruction using spectral patch-based low-rank penalty. *IEEE Transactions on Medical Imaging*, 2015, 34(3):748.

10. Zhang R, Thibault J B, Bouman C A, et al. Model-Based Iterative Reconstruction for Dual-Energy X-Ray CT Using a Joint Quadratic Likelihood Model. *IEEE Transactions on Medical Imaging*, 2014, 33(1):117-134.

11. Ducros N, Abascal J, Sixou B, et al. Regularization of nonlinear separation of spectral x-ray Projected images. *Medical Physics*, 2017, 44(9):e174.

12. I.A. Elbakri, J.A. Fessler. Statistical image reconstruction for polyenergetic X-ray computed tomography. *IEEE Transactions on Medical Imaging*, 2002, 21(2):89-99.

13. Xu Q, Mou X, Tang S, et al. Implementation of penalized-likelihood statistical reconstruction for polychromatic dual-energy CT. *Progress in Biomedical Optics and Imaging Proceedings of SPIE*, 2009, 7258:72585I-72585I-9.

14. Long Y, Fessler J A. Multi-material separation using statistical image reconstruction for spectral CT. *IEEE Transactions on Medical Imaging*, 2014, 33(8):1614-26.

15. Chen B, Zhang Z, Sidky E Y, et al. Image reconstruction and scan configurations enabled by optimization-based algorithms in multispectral CT. *Physics in Medicine & Biology*, 2017, 62(22):8763.

16. Niu T, Dong X, Petrongolo M, et al. Iterative image-domain decomposition for dual-energy CT, *Med. Phys.*, 2014, 41(4):041901.

17. Ding Q, Niu T, Zhang X, et al. Image-domain multimaterial decomposition for dual-energy CT based on prior information of material images, *Med. Phys.*, 2018, 45(8): 3614–3626.

18. Foygel Barber R, Sidky E Y,. Gilat Schmidt T, et al. An algorithm for constrained one-step inversion of spectral CT data. *Physics in Medicine & Biology*. 2016, 61: 3784–3818.

19. Schlomka JP, Roessl E, Dorscheid R, et al. Experimental feasibility of multi-energy photon-counting K-edge imaging in pre-clinical computed tomography. *Physics in Medicine and Biology*, 2008, 53(15):4031-4047.

20. Schirra C O, Roessl E, Koehler T, et al. Statistical reconstruction of material decomposed data in spectral CT. *IEEE Transactions on Medical Imaging*, 2013, 32(7):1249-1257.

21. K. Mechlem et al., Joint Statistical Iterative Material Image Reconstruction for Spectral Computed Tomography Using a Semi-Empirical Forward Model, *IEEE Transactions on Medical Imaging*, 2018, 37(1): 68-80.

22. Wei J, Han Y, Chen P. Improved contrast of materials based on multi-voltage images separation in X-ray CT. *Measurement Science & Technology*, 2016, 27(2):025402.

23. Wei J, Han Y, Chen P. Narrow-Energy-Width CT Based on Multi-voltage X-Ray Image Separation. *International Journal of Biomedical Imaging*, 2017, 2017(2):1-9.

24. Alvarez R E and MacOvski A. Energy-selective reconstructions in X-ray computerised tomography. *Physics in Medicine and Biology*, 1976, 21: 733–744.
25. Sukovic P, Clinthorne N H. Penalized weighted least-squares image reconstruction for dual energy X-ray transmission tomography. *IEEE Transactions on Medical Imaging*, 2000, 19(11): 1075–1081.
26. Lin C J. Projected gradient methods for nonnegative matrix factorization. *Neural Computation*, 2007, 19(10): 2756–2779.
27. C. H. Yan, R. T. Whalen, G. S. Beaupre, S. Y. Yen, and S. Napel, Modeling of polychromatic attenuation using computed tomography reconstructed images, *Medical Physics*. 26, 631 (1999).
28. Schweiger M, Arridge S R, Nissilä R. Gauss-Newton method for image reconstruction in diffuse optical tomography. *Physics in Medicine and Biology*, 2005, 50: 2365–2386.
29. Heinz D C, Chang C I. Fully constrained least squares linear spectral mixture analysis method for material quantification in hyperspectral imagery. *IEEE Transactions on Geoscience and Remote Sensing*, 2001, 39: 529–545.
30. Lee D D, Seung H S. Algorithms for non-negative matrix factorization. *Advances in Neural Information Processing Systems*, 2001, 13: 556–562.
31. Forsgren A, Gill P E, Wright M H. Interior methods for nonlinear optimization. *SIAM Review*, 2002, 44: 525–597.
32. Andrei N. An acceleration of gradient descent algorithm with backtracking for unconstrained optimization. *Numerical Algorithms*, 2006, 42: 63–73.
33. Hansen P C. Analysis of Discrete Ill-Posed Problems by Means of the L-Curve. *SIAM Review*, 1992, 34: 561–580.

Direct Iterative Basis Image Reconstruction Based on MAP-EM Algorithm for Spectral CT

Zhengdong Zhou

1 Introduction

The concept of spectral CT was firstly proposed by Alvarez and Macovski in 1976 [1]. Spectral CT has been attracting great research interest in the last decades. Nowadays, spectral dual-energy CT system has been available in various clinical applications, and spectral photon-counting CT system is the research focus in the field of CT imaging. Using spectral CT, virtual monoenergy images can be reconstructed to reduce the effect of beam-hardening artifacts [2]. By identifying the energy information with different energy spectra, spectral CT can also provide more material properties and is thus superior to traditional CT in material identity [3]. Spectral CT has broad application prospects in the fields of medical diagnosis, non-destructive testing and security inspection, etc.

Spectral CT has the distinctive ability of material decomposition and identification. Currently, the most widespread methods of material decomposition can be divided into two categories: two-step and one-step methods. Two-step methods include image domain [4–6] and projection-domain material decomposition methods [1, 7–9]. Image-based methods perform reconstruction first and then decomposition. The image-based methods often integrate some form of empirical beam-hardening correction but requiring knowledge of the material volumes in advance for perfect beam-hardening correction. On the other hand, projection-based methods perform decomposition first and then reconstruction. However, it is hard to achieve a robust decomposition for typical choices of basis materials (e.g., water, bone, and a high-Z contrast agent) without enough difference of the normalized attenuation profiles, thus the aberrant material line integrals maybe caused due to

Z. Zhou (✉)
State Key Laboratory of Mechanics and Control of Mechanical Structures, Nanjing University of Aeronautics and Astronautics, Nanjing, China
e-mail: zzd_msc@nuaa.edu.cn

© The Author(s), under exclusive license to Springer Nature Switzerland AG 2023
K. (Kris) Iniewski (eds.), *Advanced X-Ray Radiation Detection*,
https://doi.org/10.1007/978-3-030-92989-3_10

the inevitable statistical noise on photon counts, resulting in strong streak artifacts in the reconstructed images. In two-step methods, a loss of information is inevitable, because the first step is hard to provide a one-to-one mapping between inputs and outputs. Therefore, the inaccuracy of material decomposition is unavoidable since the second step cannot compensate for the loss of information in the first step [10]. The one-step methods, also named as "one-step inversion," refer to the direct iterative material decomposition methods, which perform decomposition and reconstruction simultaneously [11]. The one-step methods can overcome the inherent shortcomings in two-step methods and improve the accuracy of material decomposition. In 2013, Cai et al. [12] proposed a one-step material decomposition method based on Bayesian model with Huber prior knowledge, in which the conjugate gradient algorithm is adopted for optimization. Thereafter, several direct iterative projection models have been proposed one after another, and all the methods for solving the models involved complex non-convex function optimization problem [13–16]. Yong et al. [13] proposed a penalized-likelihood (PL) method to reconstruct multi-material images using a similar constraint from sinogram data, where the edge-preserving regularizers for each material are employed. And an optimization transfer method with a series of pixel-wise separable quadratic surrogate (PWSQS) functions was developed to monotonically decrease the complicated PL cost function. Foygel Barber et al. [14] proposed a primal-dual algorithm for material decomposition with one-step inversion from spectral CT projection data, where the image constraints are enforced on the basis maps during the inversion. Combined with a convex-concave optimization algorithm and a local upper bounding quadratic approximation, the algorithm is derived to generate descent steps for non-convex spectral CT data discrepancy terms. Weidinger et al. [15] proposed a dedicated statistical algorithm based on local approximations of the negative logarithmic Poisson probability function to perform a direct material decomposition for photon-counting spectral CT. The algorithm allows for parallel updates of all image pixels, which can compensate for the rather slow convergence that is intrinsic to statistical algorithms. Mechlem et al. [16] proposed an algorithm based on a semi-empirical forward model for joint statistical iterative material image reconstruction, where the semi-empirical forward model is tuned by calibration measurements. This strategy allows to model spatially varying properties of the imaging system without requiring detailed prior knowledge of the system parameters. And an efficient optimization algorithm based on separable surrogate functions is employed to accelerate convergence and reduce the reconstruction time. All the methods need to be solved iteratively, currently no analytical inversion formula is available for the material decomposition problem, let alone for one-step inversion [10].

In previous studies, we proposed methods for spectral CT image reconstruction by using maximum a posteriori expectation maximization (MAP-EM) algorithm [17, 18], which has advantages in noise robustness and image reconstruction quality. Based on previous studies, a novel and robust direct iterative basis material image reconstruction based on maximum a posteriori expectation–maximization algorithm (MAP-EM-DD) is proposed. Furthermore, by incorporating polar coordinate transformation into MAP-EM-DD, MAP-EM-PT-DD method is proposed. The iterative formulas of MAP-EM-DD and MAP-EM-PT-DD methods are derived. For readability, we present the case of dual-energy CT (DECT) and decompose on a basis of bone tissue and soft tissue, but all the derivations can be adapted to spectral photon-counting CT. The performance of the proposed methods was evaluated and compared with the image domain material decomposition method based on FBP algorithm (FBP-IDD) [19].

2 Materials and Methods

2.1 MAP-EM Statistical Reconstruction Algorithm

In spectral CT system, noise is inevitable in the projection due to the effects of incoherent scattering, pulse accumulation, and electronic noise. The statistical property of the detected data can be described by Poisson model. The MAP-EM algorithm with the Poisson model tries to find an image I to maximize the conditional probability $P(I|y)$ by a set of measured projection vector y, as follows [20]:

$$P(I|y) = P(y|I) P(I) / P(y) \tag{1}$$

where $P(I)$ is the prior knowledge of the image, y is the known measured projection. Applying Gibbs prior distribution as the prior knowledge, then $P(I)$ can be expressed as follows:

$$P(I) = \exp(-\beta \cdot U(I)) / c \tag{2}$$

where β is a regularization parameter, the constant c is an unknown normalization factor, and the function $U(I)$ denotes the total energy of I.

Applying the form of the Gibbs distribution given by Eq. (2) in the MAP-EM procedure, the iterative formula is given as follows [21]:

$$I_n^{s+1} = \frac{I_n^s}{\sum_k a_{kn} + \beta \frac{\partial U(I^s)}{\partial I_n}} \sum_k \frac{y_k a_{kn}}{\sum_{n'} a_{kn'} I_{n'}^s} \tag{3}$$

where s is the number of iteration, I_n means the nth pixel value in the reconstructed image I, y_k means the projection detected by the kth detector element, a_{kn} represents the probability that the photon passes through the nth pixel collected by the kth detector element, and $\beta \partial U(I^S)/\partial I_n$ is the term of penalty function which can suppress the noise in the reconstructed image. When β equals zero, the algorithm tends toward maximum likelihood expectation–maximization (ML-EM) algorithm.

2.2 Direct Iterative Material Decomposition Method Based on MAP-EM Algorithm

The linear attenuation coefficient of material depends on its specific properties and the incident energy of X-ray. In accordance with the basis material decomposition model, the linear attenuation coefficient of an object can be expressed as the weighted sum of the linear attenuation coefficients of certain materials [22] as follows:

$$\mu(E, l) \approx \sum_{m=1}^{n_f} x_m(l)\, \mu_m(E) \tag{4}$$

where n_f is the number of basis material, $x_m(l)$ is the basis image of the mth material, l is the path of X-ray, and $\mu_m(E)$ is the linear attenuation coefficient of the mth material at the energy E.

The discrete form of the projection function along path l at certain energy E can be expressed as follows:

$$P(E, l) = \sum_{m=1}^{n_f} \mu_m(E) B_m(l) \tag{5}$$

where $B_m(l) = \int_l x_m(l)dl$ is the projection of the basis image of the mth material along path l.

Let the continuous X-ray energy spectrum under the tube voltage u be divided into Q energy bins. The projection of the target under the tube voltage u can then be expressed as a weighted sum of the projections at each energy bin as follows:

$$
\begin{aligned}
P(u, l) &= \sum_{q=1}^{Q} w(E_q) \cdot P(E_q, l) \\
&= \sum_{q=1}^{Q} \sum_{m=1}^{n_f} w(E_q)\, \mu_m(E_q)\, B_m(l) = \sum_{m=1}^{n_f} \mu_m(u)\, B_m(l)
\end{aligned}
\tag{6}
$$

where $w(E_q) = n_p(E_q) / \sum_{q=1}^{Q} n_p(E_q)$ is the normalization coefficient of the number of photons in the qth energy bin, $n_p(E_q)$ is the number of photon in the qth energy bin, and $\mu_m(u)$ is the effective linear attenuation coefficient of a material at a given tube voltage u.

The projection of the basis image of the mth material along path l can be expressed as follows:

$$B_m(l) = A x_m(l), \, m = 1, \cdots, n_f. \tag{7}$$

Substituting Eq. (7) into Eq. (6), the discrete form of the direct iterative projection function can be obtained as follows [12]:

$$P = JRx \tag{8}$$

where $x = \left[x_1, x_2, \cdots, x_{n_f}\right]^T$ is the set of basis images, the dimension is $\mathcal{R}^{n_f N}$, N is the total number of pixels in the basis images, and J is the joint basis material matrix, which is expressed as follows:

$$J = J_1 \otimes C_K, \ J_1 = \begin{bmatrix} \mu_1(u_1) & \cdots & \mu_{nf}(u_1) \\ \vdots & \ddots & \vdots \\ \mu_1(u_{N_e}) & \cdots & \mu_{nf}(u_{N_e}) \end{bmatrix} \tag{9}$$

where N_e is the total number of energy bins, K is the total observation dimension, C_k is the identity matrix of size $K \times K$, and R is a joint projection operator expressed as follows:

$$R = C_{n_f} \otimes A \tag{10}$$

where C_{n_f} is the identity matrix with size of $n_f \times n_f$, and $\otimes$ denotes the matrix Kronecker product.

For the material decomposition of DECT with two basis materials, i.e., $N_e = 2$, $n_f = 2$, the material decomposition model can be expressed as follows:

$$P = JRx = \begin{bmatrix} \mu_1(u_L)A & \mu_2(u_L)A \\ \mu_1(u_H)A & \mu_2(u_H)A \end{bmatrix} \cdot \begin{bmatrix} x_1 \\ x_2 \end{bmatrix} \tag{11}$$

where $P = [P_1, P_2]^T$ is the set of projections at two different tube voltages, x_1 and x_2 are the basis images, and $x = [x_1, x_2]^T$ is the set of basis images.

Let $W = JR$, which denotes the transformed projection operator, then $P = Wx$. Evidently, the material decomposition problem is the same as the problem of image reconstruction. In accordance with the transformed projection operator W and the set of projection P, the basis images can be reconstructed by different kinds of reconstruction algorithms.

Let $W = \{c_{kn}\}$, the iterative formulas for the direct material decomposition using MAP-EM algorithm (MAP-EM-DD) can be derived from Eq. (3) as follows:

$$x_{1n}^{s+1} = \frac{x_{1n}^s}{\sum_k c_{kn} + \beta \frac{\partial U(x_1^s)}{\partial x_{1n}}} \sum_k \frac{y_{1k} c_{kn}}{\sum_{n'} c_{kn'} x_{1n'}^s},$$

$$x_{2n}^{s+1} = \frac{x_{2n}^s}{\sum_k c_{kn} + \beta \frac{\partial U(x_2^s)}{\partial x_{2n}}} \sum_k \frac{y_{2k} c_{kn}}{\sum_{n'} c_{kn'} x_{2n'}^s} \tag{12}$$

where x_{1n} and x_{2n} represent the nth pixel value in basis images x_1 and x_2, respectively; $U(x_1^s)$ and $U(x_2^s)$ are the energy functions of basis images x_1^s and x_2^s at sth iteration, respectively; and y_{1k} and y_{2k} are the projection detected by the kth detector element at the two different tube voltages, respectively; c_{kn} represents the weighted probability that the photon passes through the nth pixel collected by the kth detector element.

Furthermore, by applying polar coordinate transformation to replace (x_1, x_2) with (r, θ), we have $x_1 = r\cos\theta$ and $x_2 = r\sin\theta$. Equation (12) can be expressed by Eq. (13) for MAP-EM-PT-DD method as follows:

$$r_n^{s+1} = \frac{r_n^s}{\sum_k c_{kn} + \beta \frac{\partial U(r^s)}{\partial r_n}} \sum_k \frac{y_{1k} c_{kn}}{\sum_{n'} c_{kn'} x_{n'}^s},$$

$$\theta_n^{s+1} = \frac{\theta_n^s}{\sum_k c_{kn} + \beta \frac{\partial U(\theta^s)}{\partial \theta_n}} \sum_k \frac{y_{1k} c_{kn}}{\sum_{n'} c_{kn'} x_{n'}^s} \tag{13}$$

where $x_{n'}^s = \begin{bmatrix} r\cos\theta \\ r\sin\theta \end{bmatrix}_{n'}^s$, r_n and θ_n represent the value of the nth pixel on the basis images r^s and θ^s respectively, and $U(r^s)$ and $U(\theta^s)$ are the energy functions of the basis images r^s and θ^s at the sth iteration, respectively.

After r and θ are calculated, basis images x_1 and x_2 can be reconstructed by applying $x_1 = r\cos\theta$ and $x_2 = r\sin\theta$.

2.3 Simulation Setup

To evaluate the proposed methods, a simulated cylindrical phantom was developed. The cross-section related geometric parameters, and material composition of the phantom are shown in Fig. 1 and Table 1. Polyethylene (PE) and hydroxyapatite (HA) were used to simulate the soft tissue and bone tissue, and 10% salt water was used to simulate the materials around the tissues. To test the robustness of the proposed material decomposition methods, aluminum (Al), an additional mineral material, was introduced in the phantom, whose mass density was beyond the range of soft tissue and bone tissue [12]. The effective linear attenuation coefficient (ELAC) of each material is given in Table 2.

A fast KV switch dual-energy CT system simulated by Geant4 (Geometry and Tracking) [23] is shown in Fig. 2. A single-slice fan beam was used to scan the

Fig. 1 Cross-section of the cylindrical phantom

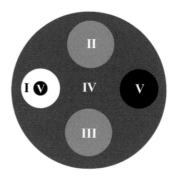

phantom, and the fan angle was set to 8.6°. The distances from the source to the center of the phantom and detector were set to 670 and 1010 mm, respectively, in the simulated X-ray CT system. The incident X-ray energy spectrum was simulated at 80 and 140 kVp with 0.8 mm-thick beryllium and 8.4 mm-thick Al filtration using Spektr software package [24]. The filtration material was used to reduce the radiation damage caused by the low-energy X-ray. The X-ray energy spectra at 80 and 140 kVp are shown in Fig. 3. In the simulation, the rotation angle was set from 0 to 2π, and the angular sampling interval was set to 1° with 2×10^6 incident X-ray photons with an energy spectra produced by the above-mentioned method.

2.4 Evaluation Metrics

To evaluate the quality of the basis images quantitatively, noise level (σ) and contrast-to-noise ratio (CNR) are used as follows:

$$\sigma = \sqrt{\frac{1}{N-1} \sum_{n=1}^{N} (z_n - \bar{z})^2} \tag{14}$$

$$\text{CNR} = \frac{|\bar{z}_c - \bar{z}_b|}{\sqrt{\sigma_b^2 + \sigma_c^2}} \tag{15}$$

where σ is the noise level, CNR is the contrast-to-noise ratio, N is the total number of pixels in the region of interest (ROI), z_n is the nth pixel value in the ROI, $\bar{z}$ is the average value of all pixels in the ROI, $\bar{z}_c$ is the average value of pixels in the region of contrast material, σ_c is the noise level of the region of contrast material, $\bar{z}_b$ is the average value of pixels in the region of background material, and σ_b is the noise level of the background material region.

Table 1 Geometric parameters and components of the cylindrical phantom in the cross-section

Number	Material	Center coordinates (mm)	Radius (mm)	Material composition
I	Aluminum	(−32.5, 0.0)	Inner 4.5, outer 12.5	Al
II	Hydroxyapatite	(30.0, 0.0)	15.0	$Ca_{10}(PO_4)_6(OH)_2$
III	Salt water	(−30.0, 0.0)	15.0	$10\%w(NaCl) + 90\%w(H_2O)$
IV	Polyethylene	(0.0, 0.0)	50.0	$(C_2H_4)_n$
V	Air	(32.5, 0.0)	12.5	$75.5\%w(N) + 23.2\%w(O) + 1.3\%w(Ar)$

Note: $w()$ denotes the mass of each component

Table 2 ELAC of each material under two scanning voltages

	ELAC				
	PE	Salt water	HA	Al	Air
80 kVp	0.2036	0.2859	0.4947	1.100	0.0003
140 kVp	0.1836	0.2340	0.3110	0.7607	0.0002

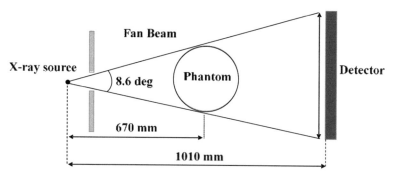

Fig. 2 Diagram of the simulated DECT system

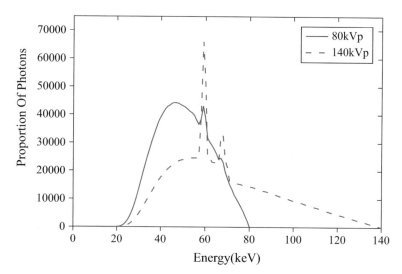

Fig. 3 Normalized energy spectrum at 80 and 140 kVp

To evaluate the accuracy of the reconstructed image quantitatively, mean square error (MSE) is used as follows:

$$MSE = \frac{\sum_{n=1}^{N} \left(z_n^{rec} - z_n^{ref}\right)^2}{N} \tag{16}$$

where Z_n^{rec} and Z_n^{ref} are the nth pixel values in the reconstructed and reference images, respectively. MSE represents the similarity between the reconstructed image and reference image. For the evaluation of the performance of the proposed basis image reconstruction algorithm, the reference image refers to the theoretical decomposition coefficient image which can be calculated by formula (18) in this paper.

To evaluate the accuracy of material decomposition, the error level of material decomposition is used as follows [25]:

$$\delta = \frac{1}{E_2 - E_1} \int_{E_1}^{E_2} \frac{|\mu(E) - (b_1\mu_1(E) + b_2\mu_2(E))|}{\mu(E)} dE \qquad (17)$$

where δ is the error level of material decomposition, and E_1 and E_2 indicate the lowest and highest photon energy to be detected, respectively. Considering the filter material, E_1 is set to 20 keV. E_2 is set to 140 keV, which should not be less than the high voltage used in DECT. $\mu_1(E)$, $\mu_2(E)$, and $\mu(E)$ are the linear attenuation coefficients of the two basis materials and the simulated phantom at energy E.

2.5   Calculation of Theoretical Decomposition Coefficients

Soft and bone tissues were selected as the two basis materials. Hydroxyapatite with a mass density of 0.74 g/cm^3 was the equivalent material of bone tissue, and PE with a mass density of 0.97 g/cm^3 was the equivalent material of soft tissue. Theoretical decomposition coefficients were calculated by the effective linear attenuation coefficients of each material at the high and low tube voltages, which can be expressed by Eq. (18). The results are shown in Table 3, in which air is omitted because its effective linear attenuation coefficient is approximately equal to zero.

$$\begin{bmatrix} b_1 \\ b_2 \end{bmatrix} = \begin{bmatrix} \mu_{1L} & \mu_{2L} \\ \mu_{1H} & \mu_{1H} \end{bmatrix}^{-1} \begin{bmatrix} \mu_L \\ \mu_H \end{bmatrix}, \qquad (18)$$

where b_1 and b_2 are the decomposition coefficients of soft and bone tissues, L represents the low tube voltage 80 kVp, H represents the high tube voltage 140 kVp, $\mu_{1L/H}$ and $\mu_{2L/H}$ are the effective linear attenuation coefficients of soft and bone tissues at the voltage of 80 and 140 kVp, respectively, and $\mu_{L/H}$ is the effective linear attenuation coefficients of the phantom at the voltage of 80 and 140 kVp. The linear attenuation coefficients of the five material insets in the phantom at the energy from 20 to 140 keV are shown in Fig. 4.

Table 3 Soft and bone decomposition coefficients of materials contained in phantom

Material	Mass density (g/cm^3)	b_1	b_2
Al	2.69	1.243	1.267
HA	0.74	0	0.740
Salt water	1.07	0.976	0.130
PE	0.97	1.000	0

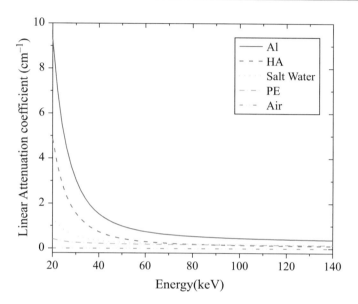

Fig. 4 Linear attenuation coefficients of phantom material [12]

3 Experimental Results

3.1 Evaluation of Regularization Parameter

The term of penalty function in Eqs. (12) and (13) can suppress the noise in the reconstructed basis images. The regularization parameter β controls the strength of the prior relative to the strength of the data in the basis image estimation. A higher β leads to a final estimation which places more emphasis on the prior and less on the data.

To evaluate the effect of parameter β on the accuracy of material decomposition and basis image quality, two metrics are used: (1) the similarity between the reconstructed and the reference images by MSE and (2) the noise level of the basis image by CNR. The smaller the MSE, the more similar is the reconstructed image to the reference image. The larger the CNR, the smaller the noise of image. For the calculation of CNR, PE is selected as the background material in the evaluation.

For each parameter β, MSE varies with the change of the iteration number and reaches the minimum at a certain number of iterations. For MAP-EM-DD method,

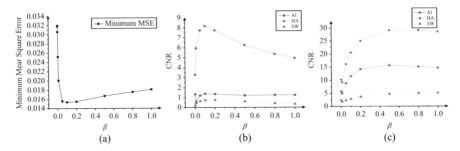

Fig. 5 Minimum MSE and CNRs of Al, HA, and salt water in basis images vs. β with MAP-EM-DD method. (**a**) Minimum MSE. (**b**) CNRs of each material in the basis image of soft tissue. (**c**) CNRs of each material in the basis image of bone tissue

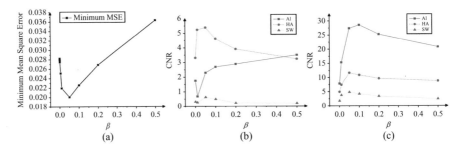

Fig. 6 Minimum MSE and CNRs of Al, HA, and salt water in basis images vs. β with MAP-EM-PT-DD method. (**a**) Minimum MSE. (**b**) CNRs of each material in the basis image of soft tissue. (**c**) CNRs of each material in the basis image of bone tissue

the minimum MSE vs. β and CNRs of each specific material in basis images of soft and bone tissue vs. β are shown in Fig. 5a–c. From Fig. 5a, as the parameter β increases, the minimum MSE decreases first and reaches the minimum at β of approximately 0.1 and then increases gradually. As shown in Fig. 5b, c, the CNRs of Al, HA, and salt water reach the maximum at β of approximately 0.1 in the basis image of soft tissue and 0.5 in the basis image of bone tissue. Considering all the trends of MSE and CNRs vs. β in Fig. 5 together, the parameter β is set to 0.2 in the experiment with MAP-EM-DD method.

With regard to MAP-EM-PT-DD method, the minimum MSE vs. β and CNRs of each specific material in the basis image of soft and bone tissues vs. β are shown in Fig. 6a–c. As shown in Fig. 6a, the minimum MSE reaches the minimum at β of approximately 0.05. As shown in Fig. 6b, c, the CNRs of Al, HA, and salt water reach the maximum at β of approximately 0.05. Therefore, the parameter β is set to 0.05 in the experiment with MAP-EM-PT-DD method.

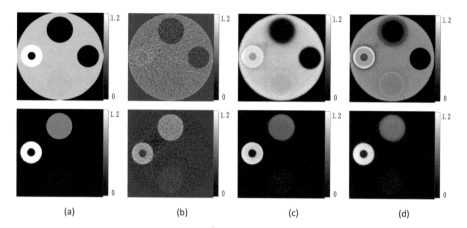

Fig. 7 Basis images reconstructed by (**a**) theoretical calculation, (**b**) FBP-IDD, (**c**) MAP-EM-DD, and (**d**) MAP-EM-PT-DD

Table 4 Noise levels of basis images

	Noise level (σ)		
	FBP-IDD	MAP-EM-DD	MAP-EM-PT-DD
Soft tissue	0.209	0.089	0.068
Bone tissue	0.077	0.028	0.022

3.2 Comparison of the Basis Images

To reconstruct the basis images with MAP-EM-DD and MAP-EM-PT-DD methods, the iteration stop condition is set as the difference of MSE between two consecutive reconstructed images to less than 1e−6. The theoretical basis images calculated by the average attenuation coefficients of each specific material in the phantom are shown in Fig. 7a. The basis images reconstructed by FBP-IDD, MAP-EM-DD, and MAP-EM-PT-DD methods are shown in Fig. 7b–d, respectively. The first row in Fig. 7 shows the basis images of soft tissue; the second row, bone tissue.

Comparison of the Noise Levels (σ)

The noise levels (σ) of basis images reconstructed by FBP-IDD, MAP-EM-DD, and MAP-EM-PT-DD methods are calculated and compared. The noise levels (σ) of basis images are shown in Table 4. Compared with FBP-IDD method, MAP-EM-DD method can reduce the noise levels (σ) of the basis images of soft tissue and bone tissue by 57.4% and 63.6%, respectively. Furthermore, compared with MAP-EM-DD method, MAP-EM-PT-DD method can reduce the noise levels (σ) of the basis images of soft tissue and bone tissue by 23.6% and 21.4%, respectively.

Comparison of the Gray Level Distributions in the Decomposition Coefficient Images

To compare the gray level distributions in the decomposition coefficient images reconstructed by the methods of FBP-IDD, MAP-EM-DD, and MAP-EM-PT-DD with theoretical gray level distribution, the red lines with arrows are plotted in the decomposition coefficient images with the above methods respectively, as illustrated in Fig. 8a, c. The gray level distribution curves through the red lines are shown in Fig. 8b, d. Figure 8b shows the curves of gray level distribution along the red line with arrows located in the theoretical basis image of soft tissue and the decomposition coefficient images of soft tissue reconstructed by the methods of FBP-IDD, MAP-EM-DD, and MAP-EM-PT-DD, respectively. Figure 8d shows the curves of gray level distribution along the red line with arrows located in the theoretical basis image of bone tissue and the decomposition coefficient image of bone tissue reconstructed by the methods of FBP-IDD, MAP-EM-DD, and MAP-EM-PT-DD, respectively.

As shown from the curves of gray level distribution in Fig. 8b, d, we can see that the curves of gray distribution along the red lines in the decomposed images of soft tissue reconstructed by FBP-IDD method have the highest fluctuation and it is quite different from the curve of theoretical gray distribution. The fluctuation of the curves of gray level distribution in the decomposed images reconstructed by MAP-EM-DD and MAP-EM-PT-DD methods is much smaller than that reconstructed by FBP-IDD. The difference between the curves of gray level distribution in the decomposed images reconstructed by MAP-EM-DD and MAP-EM-PT-DD methods is small, the trends of the curves are similar, and the curves are basically consistent to the curve of theoretical gray distribution, however the range of fluctuation of the curves of gray level distribution in the decomposition coefficient image of bone tissue is less than that of the curves of gray level distribution in the decomposition coefficient image of soft tissue. It implies that the noise level of the decomposition coefficient image of soft tissue is higher than that of the decomposition coefficient image of bone tissue. The accuracy of the decomposition coefficient images reconstructed by the direct iterative material decomposition methods, i.e., MAP-EM-DD and MAP-EM-PT-DD methods, is higher than that of the image domain material decomposition method, i.e., FBP-IDD method, and the noise level of the decomposition coefficient images reconstructed by the direct iterative material decomposition method, i.e., MAP-EM-DD and MAP-EM-PT-DD methods, is lower than that of the image domain material decomposition method, i.e., FBP-IDD method.

Comparison of Contrast Noise Ratios

Table 5 shows the CNRs of material-specific regions in basis images, where CNR_1, CNR_2, and CNR_3 represent the CNR of basis images reconstructed by the methods of FBP-IDD, MAP-EM-DD, and MAP-EM-PT-DD, respectively. As shown in Table 5, $CNR_2 > CNR_1$ and $CNR_3 > CNR_1$ in all the material-specific regions in basis

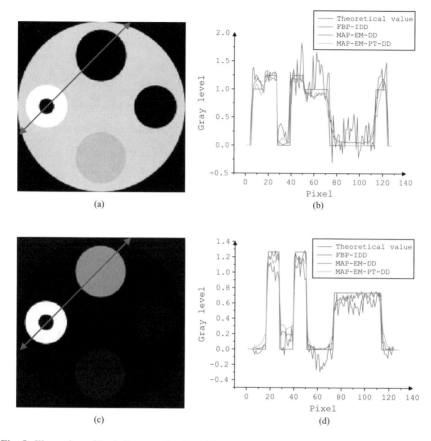

Fig. 8 Illustration of basis images of soft and bone tissue and the corresponding curves of gray level distribution along the red line with arrows located in the theoretical basis images and the basis images reconstructed by the methods of FBP-IDD, MAP-EM-DD, MAP-EM-PT-DD. (**a**) Basis image of soft tissue. (**b**) The curves of gray level distribution along the red lines located in the basis images of soft tissue. (**c**) Basis image of bone tissue. (**d**) The curves of gray level distribution along the red line located in the basis images of bone tissue

images. Compared with FBP-IDD method, MAP-EM-DD method can improve the CNRs ranging from 63.8% to 237.3%. Furthermore, $CNR_3 > CNR_2$ in the material-specific regions of Al in basis images, whereas $CNR_3 < CNR_2$ in the material-specific regions of HA in basis images, and CNR_2 and CNR_3 are almost equivalent in the region of salt water in basis images. Compared with MAP-EM-DD method, MAP-EM-PT-DD method can improve the CNRs ranging from 9.4% to 68.5% in the material-specific regions of Al in basis images and by 30.0% in the region of salt water in the basis image of bone tissue, and reduce the CNR by 18.4% in the region of salt water in the basis image of soft tissue, and reduce the CNRs ranging from 18.1% to 30.5% in the material-specific regions of HA in basis images.

Table 5 CNRs of material-specific regions in basis images

	CNR								
	Al			HA			Salt water		
	CNR_1	CNR_2	CNR_3	CNR_1	CNR_2	CNR_3	CNR_1	CNR_2	CNR_3
Soft tissue	0.832	1.363	2.297	3.229	7.754	5.386	0.301	0.762	0.622
Bone tissue	7.419	25.022	27.375	5.327	14.170	11.604	1.240	3.694	4.802

Table 6 Average decomposition coefficients of the specific materials in the basis images

	Al		HA		Salt water		PE	
	b_1	b_2	b_1	b_2	b_1	b_2	b_1	b_2
FBP-IDD	1.382	1.011	0.098	0.634	1.003	0.147	1.092	0.014
MAP-EM-DD	1.194	1.221	0.111	0.686	0.985	0.137	1.020	0.018
MAP-EM-PT-DD	1.259	1.197	0.088	0.682	0.980	0.140	1.041	0.004

Comparison of Error Levels of the Decomposition Coefficients

Table 6 and Fig. 9 show the averages and error levels of the decomposition (δ) coefficients of the specific materials by the methods of FBP-IDD, MAP-EM-DD, and MAP-EM-PT-DD. Compared with FBP-IDD method, MAP-EM-DD method can reduce the error levels of decomposition (δ) coefficients of HA, PE, salt water, and Al by 31.7%, 45.7%, 45.8%, and 62.1%, respectively. Compared with MAP-EM-DD method, MAP-EM-PT-DD method can reduce the error levels of decomposition (δ) coefficients of salt water, Al, PE, and HA by 1.9%, 17.9%, 26.9%, and 36.3%, respectively.

Comparison of the Reconstruction Time of Basis Images

The cost time for the basis image reconstruction is also critical in CT applications. In general, FBP algorithm is faster than iterative algorithms. To compare the computation performance of MAP-EM-DD and MAP-EM-PT-DD methods quantitatively, their initialization of the basis images and stop condition of iteration are set to be the same. The reconstruction of basis images is implemented on a laptop with 2.6 GHz Intel Core i7-8850H CPU and 16G RAM using Matlab 2017a (Mathworks, Inc.). The reconstruction times of basis images with the methods of FBP-IDD, MAP-EM-DD, and MAP-EM-PT-DD are 48 s, 718 s, and 617 s, respectively. FBP-IDD method takes the least reconstruction time. Compared with MAP-EM-DD method, MAP-EM-PT-DD method can reduce the reconstruction time by 14.1%.

Therefore, in the case of quick decomposition but without considering of high precision, the FBP-IDD method can be employed. In the case of considering both the accuracy and time, the MAP-EM-PT-DD method is the most suitable method. Furthermore, it is encouraged to perform the iterative procedure of both MAP-EM-

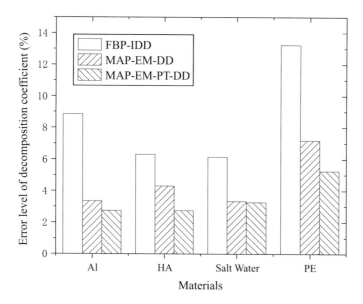

Fig. 9 Comparison of error levels of decomposition (δ) coefficients of the specific materials

DD and MAP-EM-PT-DD methods with GPU to meet the requirement of quick and accuracy basis image reconstruction.

4 Discussion

Spectral CT can provide information on material characterization and quantification by material decomposition algorithms. However, traditional methods suffer from the non-ideal effects of X-ray imaging systems, such as the noise and spectrum distortion in measurements, resulting in the decreased accuracy of material decomposition.

Our previous studies have shown that MAP-EM algorithm can improve the quality of spectral CT image effectively. Basing on this finding, we proposed a direct iterative material decomposition method based on MAP-EM algorithm (MAP-EM-DD) and an alternative method (MAP-EM-PT-DD) to improve the accuracy of material decomposition in this work. The performance of the proposed methods was evaluated and compared with FBP-IDD method using an established cylinder phantom. The effect of the regularization parameter β on the accuracy of material decomposition and quality of basis images was investigated. The metrics of noise level, CNR, and error level were calculated and compared to evaluate the performance of the proposed methods.

One-step methods perform decomposition and reconstruction simultaneously. Unlike two-step methods, one-step methods can circumvent the drawbacks of information loss, improve image quality, and reduce noise-induced bias. Considering the advantages of one-step methods and MAP-EM algorithm, MAP-EM-DD method should be able to improve the accuracy of material decomposition for spectral CT. Incorporating the polar coordinate transformation into MAP-EM-DD, MAP-EM-PT-DD method is proposed. The transformation can make the basis images more relevant during iteration, which can benefit material decomposition and basis image reconstruction.

The results demonstrate that MAP-EM-DD method can remarkably reduce the noise in the decomposed basis images and the error of material decomposition and improve the CNR of the basis images. This result is because the statistical properties of the projection and preservation of information are considered in MAP-EM-DD method, imposing the prior knowledge of the basis images on the material decomposition to suppress noise. Furthermore, compared with MAP-EM-DD method, MAP-EM-PT-DD method performs better with regard to noise level, error level, and reconstruction time, while comparable with respect to CNR. The latter method saves more than 14% of reconstruction time. This result may be contributed to the more relevant of basis images during the iteration procedure using r and θ in the polar coordinate system. However, the edge-gradient effect for MAP-EM-PT-DD method is worse than that for MAP-EM-DD method.

Although MAP-EM-DD and MAP-EM-PT-DD methods can remarkably improve the accuracy of material decomposition and basis image quality, this study bears certain limitations. Firstly, the proposed methods should be compared with other one-step methods for further performance evaluation. Secondly, the performance of the proposed methods with projection datasets from commercial spectral CT system should be validated. Thirdly, it is necessary to investigate how to reduce the edge-gradient effect in the proposed algorithms as that in the regular CT image reconstruction. Finally, the improvement of computation speed of the proposed methods is encouraged for practical use.

5 Conclusion

Two novel and robust material decomposition methods, MAP-EM-DD and MAP-EM-PT-DD, are proposed in this paper. MAP-EM-DD method outperforms FBP-IDD method with regard to noise level, CNR and error level. MAP-EM-PT-DD method outperforms MAP-EM-DD method with regard to noise level, error level, and reconstruction time. The proposed methods can evidently improve the accuracy of material decomposition and improve the quality of basis images for spectral CT. The suitable regulation parameters are also necessary to achieve high accuracy of material decomposition and high quality of basis images.

Acknowledgements I wish to thank all my coworkers in the State Key Laboratory of Mechanics and Control of Mechanical Structures and the Department of Nuclear Science and Engineering at Nanjing University of Aeronautics and Astronautics, especially my Master students, Runchao

Xin and Xuling Zhang, for their continuing efforts in researching on this exciting new X-ray imaging science and technology. I also thank to the invitation from Dr. Iniewski and the support in part by the National Natural Science Foundation of China (51575256), Key Research and Development Plan (Social Development) of Jiangsu Province (BE2017730), Key Industrial Research and Development Plan of Chongqing (cstc2017zdcy-zdzxX0007), Shanghai Aerospace Science and Technology Innovation Fund (SAST 2019-121), and the Priority Academic Program Development of Jiangsu Higher Education Institutions (PAPD).

References

1. Alvarez, R.E., Macovski, A.: Energy-selective reconstructions in X-ray computerized tomography. Phys. Med. Biol. 21(5), 733-44 (1976). https://doi.org/10.1088/0031-9155/21/5/002
2. Sidky, E.Y., Zou, Y., Pan, X.: Impact of polychromatic x-ray sources on helical, cone-beam computed tomography and dual-energy methods. Phys. Med. Biol. 49(11), 2293-2303 (2004). https://doi.org/10.1088/0031-9155/49/11/012
3. Thieme, S.F., Graute, V., Nikolaou, K., Maxien, D., Johnson, T.R.C.: Dual energy ct lung perfusion imaging—correlation with SPECT/CT. Eur. J. Radiol. 81(2), 360-365 (2010). https://doi.org/10.1016/j.ejrad.2010.11.037
4. Lambert, J.W., Sun, Y., Gould, R.G., Ohliger, M.A., Li, Z., Yeh, B.M.: An image-domain contrast material extraction method for dual-energy computed tomography. Invest. Radiol. 52(4), 245-254 (2017). https://doi.org/10.1097/RLI.0000000000000335.
5. Xie, B., Su, T., Kaftandjian, V., Niu, P., Yang, F., Robini, M., Zhu, Y., Duvauchelle, P.: Material Decomposition in X-ray Spectral CT Using Multiple Constraints in Image Domain. J. Nondestruct. Eval. 38(1), 16 (2019). https://doi.org/10.1007/s10921-018-0551-8
6. Maass, C., Baer, M., Kachelriess, M.: Image-based dual energy CT using optimized precorrection functions: a practical new approach of material decomposition in image domain. Med. Phys. 36(8), 3818–3829 (2009). https://doi.org/10.1118/1.3157235
7. Schlomka, J.P., Roessl, E., Dorscheid, R., Dill, S., Martens, G., Istel, T., Bäumer, C., Herrmann, C., Steadman, R., Zeitler, G., Livne, A., Proksa, R.: Experimental feasibility of multi-energy photon-counting K-edge imaging in pre-clinical computed tomography. Phys. Med. Biol. 53(15), 4031-4047 (2008). https://doi.org/10.1088/0031-9155/53/15/002
8. Sawatzky, A., Xu, Q., Schirra, C.O., Anastasio, M.A.: Proximal ADMM for multi-channel image reconstruction in spectral x-ray CT. IEEE Trans. Med. Imaging 33(8), 1657–1668 (2014). https://doi.org/10.1109/tmi.2014.2321098
9. Ducros, N., Abascal, J.F.P.J., Sixou, B., Rit, S., Peyrin, F.: Regularization of nonlinear decomposition of spectral x-ray projection images. Med. Phys. 44(9), e174–e187 (2017). https://doi.org/10.1002/mp.12283
10. Mory, C., Sixou, B., Si-Mohamed, S., Boussel, L., Rit, S.: Comparison of five one-step reconstruction algorithms for spectral CT. Phys. Med. Biol. 63, 235001 (2018). https://doi.org/10.1088/1361-6560/aaeaf2
11. Foygel, B.R., Sidky, E.Y., Gilat, S.T., Pan, X.: An algorithm for constrained one-step inversion of spectral CT data. Phys. Med. Biol. 61(10), 3784-3818 (2016). https://doi.org/10.1088/0031-9155/61/10/3784
12. Cai, C., Rodet, T., Legoupil, S., Mohammad-Djafari, A.: A full-spectral Bayesian reconstruction approach based on the material decomposition model applied in dual-energy computed tomography. Med. Phy. 40(11), 111916 (2013). https://doi.org/10.1118/1.4820478
13. Long Y., Fessler J.A.: Multi-Material Decomposition Using Statistical Image Reconstruction for Spectral CT. IEEE Transactions on Medical Imaging, 33(8): 1614-1626 (2014). https://doi.org/10.1109/TMI.2014.2320284.

14. Foygel Barber R., Sidky E.Y., Gilat Schmidt T., Pan X.: An algorithm for constrained one-step inversion of spectral CT data. Physics in Medicine & Biology, 61(10): 3784, (2016). https://doi.org/10.1088/0031-9155/61/10/3784.

15. Weidinger T., Buzug T.M., Flohr T., Kappler S., Stierstorfer K.: Polychromatic iterative statistical material image reconstruction for photon-counting computed tomography. International journal of biomedical imaging, 5871604 (2016). https://doi.org/10.1155/2016/5871604.

16. Mechlem, K., Ehn, S., Sellerer, T., Braig, E., Munzel, D., Pfeiffer, F., Noël P.B.: Joint statistical iterative material image reconstruction for spectral computed tomography using a semi-empirical forward model. IEEE Trans. Med. Imaging 37(1): 68-80 (2018). https://doi.org/10.1109/TMI.2017.2726687

17. Zhou, Z.D., Xin, R.C., Guan S.L., Li, J.B., Tu J.L.: Investigation of maximum a posteriori probability expectation-maximization for image-based weighting spectral X-ray CT image reconstruction. J. X-Ray Sci. Technol. 26(5): 853-864 (2018). https://doi.org/10.3233/XST-180396

18. Zhou, Z.D., Guan, S.L., Xin, R.C., Li, J.B.: Investigation of contrast-enhanced subtracted breast CT images with MAP-EM based on projection-based weighting imaging. Australas. Phys. Eng. Sci. Med. 41, 371-377 (2018). https://doi.org/10.1007/s13246-018-0634-y

19. Maaß C., Meyer E., Kachelrieß M., 2011. Exact dual energy material decomposition from inconsistent rays (MDIR). Med. Phys. 38(2), 691-700 (2011). https://doi.org/10.1118/1.3533686

20. Chen, Y., Ma, J.H., Feng, Q.J., Luo, L.M., Shi, P.C., Chen, W.F.: Nonlocal prior Bayesian tomographic reconstruction. J. MATH. IMAGING. VIS. 30(2), 133-146 (2008). https://doi.org/10.1007/s10851-007-0042-5

21. Green, P.J.: Bayesian reconstructions from emission tomography data using a modified EM algorithm. IEEE Trans. Med. Imaging 9(1), 84-93 (1990). https://doi.org/10.1109/42.52985

22. Brooks, R.A.: A quantitative theory of the Hounsfield unit and its application to dual energy scanning. J. Comput. Assist. Tomo. 1(4), 487-93 (1977). https://doi.org/10.1097/00004728-197710000-00016

23. Agostinelli S, Allison J, Amako K, et al. Geant4—a simulation toolkit. Nuclear Instruments and Methods in Physics Research Section A Accelerators Spectrometers Detectors and Associated Equipment, 2003, 506(3):250. https://doi.org/10.1016/S0168-9002(03)01368-8

24. Punnoose, J., Xu, J., Sisniega, A., Zbijewski, W., Siewerdsen, J. H.: Technical note: spektr 3.0- a computational tool for x-ray spectrum modeling and analysis. Med. Phys. 43(8), 4711-4717 (2016). https://doi.org/10.1118/1.4955438

25. Zhang, G.W., Cheng, J.P., Zhang, L., Chen Z.Q., Xing Y.X.: A practical reconstruction method for dual energy computed tomography. J. X-Ray Sci. Technol. 16(2), 67-88 (2008).

Prototypes of SiPM-GAGG Scintillator Compton Cameras

Xiaosong Yan

1 Introduction

Following the Fukushima Daiichi Nuclear Power Plant (FNPP) accident on March 11, 2011, substantial amounts of radionuclides were deposited primarily within a 20 km radius from the plant and rapidly extending up to 30 km to the northwest. In addition to nuclear accidents, the capability of producing, the possession of, and the illicit transportation of nuclear weapons or special nuclear materials (SNM) by hostile individuals, groups, and non-nuclear weapons countries, is considered to be an even more general nuclear threat to the world security and public safety.

Gamma-ray imager is an essential tool for localizing or mapping radioactivity in many fields such as nuclear incident emergency response, radioactivity decontamination, homeland security, biomedical research, etc. Widely used imagers include pinhole camera, coded-aperture camera, and Compton camera. Different from traditional gamma cameras equipped with a pinhole or coded apertures, the Compton camera utilizes the Compton scattering kinematics as electronic collimation instead of mechanical collimators. Benefiting from not using heavy mechanical collimators, Compton camera features compactness, wide field of view and broad energy range. Especially after the Fukushima accident, more and more researches focus on Compton camera and various approaches have been developed for specific applications.

Different types of detectors have been considered. Semiconductor detectors (Si, CdTe, CdZnTe, Ge, et al.) provide excellent energy resolution and have been selected for constructing Compton cameras with superior spatial resolution [1–3]. For example, room temperature CdZnTe detectors offer energy resolution of

X. Yan (✉)
Institute of Nuclear Physics and Chemistry, China Academy of Engineering Physics, Mianyang, China

Table 1 Scintillators used in Compton cameras

Scintillator	NaI:Tl	CsI:Tl	LaBr$_3$:Ce	CeBr$_3$	GAGG:Ce
Effective atomic number	51	54	45	46	51
Density (g/cm^3)	3.67	4.51	5.3	5.1	6.63
Light yield (photons MeV^{-1})	45,000	54,000	75,000	60,000	57,000
Peak wavelength (nm)	415	550	380	380	520
Energy resolution @ 662 keV	6%	5%	3%	4%	5%
Self-background	No	No	Yes	Low	No
Hygroscopic	Yes	Slightly	Yes	Yes	No

below 1% FWHM at 662 keV and Ge detectors cooled with liquid nitrogen exhibit energy resolution of better than 0.5% FWHM at 662 keV. However, they generally suffer from complicated operation, limited sensitivity (crystal size), and high cost. Scintillator detectors can provide high sensitivities and good coincidence timing resolutions. Various Compton cameras have been constructed with NaI:Tl [4, 5], CsI:Tl [6–8], LaBr$_3$:Ce [9, 10], CeBr$_3$ [11], and GAGG:Ce [12–17] scintillators. A comparison of the scintillators used in Compton camera is listed in Table 1. Compared with other scintillators, the novel GAGG:Ce scintillator offers moderate energy resolutions and has advantages of no intrinsic radioactivity and hygroscopicity [18, 19]. Thanks to recent developments, the GAGG:Ce scintillator has been widely applied in a number of applications, including medical imaging modalities such as PET and SPECT, and also the Compton camera. Conventionally, photomultiplier tubes (PMT) are used to convert optical lights into electrical signals but such systems are usually cumbersome. Benefiting from recent advances of solid-state semiconductor technology, bulky PMTs can be replaced by compact (typically within 1 mm thickness) silicon photomultipliers (SiPM, also solid-state photomultiplier, SSPM, or multi-pixel photon counter, MPPC). Scintillators coupled to thin SiPMs minimize unnecessary scattering or attenuation on gamma rays traversing them. In addition, SiPMs realize similar gain and quantum efficiency with much lower bias voltage (typically several tens of Volts). As a result, GAGG scintillators coupled to SiPMs allow developing Compton cameras that have advantages of high sensitivity, easy operation, compact size, and low cost.

This chapter aims to review the latest advances in Compton cameras based on SiPMs coupled to GAGG:Ce scintillators. Section 2 summarizes the main characteristics of Compton cameras. Section 3 introduces the most widely used reconstruction algorithms. Section 4 presents some developed prototypes for different applications.

2 Compton Camera Characteristics

2.1 Principle

A typical Compton camera (Fig. 1) consists of two position-sensitive detectors operated in time coincidence: a "scatterer" and an "absorber." When an incident photon with energy of E_γ is scattered by the scatterer with an energy deposition of E_s and then absorbed by the absorber with an energy deposition of E_a ($E_s + E_a = E_\gamma$), the Compton scattering angle (ω) can be calculated using Eq. (1). The arrival direction of the incident photon is confined on a cone with aperture angle given by the Compton scattering angle, its vertex given by the interaction point in the scatterer, and the axis given by the straight line connecting the interaction points in the scatterer and absorber. Location or distribution of the gamma-ray source can be statistically estimated by overlapping the Compton cones many times (Fig. 1), and the estimation procedure is generally called image reconstruction.

$$\cos \omega = 1 - \frac{E_s}{\gamma \left(E_\gamma - E_s\right)} \tag{1}$$

where $\gamma = E_\gamma/m_e c^2$, $m_e c^2 = 511$ keV.

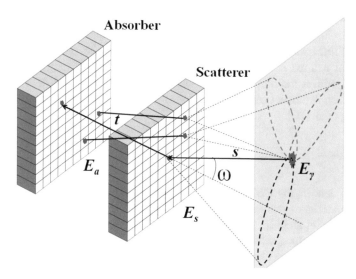

Fig. 1 Illustration of a Compton camera consists of a scatterer and an absorber

2.2 Angular Resolution

Any energy uncertainty for E_s due to either energy resolution or the Doppler broadening effect of the scatterer, as well as any measurement position uncertainty, will lead to an angular uncertainty of the source location. The overall angular uncertainty due to the three components is described as:

$$\Delta\omega^2{}_{\text{overall}} = \Delta\omega^2{}_{\text{energy-resolution}} + \Delta\omega^2{}_{\text{Doppler}} + \Delta\omega^2{}_{\text{geometry}} \qquad (2)$$

where $\Delta\omega_{\text{overall}}$ is the overall angular uncertainty, $\Delta\omega_{\text{energy}}$ comes from scatter energy measurement uncertainty and the Doppler broadening effect, $\Delta\omega_{\text{energy-resolution}}$ is the detector energy resolution contribution, $\Delta\omega_{\text{Doppler}}$ is the Doppler broadening contribution, and $\Delta\omega_{\text{geometry}}$ is the detector position resolution contribution.

Using error propagation from Eq. (1), the energy uncertainty contribution to angular uncertainty is given by

$$\begin{aligned}
\Delta\omega_{\text{energy}} &= \frac{(1+\gamma(1-\cos\omega))^2}{\gamma E_\gamma \sin\omega} \Delta E_s \\
&= \frac{(1+\gamma(1-\cos\omega))^2}{\gamma E_\gamma \sin\omega} \sqrt{\Delta E^2_{\text{energy-resolution}} + \Delta E^2_{\text{Doppler}}}
\end{aligned} \qquad (3)$$

For our SiPM-GAGG scintillator detector, the general relationship between energy resolution and the deposited energy (E) is given by $\Delta E_{\text{energy-resolution}} = 0.45\sqrt{E} + 30$ according to the measured energy resolution at several incident energies. Furthermore, it is assumed in Eq. (1) that the electron that scatters the incident gamma ray is at rest and free, but practical scattering occurs in an interaction with a moving electron bound to atom(s) composed of the scatterer. Then the Doppler effect causes dispersion of the deposited energy and is given by Matscheko et al. [20]:

$$\Delta E_{\text{Doppler}} = \frac{E_\gamma - E_s}{E_\gamma} \sqrt{E_\gamma^2 + \left(E_\gamma - E_s\right)^2 - 2E_\gamma\left(E_\gamma - E_s\right)\cos\omega}\, \frac{\Delta p_z}{m_e c} \qquad (4)$$

where Δp_z denotes the FWHM of the Compton profile which is the distribution of p_z on the material of scatterer and can be calculated based on the Compton profiles given by Biggs et al. [21].

For pixel detectors with square cross-section, the geometry contribution to the angular uncertainty can be calculated using Eq. (5) given by Ordonez et al. [22]:

$$\begin{aligned}
\Delta\omega^2{}_{\text{geometry}} &= \frac{1}{R_1^2}\left[\Delta L_s^2(1 + \alpha\cos\omega)^2 + (\Delta z_s \alpha\sin\omega)^2\right] \\
&+ \frac{\cos^2\omega}{R_2^2}\left[\Delta L_a^2\cos^2\omega + (\Delta z_a \sin\omega)^2\right]
\end{aligned} \qquad (5)$$

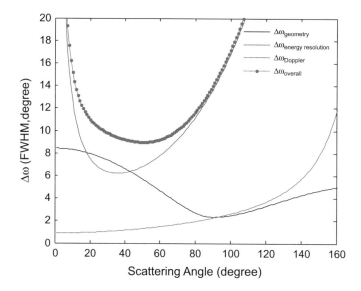

Fig. 2 Estimation of the angular uncertainty contributions for 662 keV gamma source. The Compton camera is composed of two 8 × 8 arrays of 6 mm × 6 mm × 6 mm GAGG pixels. The distance between the two arrays is 50 mm

where R_1 is the distance from the gamma-ray source to the scatterer, R_2 is the distance between scatterer and absorber, $(\Delta x_s, \Delta y_s, \Delta z_s)$ is the spatial resolution of the scattering detector, $(\Delta x_a, \Delta y_a, \Delta z_a)$ is the spatial resolution of the absorbing detector, $\Delta L_s = \Delta x_s = \Delta y_s$, $\Delta L_a = \Delta x_a = \Delta y_a$, and $\alpha = R_1 \mid \cos \omega \mid / R_2$.

For a Compton camera with scatterer and absorber both composed of a 8 × 8 array of 6 mm × 6 mm × 6 mm GAGG pixels, a 662 keV ^{137}Cs point source placed 100 mm in front of the scatterer and scatter-to-absorber distance of 50 mm, the overall angular uncertainty and its contributions are shown in Fig. 2. For the Compton camera mainly covering scattering angle between 10° and 60°, the overall angular uncertainty is between 9° and 15°. Geometry and energy resolution are the main contributions to the overall angular uncertainty.

2.3 Efficiency

The intrinsic efficiency ε of a Compton camera is generally defined as the fraction of photons entering the scatterer that undergo only one Compton scattering in the scatterer and then fully deposit the energy with photoelectric absorption in the absorber. ε for a point source can be expressed as [23]:

$$\varepsilon = \frac{1}{\Omega_1} \int_{\text{sca}} dV_1 \int_{\text{abs}} dV_2 \Delta\Omega_1 \mu_c \left(E_\gamma\right) \cdot \exp\left(-\mu_t\left(E_\gamma\right) L_1\right)$$
$$\times \frac{d\sigma}{d\Omega}\left(\theta_{12}\right) \Delta\Omega_2 \cdot \exp\left(-\mu_t\left(E_s\right) L_{12}\right) \cdot \mu_p\left(E_s\right) \tag{6}$$

where Ω_1 is the solid angle subtended by the scatterer, $\Delta\Omega_1$ is the solid angle for dV_1 which is a differential volume element, L_1 is the attenuation distance in the scatterer from the source to dV_1, L_{12} is the attenuation distance from dV_1 to dV_2, and μ_t, μ_c, and μ_p represent the total, Compton scattering, and photoelectric coefficients, respectively.

Analytical calculation of ε cannot be done unless numerical integration over the two detector volumes is performed and interaction probabilities for the initial and scattered gamma-rays are known. Instead, Monte Carlo programs such as EGS5 [24] and Geant4 [25] are usually employed to numerically compute the efficiency of Compton cameras with different parameters. The efficiency of the Compton camera described in Sect. 2.2 is calculated to be about 0.15% for a 662 keV ^{137}Cs gamma source. The efficiency can be as high as 1% with thicker and closer detectors. Generally the angular resolution would be poor when the efficiency is high and vice versa.

3 Reconstruction Algorithm

In general, Compton cameras record information of the two consecutive energy depositions, interaction positions, and timing, but this information does not reflect location or distribution of gamma-ray source. The recorded information has to be processed to uncover the demanded information, and the process is called image reconstruction. In this section, three most widely used reconstruction algorithms will be introduced, including simple back projection (SBP), filtered back projection (FBP), and maximum likelihood expectation maximization (MLEM).

3.1 Simple Back Projection

The simplest way to produce a Compton image is to back project the cones to the imaging plane as shown in Fig. 1 and the method is called simple back projection (SBP) algorithm. s is the image vector decided by Compton scattering position and each pixel position in the image plane. t is the interaction vector which is the axis of the Compton cone. SBP can be expressed as

$$f(s) = \sum_{i=0}^{N} \delta\left(\langle s, t_i \rangle, \cos\omega_i\right) \tag{7}$$

where i denotes the ith Compton scattering event, N is the total number of events used for reconstruction, and $<s,t_i>$ denotes the inner product of vector s and t_i. In practice, the δ function can be approximated with a Gaussian function of which the σ value is estimated using Eq. (2).

Images created using SBP make no correction for the variations of the sensitivity of the imager with position and suffer from chance intersections combined with measurement errors and background interferences. Since this method is simple and fast, it might be useful in some special applications which require fast response rather than promise angular resolution, such as source probing. Imaging of extended shapes requires a more advanced algorithm.

3.2 Filtered Back Projection

In 2000, Parra presented an analytical inversion algorithm based on spherical harmonics expansion for the complete date set of all possible scattering angles according to Klein–Nishina formula [26]. As an improvement, Tomitani and Hirasawa proposed a reconstruction algorithm for limited angle Compton data set based on the Legendre polynomial (P_n) expansion [27, 28]:

$$f(s) = \sum_{i=0}^{N} \sum_{n=0}^{N\,MAX} \frac{2n+1}{4\pi} \frac{W_n}{H_n} P_n(\cos \omega_i) P_n(\langle s, t_i \rangle) \tag{8}$$

where W_n is an appropriate filter function and H_n is expressed as:

$$H_n = \int_{\theta_1}^{\theta_2} d\cos\theta \, \sigma(\theta) P_n^2(\cos\theta) \tag{9}$$

where $\sigma(\theta)$ is the Klein–Nishina differential scattering cross-section formula

$$\sigma(\theta) = \frac{r_0^2}{2} \frac{1+\cos^2\theta}{\{1+\gamma(1-\cos\theta)\}^2} \left[1 + \frac{\gamma^2(1-\cos\theta)^2}{(1+\cos^2\theta)\{1+\gamma(1-\cos\theta)\}}\right] \tag{10}$$

Because the image is obtained by transforming a given source distribution into spherical harmonics space, filtering and then transforming back into the angular space, this method is named as filtered back projection (FBP). Compared with SBP, FBP offers a better angular resolution with slightly larger computation time. Since FBP algorithm can also be executed event by event, it is easy to implement this method in real time imaging.

3.3 Maximum Likelihood Expectation Maximization

The maximum likelihood expectation maximization (MLEM) is a widely used and accepted iterative algorithm in the field of gamma-ray imaging. Different from SBP and FBP, which is a linear superposition of the Compton cones, MLEM uses the combined knowledge of the entire group of the events to calculate and converge to a source distribution with the maximum likelihood given the measured data set and a modeling of the detector system. The convergence is assured as long as the total number of events is larger than the pixel number in the image plane. MLEM is performed using Eq. (11) [29] below, where λ_j^n represents the calculated amplitude of pixel j in the image at the nth iteration. To start the process, usually a SBP image is created as the initial image. The main summation is over all recorded event sequences $i = 1, N$ which had a reconstruction cone intersecting pixel j. The weighted likelihood that event I originated from pixel j is t_{ij}. The secondary summation is over all pixels k intersected by the reconstruction cone of event i.

$$\lambda_j^n = \frac{\lambda_j^{n-1}}{s_j} \sum_{i=1}^N \frac{Y_i t_{ij}}{\sum_k t_{ik} \lambda_k^{n-1}} \tag{11}$$

Pixel weights are calculated from Eq. (12), where Δz is the attenuation thickness of absorber that a scattered photon would travel through. The interaction cross-section (σ_t) is at energy E_a, R_{12} is the distance between the first and second interaction positions, and θ_{12} is the polar angle of the vector pointing from the scatter position to the absorption position.

$$t_{ij} = \sigma(\omega) \frac{1 - \exp(-\sigma_t \Delta z)}{R_{12}^2 \cos \theta_{12}} \tag{12}$$

S_j is the sensitivity function that represents the probability that a gamma-ray originating from pixel j is detected anywhere in the Compton camera and can be calculated analytically before imaging. The formula for the calculation of the sensitivity function is shown in Eq. (13). It is calculated as the sum of the probability that a photon is emitted in the direction of the scatterer times the probability that a photon survives the distance R_{01} (through air) from the image pixel to each of the scatterer pixels. This is then multiplied by the sum of the probabilities that the photon will scatter in the thickness of scatterer (ΔZ_{sca}) as seen from the image pixel. Next is the probability that the scattered photon will survive the distance R_{12} (through air) to each of the absorber pixels multiplied by the probability of scattering (the Klein–Nishina differential cross-section) at the angle between the vectors formed by source to scatter position and the scatter to absorption position followed by the probability that the scattered photon will interact in the thickness of absorber (ΔZ_{abs}) seen from the scattering position times the solid angle covered by the absorber crystal $(A_{abs} \cos \theta_{12} / R_{12}^2)$.

$$S_j = \sum_{sca} \left[\exp\left(-R_{01}/\Lambda_{air}\right) \left(1 - \exp\left(-\Delta Z_{sca}/\Lambda_{sca}\right)\right) \frac{A_{sca} \cos\theta_{01}}{4\pi R_{01}^2} \right.$$
$$\left. \times \sum_{abs} \left[\exp\left(-R_{12}/\Lambda_{air}\right) \left(1 - \exp\left(-\Delta Z_{abs}/\Lambda_{abs}\right)\right) \sigma\left(\omega\right) \frac{A_{abs} \cos\theta_{12}}{R_{12}^2} \right] \right] \tag{13}$$

Using this equation, iterations are performed until either the maximum likelihood is achieved (with image convergence guaranteed) or the iterations are terminated based on a designated stopping criterion. Generally, MLEM reconstruction algorithm can achieve the best angular resolution among the three discussed methods. However, because of the iterative calculation, the computation cost for this method is much higher. As a result, it is difficult to apply this method for applications when real time imaging is demanded.

4 Prototypes of SiPM-GAGG Scintillator Compton Cameras

Thanks to recent developments of the GAGG:Ce scintillator crystal and SiPM advancements, lots of SiPM-GAGG scintillator Compton cameras have been developed especially after the Fukushima disaster. As energy-position-sensitive imaging systems, Compton cameras typically have several tens or hundreds of readout channels. Besides the similar constituent of SiPM and GAGG:Ce scintillator, different Compton cameras are developed with individual readout electronics [13, 14, 30], charge division resistive/capacitive networks [12, 15–17]. Generally, with individual readout electronics it is possible to keep inherent characteristics of each detector channels and promising for developing high-performance imagers. Particularly, the ASIC (Applied Specific Integrated Circuit)-based individual readout electronics allow developing very compact and high-performance imagers. However, this approach suffers from long development cycle, high cost, and low flexibility. As a result, various resistive/capacitive network electronics [31–35] have been proposed to decrease the large number of the output channels and most of the scintillator Compton cameras are developed with resistive/capacitive network electronics.

4.1 Prototypes with Individual Readout Electronics

J. Jiang et al. developed a Compton camera prototype (Fig. 3a) and the prototype was mounted on an unmanned helicopter for aerial radiation monitoring of contaminated areas in Fukushima [13]. 16 GAGG crystals ($10 \times 10 \times 5$ mm^3) were coupled to a 4×4 array of SiPMs (KETEK PM6660) as the scatterer and other 16 GAGG crystals ($10 \times 10 \times 10$ mm^3) were coupled to a 4×4 array of avalanched photodiode (APD, Hamamatsu S8664-55) as the absorber. Discrete 16 individual readout circuits were used to read out signals from SiPMs and two ASIC chips of individual readout circuits were employed to read out signals from APDs. In the

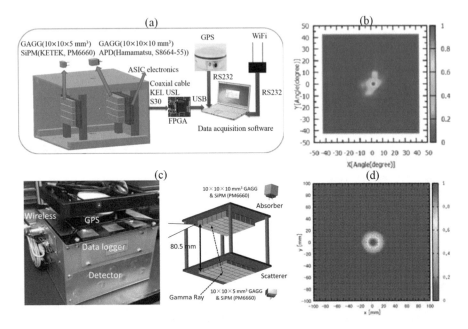

Fig. 3 Prototypes of Compton camera with individual readout. (**a**) Design of the prototype with two 4 × 4 arrays. (**b**) Point source image while the ^{137}Cs source was located at position (0°, 0°). (**c**) The improved prototype with two 8 × 8 arrays. (**d**) Point source image while the ^{137}Cs source was located at position (0°, 0°)

readout circuits, the pulse height of each detection event was linearly converted into digital pulse with corresponding time width [36]. A data-acquisition (DAQ) field-programmable gate array (FPGA, Altera Cyclone IV) was used to count the width of the digital signals and perform coincidence measurements. The detector realized an energy resolution (FWHM) of 6.5% at the ^{137}Cs 662 keV photopeak. The distance between the two detector arrays was set to be 32.5 mm for taking ^{137}Cs images at varied locations and the FBP algorithm was applied for reconstructing images. The camera realized the best angular resolution of 14° (Fig. 3b) when the source was located right in front of the camera. With this angular resolution, the detector system is expected to achieve a spatial resolution of about 2.4 m (FWHM) when the Compton camera is working at an altitude of 10 m.

The prototype was then improved to a larger size for detecting cesium radiation hot spots and confirming the decontamination effect in Fukushima (Fig. 3c) [14]. The GAGG scintillators were assembled into 8 × 8 arrays measuring 10 mm × 10 mm × 5 mm and 10 mm × 10 mm × 10 mm to provide scatterer and absorber layer. Each array was processed with a 91 × 112 mm^2 readout board implementing 64 channels of dToT-based individual readout electronics [37]. The angular resolution was improved from approximately 14° to less than 10° (Fig. 3d) on the expansion of the interlayer distance from 32.5 to 80.5 mm.

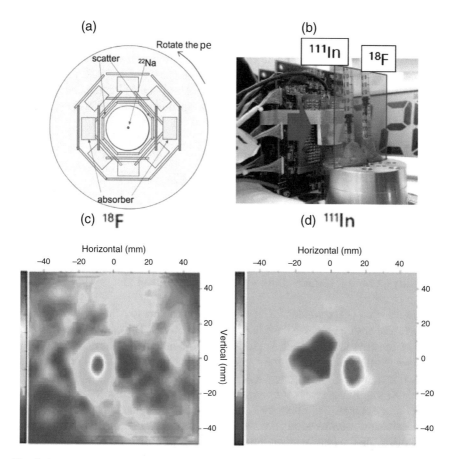

Fig. 4 A prototype of PET-Compton system developed by K. Shimazoe et al. (**a**) Setup for positron emission tomography (PET) imaging using PET–Compton hybrid modules. (**b**) Experimental setup for evaluation of simultaneous imaging using positron emission tomography (PET)–Compton hybrid modules. (**c**) Compton image of ^{18}F. (**d**) Compton image of ^{111}In

K. Shimazoe et al. combined the Compton camera with positron emission tomography (PET) system for simultaneous PET and Compton imaging (Fig. 4a) [30]. GAGG arrays with thicknesses of 1.5 and 10 mm were chosen for scatterers and absorbers, respectively. The pixel size is 2.5 mm × 2.5 mm, and the pitch size is 3.25 mm. Each crystal is separated by $BaSO_4$ reflectors. The distance between the absorber and scatterer was set to be 30 mm to form a single Compton and PET camera module. The fabricated scintillator array was coupled to an 8 × 8 pixel TSV-SiPM array (Hamamatsu MPPC S13361-3050). Each array was processed with a readout board implementing 64 channels of dToT-based individual readout electronics and the digital ToT signals were fed to a FPGA (Xilinx Kintex 7 XC7K70T)-based DAQ system. Figure 4b shows the experimental setup for the simultaneous PET and Compton imaging of ^{111}InCl$_3$ and ^{18}F-FDG housed in two

10-ml syringes. The source strengths of ^{111}In and ^{18}F were adjusted to be 1 MBq in total. Image reconstruction was achieved via FBP and MLEM for Compton and PET imaging, respectively. Figure 4c, d show the Compton imaging results of ^{18}F using gamma-energy of 511 keV and ^{111}In using gamma-energy of 245 keV, respectively. The location of each radioisotope is clearly visualized in the simultaneous imaging experiment. The author emphasized that the spatial resolution will be further improved via adoption of higher-resolution pixel detectors.

4.2 Prototypes with Resistive Network Electronics

The resistive network is the simplest method and has been applied in developing Compton cameras by many groups.

A Handy Compton Camera Developed by J. Takaoka et al.

J. Takaoka et al. [12] developed a handy Compton camera to help identify radiation hotspots and ensure effective decontamination operation. The camera was based on position-sensitive GAGG scintillators coupled to a monolithic MPPC array. Figure 5a shows a photo of the first prototype module, measuring $8.5 \times 14 \times 16$ cm^3 in size and weighting 1.5 kg. They adopted a 50×50 array of $1 \times 1 \times 10$ mm thick Ce:GAGG scintillator arrays for both the scatterer and the absorber, and four 8×8 MPPC array boards. To reduce the number of output signal channels, a resistive charge division network similar with the one shown in Fig. 5b [38] was applied for multichannel signal readout. Energy and position of a detection event were calculated by using Eqs. (14) and (15). The distance between the scatterer and the absorber was set to 15 mm. A fisheye visible light camera was equipped in order to take an optical image to be superimposed on the gamma-ray image of 180° field of vision (FOV). The camera achieved angular resolution of about 14° (FWHM with MLEM reconstruction algorithm) for 662 keV gamma rays. An integration time of 30 s is sufficient to reconstruct the image for a weak isotope when the corresponding radiation dose is about 6 μSv/h.

Furthermore, they also exploited the concept of a novel DOI (depth of interaction)-Compton camera with two identical 1 cm cubic 3D position-sensitive MPPC-GAGG detector modules. The camera realized a sensitivity of about 1% for 662 keV gamma rays, which is about 50 times superior to other cameras being tested in Fukushima. Each GAGG cubic consisted of five layers of the 5×5 matrix of 2 mm GAGG cubic crystals. Double-sided MPPC arrays consisting of 4×4 channels of 3 mm $\times$ 3 mm pixels were coupled to both ends of a GAGG scintillator block. A BaSO$_4$ reflector (0.2 mm thick in the 2D direction) divides each pixel of the crystal block (2 mm cubic), whereas it is separated by a layer of air (about 10 μm-thick) that forms naturally by laying on top of each other. Figure 5d shows

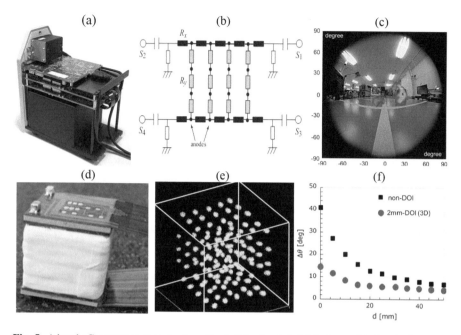

Fig. 5 A handy Compton camera developed by J. Takaoka et al. (**a**) A photo of the handy Compton camera prototype. (**b**) A schematic diagram of the resistive network. (**c**) Reconstructed image of a ^{137}Cs isotope, taken with a prototype Compton camera. (**d**) Photo of 3D position-sensitive crystal array. (**e**) 3D position histogram of a crystal block measured with 662 keV gamma rays. (**f**) Geant4 simulation results of the angular resolution as a function of distance d for DOI and non-DOI configurations

a photo of the DOI module. By measuring the pulse-height ratio of MPPC-arrays coupled at both ends of a Ce:GAGG scintillator block, the DOI is obtained for incident gamma rays as well as the usual 2D positions. Figure 5e shows the position response of a performance test for a 5 × 5 × 5 matrix of 2 mm cubic GAGG crystal pixels measured with 662 keV fully absorbed gamma rays. Each crystal block is clearly distinguished, working as a high-resolution 3D position-sensitive detector. The two modules were stacked together with a distance of 4 mm to form a Compton camera. The average energy resolution was measured 10% for 662 keV gamma rays. The DOI-Compton camera prototype achieved angular resolution of about 10° (FWHM with MLEM reconstruction algorithm) for 662 keV gamma rays. Figure 5f shows Geant4 simulation of results of the angular resolutions as a function of d (distance between the scatterer and the absorber) for the DOI (2 mm resolution) and non-DOI configurations, showing a significant improvement of the DOI-Compton cameras.

They also proposed a new concept for the stereo measurement of gamma rays by using two Compton cameras, thus enabling the 3-D positional measurement of radioactive isotopes [39]. According to the simulation results, they ensured that the source position of a single gamma-ray source could be determined typically to within 2 m accuracy and they also confirmed that more than two sources can be clearly separated by event selection.

A Prototype of Compton Camera Developed by J. Zhang et al.

J. Zhang et al. developed a prototype of SiPM-GAGG Compton camera for radioactive materials detection [16]. The factual picture of this system is shown in Fig. 6a. The SiPM array was an 8 × 8 pixel array made by arranging 64 MicroFC-60035 SiPMs close to each other. The gap between two adjacent pixels was 0.2 mm, and the total area of SiPM array was 57.4×57.4 mm^2. The Ce:GAGG array was designed as a 23 × 23 array of 2.2×2.2 mm^2 pixels. Each pixel was divided with a 0.1 mm thick reflective BaSO$_4$ layer. The thickness of scatterer was 5 mm and the thickness of absorber was 10 mm. The distance of between the scatterer and absorber was set to 25 mm to balance image angular resolution and detection efficiency.

For an 8 × 8 SiPM array detector, 64 anode signals and one summed fast output signal were acquired by an analog circuit, which using symmetric charge division circuit (SCDC) and impedance bridge circuit (IBC) to reduce the number of output signals. In this design, the 8 × 8 channel signals were divided into 8 (x direction) + 8 (y direction) by the equivalent resistance network of SCDC. Every 8 signals of one direction were firstly fed into the amplifier (AMP), and be reduced to 2 charge signals by IBC. After filtered by the filter circuits (FC) [16], a total of 4 quasi-Gaussian charge signals (x^+, x^-, y^+, y^-) were fed into digital circuit. Energy and position of a detection event were calculated by using Eqs. (14) and (15). The schematic diagram of SCDC, IBC, and FC is shown in Fig. 6b. A 2D position histogram was measured with a ^{137}Cs source and the flood image result is shown in Fig. 6c. The ^{137}Cs energy spectrum of single pixel was measured and the average energy resolution (FWHM) is $8.5 \pm 0.4\%$, as shown in Fig. 6d.

$$E = x^+ + x^- + y^+ + y^- \tag{14}$$

$$(X, Y) = \left(\frac{x^+}{x^+ + x^-}, \frac{y^+}{y^+ + y^-} \right) \tag{15}$$

The capability of this prototype system for single point source imaging was evaluated in laboratory. A ^{137}Cs point source at the center of FOV can be imaged within 20 s where the radiation dose at the camera position was 1.0 μSv/h, as shown in Fig. 6e. The ARM (FWHM) of the reconstructed point source was about 7°. In order to further evaluate prototype system's coupling resolution of position and energy, two γ point sources with the interval of 25° were used in laboratory. One of them was a 5 μCi ^{22}Na point source and was placed at the center of FOV with the distance of 100 mm. Another was a 25 μCi ^{137}Cs point source and was placed at (90°, 115°) with the distance of 350 mm. The measured imaging result is shown

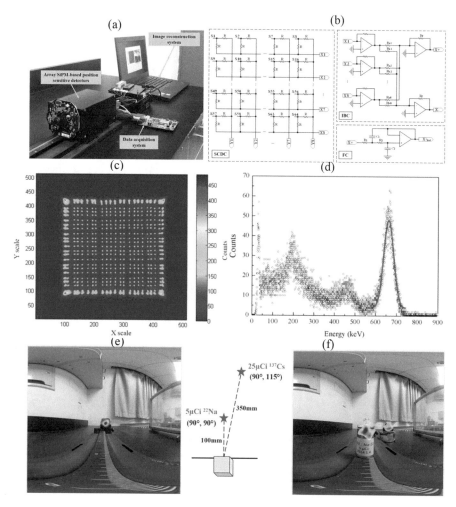

Fig. 6 A prototype of Compton camera developed by J. Zhang et al. (**a**) The prototype system. (**b**) Schematic diagram of symmetric charge division circuit

in Fig. 6f. It is confirmed that two point sources with different γ-ray energy can be separated clearly by this Compton camera prototype system. The simple imaging experiments suggested that this Compton camera offers capabilities for applications like source term investigation and radioactive materials detection.

4.3 A Prototype with Capacitive Multiplexing Readout

S. Jiang et al. developed a Compton camera with capacitive multiplexing readout [17]. A 4 × 4 array of 6 mm × 6 mm × 3 mm Ce:GAGG scintillators was used as the scatterer and another 4 × 4 array of 6 mm × 6 mm × 6 mm Ce:GAGG scintillators was used as the absorber. Each scintillator array was directly coupled to a 4 × 4 array of SenSL MicroFC-60035 SiPMs. The gap between two adjacent SiPM pixels was 1.2 mm. The pixelated scintillators were polished and optically isolated by 1.2 mm thick $BaSO_4$ reflectors and the outside of the scintillator array was covered by $BaSO_4$ reflectors. The whole detector array was encapsulated in a black box to prevent the ambient light from reaching the SiPMs. A detector module is shown in Fig. 7a.

Sixteen channels of a detector module were read out with capacitive multiplexing circuit (Fig. 7b). Each standard anode signal of the SiPM is split and transferred to one or two (with 1/3 and 2/3 branching ratio) output channels depending on the position of the SiPM. Sixteen anode signals are encoded to four position-weighted signals that are sent to four fast preamplifiers (P, Q, R, and S) powered with ±3.3 V (V_s). The energy (E) and position (X, Y) of a detection event are calculated as:

$$E = P + Q + R + S \tag{16}$$

$$(X, Y) = \left(\frac{Q + S - P - R}{P + Q + R + S}, \frac{P + Q - R - S}{P + Q + R + S} \right) \tag{17}$$

Signals from the fast preamplifiers were collected and processed by a CAEN 5725 digitizer. The measured flood image of the scatterer and ^{137}Cs energy spectra of the scatterer pixels irradiated with a ^{137}Cs source are shown in Fig. 7c and d, respectively. The energy resolutions of the pixels were between 6.3% and 7.5%.

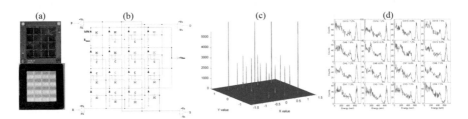

Fig. 7 Principle and results of the scintillator module with capacitive multiplexing readout. (**a**) A 4 × 4 SiPM array and Ce:GAGG scintillator array. (**b**) Schematic diagram of the 16:4 capacitive multiplexing circuit. (**c**) 2D flood image of 16 pixels measured with a ^{137}Cs source. (**d**) The measured ^{137}Cs energy spectra of 16 pixels

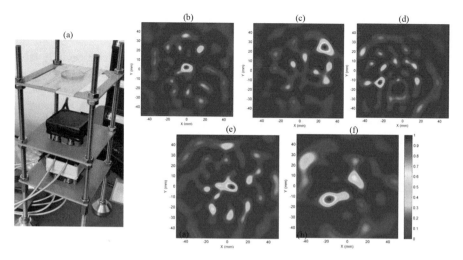

Fig. 8 The built prototype and imaging results with the FBP reconstruction algorithm. (**a**) A photo of the built prototype. (**b**) Reconstructed image of a ^{137}Cs point source located at (0, 0) mm. (**c**) Reconstructed image of a ^{137}Cs point source located at (25, 25) mm. (**d**) Reconstructed image of a ^{137}Cs point source located at (−15, −25) mm. (**e**) Reconstructed image of a ^{137}Cs point source located at (0, 0) mm. (**f**) Reconstructed image of a ^{22}Na point source located at (−15, −20) mm

Two detector modules are placed parallel with a fixed distance of 50 mm to build a Compton camera (Fig. 8a) and the FBP algorithm was adopted for imaging reconstruction. Firstly, a ^{137}Cs point source was placed at 50 cm in front of the camera. Efficiency of the prototype for the ^{137}Cs point source located right in front was measured to be 0.09%. Figure 8b–d show the obtained images when the ^{137}Cs source was placed at (0, 0) mm, (25, 25) mm, and (−15, −25) mm. The source was localized at (2.2, −0.8) mm with angular resolution of 7.2°, (24.8, 26.2) mm with angular resolution of 7.3°, and (−12.3, −24.3) mm with angular resolution of 7.0° for the three positions, respectively. Then the ^{137}Cs source and the ^{22}Na source were placed at 50 mm in front of the scatterer plane simultaneously for performing isotopic Compton imaging. The ^{137}Cs source was placed at (0, 0) mm, and the ^{22}Na source was located at (−15, −20) mm. The gamma-ray images were reconstructed separately with $E_\gamma = 662$ keV for ^{137}Cs and $E_\gamma = 511$ keV for ^{22}Na. The ^{137}Cs source was localized at (1.0, 2.5) mm with angular resolution of 8.2°(Fig. 8e), and the ^{22}Na source was localized at (−12.5, −17.5) mm with angular resolution of 8.8° (Fig. 8f).

The prototype system is demonstrated to be capable of identifying and localizing multi-gamma sources with good energy and spatial resolutions. The author plans to extend the prototype with larger scintillator arrays readout with 64-4 capacitive readout electronics [40] and equip a fisheye camera for practical applications.

5 Conclusion

The development of Compton cameras has gained renewed interest especially after the 2011 Fukushima nuclear accident. Compton cameras based on scintillators have advantages of easier operation, higher sensitivity, and lower cost compared with semiconductor based Compton cameras. Many SiPM-GAGG Compton camera prototypes have been developed with individual readout electronics, charge sharing resistive/capacitive multiplexing readout electronics. The SiPM-GAGG Compton cameras have been applied in radiation hotspot identification in Fukushima, and are to be applied in homeland security, emergency response, consequence management, medical imaging and astronomy research, etc.

References

1. S. Takeda, A. Harayama, Y. Ichinohe, *et al.*, "A portable Si/CdTe Compton camera and its applications to the visualization of radioactive substances", *Nucl. Instrum. Meth. A,* vol. 787, pp. 207-211, 2015.
2. M. Galloway, A. Zoglauer, M. Amman, *et al.*, "Simulation and detector response for the High Efficiency Multimode Imager", *Nucl. Instrum. Meth. A,* vol. 652, *no.* 1, pp. 641-645, 2011.
3. L. Mihailescu, K. M. Vetter, M. T. Burks, *et al.*, "SPEIR: A Ge Compton camera", *Nucl. Instrum. Meth. A,* vol. 570, pp. 89-100, 2007.
4. A. M. L. MacLeod, P. J. Boyle, P. R. B. Saull, *et al.*, "Development of a Compton imager based on scintillator bars", *2011 IEEE Nuclear Science Symposium Conference Record,* 2011.
5. A. M. L. MacLeoda, P. J. Boylea, D. S. Hannaa, *et al.*, "Development of a Compton imager based on bars of scintillator", *Nucl. Instrum. Meth. A,* vol. 767, pp. 397-406, 2014.
6. P. R. B. Saulla, L. E. Sinclairb, H. C. J. Seywerdb, *et al.*, "First demonstration of a Compton gamma imager based on silicon photomultipliers", *Nucl. Instrum. Meth. A,* vol. 679, pp. 89-96, 2012.
7. M. Kagayaa, H. Katagiri, R. Enomoto, *et al.*, "Development of a low-cost-high-sensitivity Compton camera using CsI (Tl) scintillators (γI)", *Nucl. Instrum. Meth. A,* vol. 804, pp. 25-32, 2015.
8. L. Sinclair, P. Saull, D. Hanna, *et al.*, "Silicon Photomultiplier-Based Compton Telescope for Safety and Security (SCoTSS)", *IEEE Trans. Nucl. Sci.,* vol. 61, *no.* 5, pp. 2745-2752, 2014.
9. G. Llosá, J. Cabello, S. Callier, *et al.*, "First Compton telescope prototype based on continuous LaBr₃-SiPM detectors", *Nucl. Instrum. Meth. A,* vol. 718, pp. 130-133, 2013.
10. G. Llosa, J. Cabello, J. E. Gillam, *et al.*, "Second LaBr₃ Compton telescope prototype", presented at the 2013 3rd International Conference on Advancements in Nuclear Instrumentation, Measurement Methods and Their Applications, 2013.
11. A. Iltis, H. Snoussi, L. Rodrigues de Magalhaes, *et al.*, "Temporal Imaging CeBr3 Compton Camera: A New Concept for Nuclear Decommissioning and Nuclear Waste Management", presented at the Advancements in Nuclear Instrumentation Measurement Methods and their Applications, 2018.
12. J. Kataoka, A. Kishimoto, T. Nishiyama, *et al.*, "Handy Compton camera using 3D position-sensitive scintillators coupled with large-area monolithic MPPC arrays", *Nucl. Instrum. Meth. A,* vol. 732, pp. 403-407, 2013.
13. J. Jiang, K. Shimazoe, Y. Nakamura, *et al.*, "A prototype of aerial radiation monitoring system using an unmanned helicopter mounting a GAGG scintillator Compton camera", *J Nucl Sci Technol,* vol. 53, *no.* 7, pp. 1067-1075, 2016.

14. Y. Shikaze, Y. Nishizawa, Y. Sanada, *et al.*, "Field test around Fukushima Daiichi nuclear power plant site using improved Ce: Gd3 (Al, Ga) 5O12 scintillator Compton camera mounted on an unmanned helicopter", *J Nucl Sci Technol*, pp. 1-12, 2016.
15. A. Kishimoto, J. Kataoka, A. Koide, *et al.*, "Development of a compact scintillator-based high-resolution Compton camera for molecular imaging", *Nucl. Instrum. Meth. A*, vol. 845, pp. 656-659, 2017.
16. J.-P. Zhang, X.-z. Liang, J.-l. Cai, *et al.*, "Prototype of an array SiPM-based scintillator Compton camera for radioactive materials detection", *Radiation Detection Technology and Methods*, vol. 3, *no.* 17, pp. 1-12, 2019.
17. S. Jiang, J. Lu, S. Meng, *et al.*, "A prototype of SiPM-based scintillator Compton camera with capacitive multiplexing readout", *J Instrum*, vol. 16, *no.* 01, pp. P01027-P01027, 2021.
18. K. Kamada, T. Yanagida, T. Endo, *et al.*, "2 inch diameter single crystal growth and scintillation properties of $Ce:Gd_3Al_2Ga_3O_{12}$", *J. Cryst. Growth*, vol. 352, *no.* 1, pp. 88-90, 2012.
19. J. Iwanowska, L. Swiderski, T. Szczesniak, *et al.*, "Performance of cerium-doped $Gd_3Al_2Ga_3O_{12}$(GAGG:Ce) scintillator in gamma-ray spectrometry", *Nucl. Instrum. Meth. A*, vol. 712, pp. 34-40, 2013.
20. G. Matscheko, G. A. Carlsson, and R. Ribberfors, "Compton spectroscopy in the diagnostic X-ray energy range: II. Effects of scattering material and energy resolution", *Phys Med Biol*, vol. 34, *no.* 2, pp. 199-208, 1989.
21. F. Biggs, L. B. Mendelsohn, and J. B. Mann, "Hartree—Fock Compton profiles for the elements", *Atomic data and nuclear data tables*, vol. 16, pp. 201-309, 1975.
22. C. E. Ordonez, W. Chang, and A. Bolozdynya, "Angular uncertainties due to geometry and spatial resolution in Compton cameras", *IEEE Trans. Nucl. Sci.*, vol. 46, *no.* 4, pp. 1142-1147, 1999.
23. Y. F. Du, Z. He, G. F. Knoll, *et al.*, "Evaluation of a Compton scattering camera using 3-D position sensitive CdZnTe detectors", *Nucl. Instrum. Meth. A*, vol. 457, pp. 203-211, 2001.
24. H. Hirayama, Y. Namito, W. R. Nelson, *et al.*, "The EGS5 code system", United States. Department of Energy, 2005.
25. S. Agostinelli, J. Allison, K. Amako, *et al.*, "Geant4-a simulation toolkit", *Nucl. Instrum. Meth. A*, vol. 506, *no.* 3, pp. 250-303, 2003.
26. L. C. Parra, "Reconstruction of cone-beam projections from Compton scattered data", *IEEE Trans. Nucl. Sci.*, vol. 47, *no.* 4, pp. 1543-1550, 2000.
27. T. Tomitani and M. Hirasawa, "Image reconstruction from limited angle Compton camera data", *Phys Med Biol*, vol. 47, *no.* 12, pp. 2129-2145, 2002.
28. T. Tomitani and M. Hirasawa, "Analytical image reconstruction of cone-beam projections from limited-angle Compton camera data", *IEEE Trans. Nucl. Sci.*, vol. 50, *no.* 5, pp. 1602-1608, 2003.
29. J. P. Sullivan, S. R. Tornga, and M. W. Rawool-Sullivan, "Extended radiation source imaging with a prototype Compton imager", *Appl Radiat Isotopes*, vol. 67, pp. 617-624, 2009.
30. K. Shimazoe, M. Yoshino, Y. Ohshima, *et al.*, "Development of simultaneous PET and Compton imaging using GAGG-SiPM based pixel detectors", *Nucl. Instrum. Meth. A*, vol. 954, *no.* 1, p. 161499, 2020.
31. T. Y. Song, H. Wu, S. Komarov, *et al.*, "A sub-millimeter resolution PET detector module using a multi-pixel photon counter array", *Phys Med Biol*, vol. 55, pp. 2573-2587, 2010.
32. D. Stratos, G. Maria, F. Eleftherios, *et al.*, "Comparison of three resistor network division circuits for the readout of 4 × 4 pixel SiPM arrays", *Nucl. Instrum. Meth. A*, vol. 702, pp. 121-125, 2013.
33. Z. Wang, X. Sun, K. Lou, *et al.*, "Design, development and evaluation of a resistor-based multiplexing circuit for a 20 × 20 SiPM array", *Nucl. Instrum. Meth. A*, vol. 816, pp. 40-46, 2016.
34. H.-J. Choe, Y. Choi, W. Hu, *et al.*, "Development of capacitive multiplexing circuit for SiPM-based time-of-flight (TOF) PET detector", *Phys Med Biol*, vol. 62, *no.* 7, pp. N120-N133, 2017.

35. X. Sun, K. Lou, and Y. Shao, "Capacitor based multiplexing circuit for silicon photomultiplier array readout", in *2014 IEEE Nuclear Science Symposium and Medical Imaging Conference (NSS/MIC)*, Seattle, US, 2014.
36. K. Shimazoe, H. Takahashi, B. Shi, *et al.*, "Dynamic Time Over Threshold Method", *IEEE Trans. Nucl. Sci.,* vol. 59, *no.* 6, pp. 3213-3217, 2012.
37. Y. Nakamura, K. Shimazoe, and H. Takahashi, "Silicon Photomultiplier-Based Multi-Channel Gamma Ray Detector Using the Dynamic Time-Over-Threshold Method", *J Instrum,* 2016.
38. T. Nakamori, T. Kato, J. Kataoka, *et al.*, "Development of a gamma-ray imager using a large area monolithic 4×4 MPPC array for a future PET scanner", *J Instrum,* vol. 7, *no.* 01, p. C01083, 2012.
39. K. Takeuchi, J. Kataoka, T. Nishiyama, *et al.*, ""Stereo Compton cameras" for the 3-D localization of radioisotopes", *Nucl. Instrum. Meth. A,* 2014.
40. H.-j. Choe, Y. Choi, D. J. Kwak, *et al.*, "Prototype time-of-flight PET utilizing capacitive multiplexing readout", *Nucl. Instrum. Meth. A,* vol. 921, pp. 43-49, 2019.

Radioactive Source Localization Method for the Partially Coded Field-of-View of Coded-Aperture Imaging in Nuclear Security Applications

Qi Liu, Yi Cheng, Xianguo Tuo, and Yongliang Yang

1 Introduction

As an important part of the clean energy that contributes to attaining carbon neutrality, nuclear power will continue to play a key role in the world's low-carbon energy structure. At the end of 2019, 443 nuclear reactors were operational and 54 reactors were under construction [1]. The increasing number of nuclear facilities inevitably produces radioactive waste that may be obtained by terrorists or criminal organizations for non-peaceful use. As reported by the International Atomic Energy Agency (IAEA) Incident and Trafficking Database (ITDB), 3686 incidents of nuclear and other radioactive materials out of regulatory control were confirmed between 1993 and 2019. Of these incidents, 290 were connected with trafficking or malicious use including high enriched uranium, plutonium, and plutonium beryllium neutron sources [2]. Therefore, the detection, monitoring, and accurate localization of radioactive sources are critical to nuclear safety, homeland security, and public health.

Gamma-ray imaging technique can simultaneously provide the spatial distribution of radioactive materials and the relative intensities of hotspots and is well suited for the remote localization of radioactive sources in nuclear security applications such as decontamination and decommissioning of nuclear facilities, nuclear waste management, searching for lost radioactive sources, and radiation protection in nuclear power plants. Various imaging techniques have been successfully applied to nuclear security applications, including pinhole camera [3–5], coded-aperture cam-

Q. Liu (✉) · Y. Cheng · X. Tuo
School of Physics and Electronic Engineering, Sichuan University of Science and Engineering, Zigong, China

Y. Yang
CNNP Nuclear Power Operations Management Co., Ltd., Haiyan, China

eras [6–10], Compton cameras [11–13], rotational modulation collimator (RMC) systems [14, 15], and time-encoded imaging systems [16–19]. Coded-aperture and Compton imaging systems have received the most development and are best suited for field use because of their excellent performance and portability. Compton cameras can provide a 4π field-of-view (FOV) but the angular resolution is typically worse than $10°$, and the performance is degraded for low-energy photons (<250 keV) because of their dependence on the Compton scattering kinematics. Coded-aperture cameras rely on the spatial modulation technique and their angular resolution is typically better than $6°$ due to the mechanical collimation using a heavy metal with a specific pattern. However, the FOV of such systems is limited, and the sources out of the fully-coded field-of-view (FCFOV) are likely to cause false hotspots due to the partial encoding effect that will be described in Sect. 2. A false localization of radioactive sources may lead personnel to accidentally receive excessive exposure to radiation and cause harm to health in nuclear security applications.

In this chapter, we will briefly introduce the theory of the coded-aperture imaging technique in Sect. 2.1. The principle of partial encoding effect is described and its influence is experimentally studied with a gamma source and a real coded-aperture camera. A brief review of existing methods to address the above issue is presented in Sect. 2.2. A deep neural network-based method to identify the false hotspots in reconstructed images and locate the actual locations of radioactive sources is proposed in Sect. 3. Experimental results of different source locations and a robot assisted method are presented in Sect. 4.

2 Background

Coded-aperture imaging is derived from the pinhole imaging technique that can directly project the radiation scene within the FOV of the imaging system into a position-sensitive detector through a single hole of the mask, while the detector is surrounded by thick shielding to decrease background noise and keep sufficient signal-to-noise ratio (SNR). The angular resolution of a pinhole camera is inversely proportional to the size of the hole. A better angular resolution can be obtained by narrowing the hole size, but the portion of incident photons will significantly decrease at the same time. Therefore, the trade-off between the spatial resolution and SNR must be considered during the mask design phase. In order to address the above issue, the coded-aperture imaging technique replaces the single hole with multi-holes to increase the SNR while maintaining the same angular resolution as the single hole case.

2.1 Theory of Coded-Aperture Imaging

Coded-aperture imaging technique for radiation detection was firstly proposed in X-ray astronomical applications by Ables and Dicke, independently [20, 21]. Coded-aperture cameras are comprised of a position-sensitive detector that can be represented by a matrix D, and a coded mask A containing opaque elements filled with heavy metal such as tungsten and transparent elements that permit the incident photons to pass through. In the early works, these holes were randomly opened on the mask and a point source can project multiple shadows on the position-sensitive detector. The counting distribution becomes more complicated than the single hole case, and a source position corresponds to a specific pattern rather than a small area of the detector. Therefore, the image formation is divided into an encoding and reconstruction process, respectively.

During the encoding phase, the photons emitted from the radioactive sources S within the FOV of imaging systems were modulated by mask A and detected by the detector D, which can be represented by:

$$D = S * A + B \qquad (1)$$

where the symbol $*$ indicates the correlation operator and B represents all the events that are not modulated by the mask.

The reconstructed distribution of radioactive sources $\hat{S}$ can be calculated through a cross-correlation between the matrix D and a decoding matrix G as follows:

$$\hat{S} = D * G = S * A * G + B * G \qquad (2)$$

The radioactive distribution $\hat{S}$ will be perfectly reconstructed if $A * G$ is the Kronecker delta function, except for the last noise term $B * G$ including high energy cosmic ray, electronic noise, as well as the photons that are not modulated by the mask. This is the so-called cross-correlation (CC) reconstruction method. However, $A * G$ is not a delta function for the random mask pattern in general, and the reconstructed images appear an inherent noise. Since the random distribution of the transparent elements from the mask, the unique optimization is required for each specific application and such masks are well suited for the applications that required a special shape such as the rotating cylindrical mask of a time-encoded imaging system [19]. The above issues were addressed and a number of portable coded-aperture cameras have been developed since the uniformly redundant arrays (URA) were proposed by Fenimore and Canon [22]. URA-based coded masks typically have around ~50% open fraction and satisfy the condition that $A * G$ is an ideal delta function without sidelobes. However, the shape of a URA mask must be rectangular and the difference between the vertical and horizontal dimensions must equal two. Modified uniformly redundant arrays (MURAs) which are derived from quadratic residues rather than pseudo-noise (PN) sequences were proposed to address the above issue. The geometric configurations of MURAs can be linear, hexagonal, and square. The square masks can be constructed by a 2×2 mosaic of the basic pattern, and the detector size only equal to the basic pattern is required to fully cover the projection pattern. The only limitation is that the mask's dimension in each direction must be a prime p so that the mask has a total of p^2 elements. Besides the MURAs, the pseudo-noise product (PNP) arrays can also provide a perfectly square geometry, but with only a 25% open fraction.

Angular resolution $\delta\theta$ and FOV were determined by the system geometry such as the mask pattern, the element size of the mask and position-sensitive detector, the mask-to-detector distance (MDD or focal length), and were given by

$$\delta\theta = \arctan\left(\frac{w_m}{\text{MDD}}\right) \tag{3}$$

$$\text{FOV} = 2\arctan\left(\frac{m-d}{2\text{MDD}}\right) \tag{4}$$

where w_m is the pinhole size, m and d are the sizes of the coded mask and position-sensitive detector, respectively. Variable zooms can be implemented by changing the MDD of coded-aperture cameras. A better angular resolution can be obtained by increasing the MDD once the hardware dimensions of a coded-aperture camera have been determined. However, the cost is that the FOV is further reduced. The FOV of a coded-aperture camera is typically smaller than $50°$ due to the two-plane geometric configuration composed of the mask and detector [23]. The limited FOV not only affects the detectable area in field applications, but the reconstructed results may be interfered with by the partial encoding effect that will be described in Sect. 2.2.

2.2 Partial Encoding Effect

The modern coded-aperture imaging systems typically use the mask with a 2×2 mosaic of the basic pattern and expand the array dimensions from $p \times p$ to $(2p - 1) \times (2p - 1)$. Compared to the box geometry that the mask and detector have the same size, this 2×2 geometric arrangement is constructed so that each source position within the FCFOV of the imaging system can cast a unique shadow pattern on the position-sensitive detector while providing a relatively larger FCFOV. However, the cyclic permutation feature of the mask poses difficulties with image reconstruction when radioactive sources are located out of the FCFOV of the coded-aperture camera.

Coded-aperture cameras are designed specifically for the localization of radioactive sources within the FCFOV such as the "Source 1" shown in Fig. 1a. All the detected photons emitted from the "Source 1" are modulated by the mask so that the hotspot can be accurately reconstructed by using the cross-correlation method [24]. In medical applications, the imaging process satisfies this condition because the radioactive materials are placed in front of the imaging systems. However, the

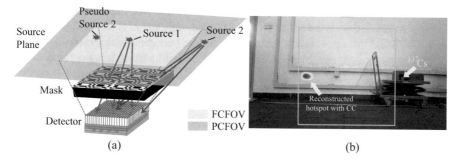

Fig. 1 (**a**) Schematic diagram of the partial encoding effect and (**b**) the reconstructed radiation image using the cross-correlation (CC) method superimposed on the optical image. The delimitation of the FCFOV was represented by the green square

approximate locations of radioactive sources are often unknown in nuclear security applications, and there is a possibility that the radioactive sources are located at the partially coded field-of-view (PCFOV) of the imaging systems such as the "Source 2" shown in Fig. 1a. The emitted photons are partially modulated by the mask while the rest directly interacts with the detector. The resulting shadow cast on the detector is partially identical to the shadow pattern of a source within the FCFOV due to the cyclic permutation feature of the MURA-based masks, which results in false hotspots (also called ghosts) in reconstructed images, as shown in Fig. 1a. This phenomenon is called the partial encoding effect and is experimentally illustrated in Fig. 1b. A ^{137}Cs source with an activity of 37 MBq was placed on a mobile platform with a 2 m source-to-mask distance (SMD). The position of the ^{137}Cs source is out of the FCFOV and on the right side of the imaging system. The result of the reconstructed radiation image using the cross-correlation method superimposes the optical image caught by the optical camera is shown in Fig. 1b. A false reconstructed hotspot was presented on the opposite side of the actual source position due to the partial encoding effect. Even worse, if the position of a radioactive source is near the boundary between the FCFOV and PCFOV of the imaging system, false hotspots will appear on both sides of the reconstructed images. This issue not only interferes with the accurate localization of radioactive sources but also affects the operator's judgment of the actual source number in the radiological scene. A false source localization poses a threat to public safety and personnel's health in nuclear security applications.

There have been some researches to address the above common issues of coded-aperture cameras. A panorama coded-aperture gamma camera was developed by combining six detector modules into a single hexagonal device to achieve a 360° horizontal FOV [25], but this method exists interference problems that radioactive sources within the FCFOV of a detector module are likely to be in the PCFOV of its adjacent detector module. Although the thick lead shielding were placed between the detector modules, the partial encoding effect cannot be eliminated because the contributions of the false reconstructed hotspots (ghosts) were derived from the photons that were modulated by the mask rather than the photons that directly interacted with the detector. Furthermore, only the horizontal direction was considered and this approach was not suited for the vertical and diagonal directions. A multimode imager combining Compton with coded-aperture imaging techniques was designed based on the active mask configuration for nuclear threat detection [26]. The imager consists of 96, 1 cm^3 CdZnTe (CZT) detector elements arranged in a two-plane geometric configuration in which the front plane uses a random mask pattern with 32 detector elements to serve simultaneously as a coded mask in coded-aperture modality and a Compton scatter layer in Compton imaging modality, and the back plane is filled with 64 detector elements. However, an FCFOV is only available for sources that are directly on-axis because its mask size is the same as the detector plane in coded-aperture modality, which makes the accurate localization for the off-axis sources difficult. To address the problems of imaging artifacts and the coded-aperture modality suffers from a limited FOV, a spherical active coded-aperture imaging system was proposed by rearranging the detector elements into a 14-cm diameter sphere with 192 available detector locations rather than a planar mask-detector configuration, resulting in a 4π FOV for both imaging modalities [27]. The random mask pattern was chosen and optimized based on the "Great Deluge algorithm" to determine the optimal number and configuration of detector elements on the sphere and the resulting optimized mask contains a total of 104 detector elements. But the coded-aperture modality can only be used in the imaging energy range of 50 ~ 400 keV and the angular resolution is approximately 10° [28]. A hybrid gamma camera combining both the coded-aperture and Compton imaging modalities was developed based on a single Timpex3 chip hybridized with a 1 mm thick cadmium telluride semiconductor, which takes advantage of the high-resolution of the coded-aperture imaging and the wide FOV of the Compton imaging technique [29]. However, the problem of the partial encoding effect still exists for the coded-aperture modality, and an additional measurement using the Compton imaging modality is required to identify the ghosts within the reconstructed image. An iterative reconstruction algorithm based on compressed sensing, with system matrices considering the PCFOV can address the above issues [30]. The choice of the source pixel size also presents a trade-off between the angular resolution and reconstruction time, and it is a waste of computing resources if no radioactive source is located at the PCFOV of the imaging system but the system matrices containing both the FCFOV and PCFOV were used.

3   Methodology

In this section, we propose a deep neural network-based method to identify and locate the radioactive sources within the PCFOV of a coded-aperture camera. Instead of placing thick shielding around the detector to attenuate the photons that were not modulated by the mask, the counting distribution of the shadow cast on the position-sensitive detector for the radioactive source within the PCFOV has specific features and was utilized in this work. The features of the raw projection data were extracted by using the deep neural network model to identify the approximate source position, which converts the task of source localization into an image recognition problem. The reconstructed images containing false hotspots were reconstructed using the conventional cross-correlation method, and these false hotspots still contain useful information of source positions that can be extracted. The resulting source position can be obtained by combining the approximate source position and the information of the false hotspots and using a regression model.

3.1 Configuration of the Coded-Aperture Camera

The coded-aperture camera adopted for this experiment is shown in Fig. 2. The front plane of the camera is equipped with a MURA rank-19 mask based on a 2×2 cyclic replication of the basic pattern with an 8 mm thick tungsten-copper alloy material, resulting in 37×37 pixels with dimensions of $88.8 \times 88.8 \times 8$ mm^3. With consideration of the mask parameters, the detector module was composed of a 19×19 pixelated bismuth germanate (BGO) scintillation crystal with a pixel size of $2.4 \times 2.4 \times 10$ mm^3 coupled to a Hamamatsu H8500 position-sensitive photomultiplier tube (PSPMT).

Compared to the semiconductor detector such as the cadmium telluride (CdTe) and cadmium zinc telluride (CdZnTe), BGO crystal has several advantages such as low cost, relatively high density (7130 kg/m^3), no intrinsic radioactivity, non-hygroscopic, and high absorption efficiency, make it well suited for the gamma-ray imaging applications in the energy range from 50 to 1500 keV. The 8×8 multi-anode outputs of the H8500 were connected to a resistive network with the discretized positioning circuit (DPC) [31], reducing the 64-pixel outputs to

(a)   (b)

Fig. 2 Photographs of (**a**) the detector module containing a 19×19 pixelated BGO scintillation crystal array coupled to a Hamamatsu H8500 PSPMT, and (**b**) the coded-aperture camera prototype

four output signals. Compared to the approach required that each channel has its amplifier to read independently, this signal processing method can effectively simplify the readout circuits. After digitizing the four signals with a 40 MSPS sampling frequency and 12-bit resolution, the incident position and energy can be extracted from the input waveform.

3.2 Data Collection

As a data-driven non-model framework, a deep neural network requires an advanced network architecture to enhance the expressive ability and adequate training data to improve the generalization performance. Data collection is the first process and plays a key role in deep learning applications. In general, training samples can be obtained with the data synthesis [32, 33], Monte Carlo simulations [34, 35], and data acquisition using a real imaging system with experimental measurements [36]. Data synthesis is an efficient approach to obtain a large number of training samples by using the simple geometrical model of the imaging system or analytical calculation of the detector response such as the linear attenuation. However, it neglects the physical processes during the data acquisition such as the inter-crystal scatter and crystal penetration, which leads to obvious discrepancies between the distributions of the real data and synthetic data. The most accurate way of acquiring training samples would be derived directly from the experimental measurements using the real imaging system and radioactive sources, all the physical effects such as the electronic noise, scattering and transmission of the mask, and scattering both within the detector and in the surrounding materials can be included. The experimental setup can be tedious due to the requirement of the precise positioning device to move the radioactive sources to different positions, which increases the cost and complexity of the measurements. The distribution of the projection data will also change if the hardware modules are replaced with a new one, which leads to discrepancies between the training data and the real data. Furthermore, the counting rate in nuclear security applications is much lower than that of the medical applications, resulting in a long acquisition time to satisfy the counting statistics of the projection data, which makes the experimental measurements unpractical. An alternative to the experimental method is to simulate the imaging process using the Monte Carlo simulations. A variety of software packages such as the MCNP, Geant4, GATE have been developed and based on the repeated random sampling of probability models of the gamma-ray transport through the matter, which offers the possibility of simulating the physics processes of a gamma camera very precisely. In this chapter, the simulation strategy was used to obtain the training samples because it can achieve a good balance between the precision and the acquisition time of the training data, only at the cost of the computing resources.

The counting distribution of the shadow cast on the detector has different response characteristics when radioactive sources are located at different positions. Therefore, the FOV including both the FCFOV and PCFOV was divided into nine

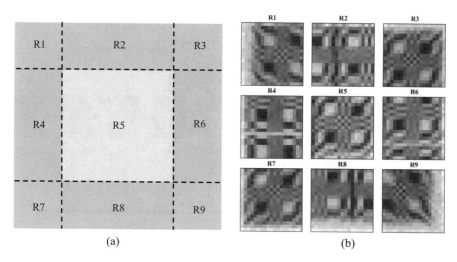

R1	R2	R3
R4	R5	R6
R7	R8	R9

(a)   (b)

Fig. 3 (**a**) Schematic diagram of the partitioned FOV of the gamma camera used in this work, the R5 is identical to the FCFOV (green region) and others are located at the PCFOV (purple regions), and (**b**) the projection data randomly chosen from each region

regions from R1 to R9 (from left to right, top to bottom) in which the R5 is equal to the FCFOV and others belong to the PCFOV, as shown in Fig. 3a. The method introduced in this chapter was developed for nuclear security applications so that only the far-field geometry was considered. All radioactive sources were simulated on a source plane with a 2 m source-to-mask distance to satisfy the far-field condition so that the detected photons can be considered to be a parallel incident. A 59 mm mask-to-detector distance was set to provide a 4618×4618 mm^2 PCFOV and a 1544×1544 mm^2 FCFOV, respectively. The position information of the radioactive sources contained in the projection data is proportion to the ratio of the photons modulated by the mask. With the above considerations, the range of each region was determined according to whether nearly half of the incident photons can be modulated. The dimensions of the diagonal regions (R1, R3, R7, R9) and the axial regions (R2, R4, R6, R8) were set to 328×328 mm^2 and 1544×628 mm^2, respectively.

The acquired dataset was comprised of 17,100 projection data and randomly split into a training dataset for network training with the size of 15,300, and a validation dataset for hyper-parameter tuning with the size of 1800. The two datasets were kept separate throughout the study. In each simulation, a total of 2.5×10^6 662 keV photons were emitted from a point source, resulting in around 180×10^3 average counts and 4.5% relative standard deviation. The computing time of each simulation for the given simulated conditions was less than 3 min on a workstation (OS: Ubuntu 18.04.3, RAM: 32GB, CPU: 2.6 GHz), resulting in a total of 35 days with a single-core computer to complete the data acquisition. The data of each region has the same size to avoid the class-imbalanced problems that may cause the network to implicitly optimize the predictions based on the most abundant class in the dataset.

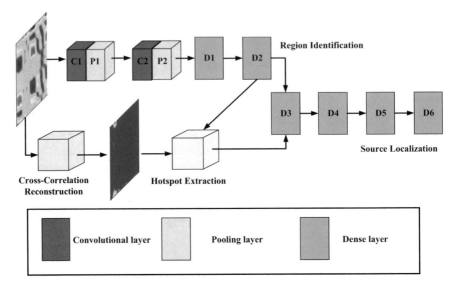

Fig. 4 The architecture of the proposed deep neural network model to identify and locate the radioactive sources within both the FCFOV and PCFOV

The projection data randomly chosen from each region is shown in Fig. 3b. It can be seen that the image edges present obvious rows or column pixels with high counts when the source position deviates from the center of the FOV, and it is difficult to distinguish between the projection data if the source position is close to the boundary between the FCFOV and PCFOV.

3.3 Deep Neural Network Architecture

The deep learning technique has achieved great successes in computer vision, image recognition because the deep neural networks were used and trained with big data. A deep neural network based on the convolutional neural network (CNN) architecture was proposed in this chapter to identify and locate the radioactive sources within both the FCFOV and PCFOV. The network is comprised of the module of region identification, hotspot extraction, and source localization, corresponding to the top left, bottom left, and right parts in Fig. 4, respectively.

The module of region identification converts the task of identifying source regions to an image recognition problem. The projection data of the used gamma camera have a single-channel, 2-dimensional structure with 19×19 pixels, which are similar to the images of handwritten digits from the MNIST database so that the CNNs are well suited for their feature extraction. Two convolutional layers were used to convolve the input data with the learned filters, while each is followed by a 2×2 max-pooling operation to reduce the number of parameters and a rectified linear unit (ReLU) activation function to enhance the expressive capability. Batch normalization was implemented to improve the convergence and stability of the training process. The extracted feature maps were reshaped into a vector as the input of a dense layer, the probabilities of each source region were calculated by using a softmax function, as follows:

$$p_j = \frac{e^{x_j}}{\sum_{i=1}^{n} e^{x_i}} \tag{5}$$

where p_j is the calculated probability of the jth source region, x_j is net input of the jth source region, and n is the number of the source region. The region identification result is the probability with the maximum value.

The reconstructed images can be obtained using the cross-correlation method with a known decoding matrix, where the contained false hotspots are directly related to the positions of the radioactive sources and are utilized in our method. After the source region was identified by the module of region identification, the hotspot was searched within a region of interest (ROI) chosen on the opposite half area of the source region in the reconstructed images, and its position information was determined by the central values of the Gaussian fitted line profiles along the X and Y axes.

The results of the region identification and extracted hotspot information were combined and input to the module of source localization because both the above two features are associated with the source position. Three fully connected layers were used to predict the position of the radioactive source, and the first two layers were followed by the ReLU activation function. The one-hot encoding representation was used to convert the indices of the identified source region to binary vectors and avoid the network assuming a natural ordering between different source regions that may cause poor prediction performance. More specifics on our network are as follows:

0. Input layer: To prevent the data size from decreasing rapidly after the convolution and pooling operations, zero-padding is used to expand the size of the input data from 19×19 to 23×23. The pixel values are normalized to a range of [0, 1], which brings different features onto the same scale.
1. C1 layer: The first convolutional layer operates upon the input projection data with 32 filters 3×3 in size, resulting in 32 feature maps 21×21 in size.
2. P1 layer: Max-pooling with stride 2 for down-sampling, producing 32 feature maps with the size of 11×11.
3. C2 layer: A total of 32 feature maps with the size of 13×13 are obtained with 32 filters 3×3 in size.
4. P2 layer: Similar to the P1 layer, a 2×2 pooling window with stride 2 is used to obtain 32 feature maps of 7×7.
5. D1 layer: The feature maps extracted from all channels are then flattened and connected to a dense layer consisting of 1568 neurons.
6. D2 layer: A dense layer with nine output nodes representing the R1 to R9 source regions.

7. D3 layer: A dense with 11 neurons representing the combination of the identified source region and extracted hotspot information.
8. D4 layer: A dense layer with 512 ReLU activations neurons.
9. D5 layer: A dense layer with 1024 ReLU activations neurons.
10. D6 layer: A dense layer with two nodes, representing the source position.

3.4 Training Procedure

The region identification and localization of the radioactive sources are classification and regression tasks, respectively. The two modules were trained and optimized individually to simplify the learning process. For the classification task, the objection function to be minimized during the training phase is the cross-entropy loss, defined

$$l_{cross\,entropy} = -\frac{1}{N} \sum_{n=1}^{N} log \frac{e^{x_{n,y_n}}}{\sum_{c=1}^{C} e^{x_{n,c}}} \qquad (6)$$

where N is the number of the training samples, C is the number of source regions, x is the input, and y denotes the corresponding label. The network was implemented in the Pytorch deep learning framework, optimized using the Adam optimizer with a learning rate of 1.43×10^{-5}, batch size of 16, and decay rates for moving averages of the gradient and its square $\beta_1 = 0.9$ and $\beta_1 = 0.999$. A random search technique was performed over the pre-defined hyper-parameter space, summarized in Table 1.

The mean squared error (MSE) between the network output and the label was used as the loss function for the regression task, as follows:

$$l_{mse} = -\frac{1}{N} \sum_{n=1}^{N} (x_n - y_n)^2 \qquad (7)$$

The same optimizer was used as the classification task with a learning rate of 2.65×10^{-6} and a batch size of 16. Table 2 presents the pre-defined hyper-parameter space used for a random search.

Table 1 Hyper-parameter space for the classification task

Hyper-parameter	Values
Learning rate	$10^{uniform[-5, -4]}$
Filter number of Conv1 layer	16, 32
Filter number of Conv2 layer	32, 64
Kernel size	$3 \times 3, 5 \times 5$
Batch size	16, 32, 64

Table 2 Hyper-parameter space for the regression task

Hyper-parameter	Values
Learning rate	$10^{\text{uniform}[-7,\ -5]}$
Dense4 layer size	512, 1024, 2048, 4096
Dense5 layer size	512, 1024, 2048
Batch size	16, 32, 64

4 Experiments

To validate the feasibility of the proposed deep neural network trained with the simulated data in real-world scenarios, we conducted experiments with the real gamma camera introduced in Sect. 3.1 and exported the optimized neural network model to a Torch Script that can be loaded and executed purely from C++, with no dependency on Python. The Torch Script was deployed on a laptop with Intel Core i7-6700HQ, 16 GB RAM, and Nvidia GeForce GTX 960M. The computation time for both the classification and regression models to perform the prediction is less than 10 ms. In addition, a robot assisted approach combined with the above models was presented to locate the radioactive sources and reduce the dose received by operators in nuclear security applications.

4.1 Different Source Locations

We first conducted an experiment using a 37 MBq ^{137}Cs point source placed on a mobile platform shown in Fig. 5a. A total of 23 source positions with a 2 m source-to-mask distance were set by moving the mobile platform and controlling its height, as shown in Fig. 5b. The detailed experimental setups are summarized in Table 3.

At each source position, the projection data were acquired at three different count levels, with ~180 × 10^3, ~36 × 10^3, and ~18 × 10^3, respectively. The acquisition time of the 180 × 10^3 projection data range from ~600 s to ~1000 s because the photons emitted from sources suffer more attenuation from the off-axis positions than the central positions.

The choices of these test positions were based on two considerations, the 19 positions roughly distributed from R1 to R6 were used to evaluate the generalization performance of the model, while the four adjacent positions located at the upper left corner of the FCFOV verified the feasibility of the model in the critical conditions.

Source regions of all the acquired data were correctly identified and the positions were accurately located in the case of the 180 × 10^3 count level. For illustration purposes, the positioning results of the deep neural network were superimposed with the captured optical image, and some typical results were presented in Fig. 6. As the count level decreased to 36 × 10^3 and 18 × 10^3, the number of the incorrectly identified and located data increased to one and five, respectively, due to the reduced counting statistics fluctuations.

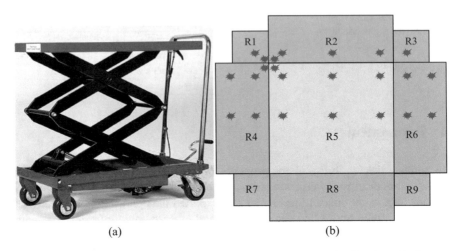

(a)   (b)

Fig. 5 (**a**) The mobile platform used for placing radioactive sources, and (**b**) schematic diagram of the experimental setup with different source positions

Table 3 Experimental setup for 23 source positions in different regions

Region	Source position (mm)	Region	Source position (mm)
R1	$(-780, 780)$, $(-900, 900)$	R2	$(-750, 780)$, $(-600, 900)$, $(0, 900)$, $(600, 900)$
R3	$(900, 900)$	R4	$(-900, 0)$, $(-1200, 0)$, $(-900, 600)$, $(-1200, 600)$, $(-780, 750)$
R5	$(-600, 0)$, $(0, 0)$, $(600, 0)$ $(-600, 600)$, $(0, 600)$, $(600, 600)$, $(-750, 750)$	R6	$(900, 0)$, $(1200, 0)$, $(900, 600)$, $(1200, 600)$

4.2 Source Localizations Assisted with a Robotic Platform

A second experiment was performed by equipping the gamma camera with a robotic platform, and the photograph of the gamma camera mounted on the robot is presented in Fig. 7a. Four Mecanum wheels were used for the robot to achieve omnidirectional maneuverability by setting different directions and velocities for each wheel independently, and the motion states were monitored through a nine-axis MEMS chip and encoders controlled by a Cortex-M4 microcontroller. Such hardware configurations allow the system to achieve a horizontal 360° source localization capability by rotating the wheels shown in Fig. 7b. Figure 7c presents the developed software to control the robotic platform and gamma camera using the C++ language in the Microsoft Visual Studio environment and the user interface (UI) was designed with Qt 5.12.

A 37 MBq ^{137}Cs point source was placed on the wall with a ~1.5 m source-to-mask distance, and the relative position between the source and the robotic platform is presented in Fig. 8a. Although the robot assisted approach with the conventional

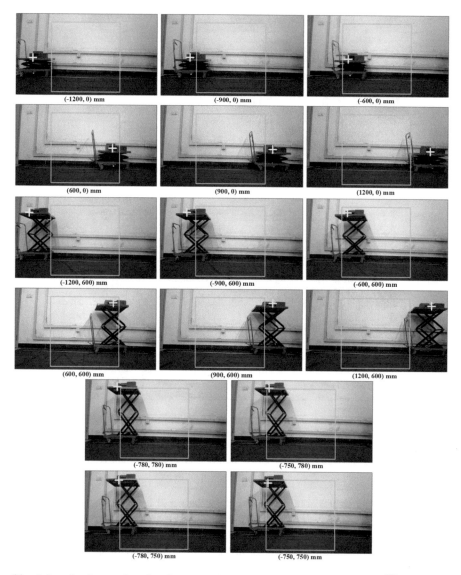

Fig. 6 Localization results of real test acquisition of different positions with a ^{137}Cs source, 180×10^3 count level. The predicted source position is at the center of the yellow cross cursor

reconstruction methods can also provide a horizontal 360° source localization capability, the result of a single measurement is affected by the partial encoding effect in the case of the geometrical configuration shown in the red rectangle in Fig. 8a. The FCFOV of the gamma camera is ~40° so that the source is located in the R4 region. Around 200×10^3 events were acquired with 450 s measurement time. The source region was correctly identified and the positioning result superimposed

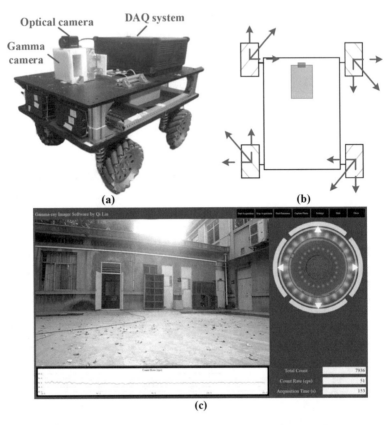

Fig. 7 (**a**) Photograph of the gamma camera mounted on the robotic platform, (**b**) the schematic diagram of the rotational motion by setting different directions and velocities for the four Mecanum wheels, and (**c**) the developed software used to control the robot and gamma camera

with a panoramic image obtained by stitching multiple optical images is shown in Fig. 8b. The speed-up robust features (SURF) algorithm was used to extract features between the adjacent images, and their homography matrix was obtained by using the RANSAC algorithm. It can be seen that the stitching method combined with the motion of the robotic platform can provide the entire scene information with a single image, addressing the issue when a radioactive source exceeds the FOV of the optical camera.

5 Conclusion

In this chapter, a deep neural network-based approach was introduced to address the partial encoding effect in coded-aperture imaging for nuclear security applications. The identification and localization of the radioactive sources within the PCFOV

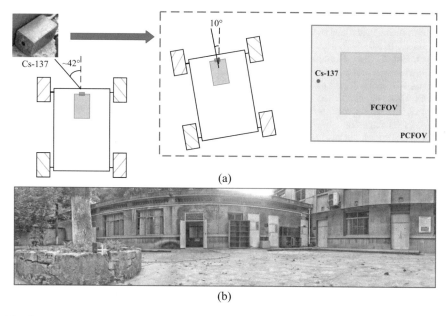

(a)

(b)

Fig. 8 (**a**) Schematic diagram of the experimental setup, and (**b**) the stitched optical image superimposed the localization result of the model labeled with the yellow cross cursor. The inconsistency of the vertical position between the schematic diagram and the located result is due to the parallax error between the gamma camera and the optical camera in this experimental setup

of a coded-aperture camera in nuclear security applications were converted to a classification and regression task, respectively. The models were trained with simulated data and we experimentally demonstrated its feasibility by setting 23 different source positions with three count levels. The results show that all the source positions can be correctly identified and accurately located at 180×10^3 count level, especially the 4 adjacent positions demonstrate the model performance in critical conditions that the sources were located at the boundary between the FCFOV and PCFOV. A robot assisted approach was introduced by mounting the gamma camera on a robotic platform based on four Mecanum wheels, and a horizontal 360° source localization capability can be achieved by setting the corresponding directions and velocities for each wheel. Our future work is focused on extending the model for complex situations such as multiple sources and extended sources.

References

1. IAEA, *Energy, Electricity and Nuclear Power Estimates for the Period up to 2050*. Vienna: INTERNATIONAL ATOMIC ENERGY AGENCY, 2020.
2. IAEA, "IAEA Incident and Trafficking Database (ITDB) Fact Sheet 2020," *IAEA Incid. Traffick. Database*, 2020.
3. O. Gal and F. Jean, "The cartogam portable gamma imaging system," *IEEE Trans. Nucl. Sci.*, vol. 47, no. 3 PART 2, 2000, doi: https://doi.org/10.1109/23.856725.

4. L. Caballero *et al.*, "Gamma-ray imaging system for real-time measurements in nuclear waste characterisation," *J. Instrum.*, vol. 13, no. 3, 2018, doi: https://doi.org/10.1088/1748-0221/13/03/P03016.

5. O. Gal, C. Izac, F. Lainé, and A. Nguyen, "Cartogam: A portable gamma camera," *Nucl. Instruments Methods Phys. Res. Sect. A Accel. Spectrometers, Detect. Assoc. Equip.*, vol. 387, no. 1–2, 1997, doi: https://doi.org/10.1016/S0168-9002(96)01013-3.

6. M. Gmar *et al.*, "Development of coded-aperture imaging with a compact gamma camera," in *IEEE Transactions on Nuclear Science*, 2004, vol. 51, no. 4 I, doi: https://doi.org/10.1109/TNS.2004.832608.

7. O. Gal *et al.*, "Development of a portable gamma camera with coded aperture," *Nucl. Instruments Methods Phys. Res. Sect. A Accel. Spectrometers, Detect. Assoc. Equip.*, vol. 563, no. 1, pp. 233–237, 2006, doi: https://doi.org/10.1016/j.nima.2006.01.119.

8. F. Carrel *et al.*, "GAMPIX: A new gamma imaging system for radiological safety and homeland security purposes," *IEEE Nucl. Sci. Symp. Conf. Rec.*, pp. 4739–4744, 2011, doi: https://doi.org/10.1109/NSSMIC.2011.6154706.

9. M. Jeong and M. Hammig, "Development of hand-held coded-aperture gamma ray imaging system based on GAGG(Ce) scintillator coupled with SiPM array," *Nucl. Eng. Technol.*, vol. 52, no. 11, 2020, doi: https://doi.org/10.1016/j.net.2020.04.009.

10. Q. Liu *et al.*, "Image reconstruction using multi-energy system matrices with a scintillator-based gamma camera for nuclear security applications," *Appl. Radiat. Isot.*, vol. 163, Sep. 2020, doi: https://doi.org/10.1016/j.apradiso.2020.109217.

11. C. G. Wahl *et al.*, "The Polaris-H imaging spectrometer," *Nucl. Instruments Methods Phys. Res. Sect. A Accel. Spectrometers, Detect. Assoc. Equip.*, vol. 784, pp. 377–381, 2015, doi: https://doi.org/10.1016/j.nima.2014.12.110.

12. L. Sinclair, P. Saull, D. Hanna, H. Seywerd, A. MacLeod, and P. Boyle, "Silicon photomultiplier-based Compton Telescope for Safety and Security (SCoTSS)," *IEEE Trans. Nucl. Sci.*, vol. 61, no. 5, pp. 2745–2752, 2014, doi: https://doi.org/10.1109/TNS.2014.2356412.

13. A. M. L. MacLeod, P. J. Boyle, D. S. Hanna, P. R. B. Saull, L. E. Sinclair, and H. C. J. Seywerd, "Development of a Compton imager based on bars of scintillator," *Nucl. Instruments Methods Phys. Res. Sect. A Accel. Spectrometers, Detect. Assoc. Equip.*, vol. 767, 2014, doi: https://doi.org/10.1016/j.nima.2014.08.012.

14. B. R. Kowash, D. K. Wehe, and J. A. Fessler, "A rotating modulation imager for locating mid-range point sources," *Nucl. Instruments Methods Phys. Res. Sect. A Accel. Spectrometers, Detect. Assoc. Equip.*, vol. 602, no. 2, pp. 477–483, 2009, doi: https://doi.org/10.1016/j.nima.2008.12.233.

15. B. R. Kowash and D. K. Wehe, "A unified near- and far-field imaging model for rotating modulation collimators," *Nucl. Instruments Methods Phys. Res. Sect. A Accel. Spectrometers, Detect. Assoc. Equip.*, vol. 637, no. 1, pp. 178–184, 2011, doi: https://doi.org/10.1016/j.nima.2010.12.037.

16. S. T. Brown, D. Goodman, J. Chu, B. Williams, M. R. Williamson, and Z. He, "Time-Encoded Gamma-Ray Imaging Using a 3-D Position-Sensitive CdZnTe Detector Array," *IEEE Trans. Nucl. Sci.*, vol. 67, no. 2, pp. 464–472, 2020, doi: https://doi.org/10.1109/TNS.2019.2953182.

17. X. Liang *et al.*, "Self-supporting design of a time-encoded aperture, gamma-neutron imaging system," *Nucl. Instruments Methods Phys. Res. Sect. A Accel. Spectrometers, Detect. Assoc. Equip.*, vol. 951, 2020, doi: https://doi.org/10.1016/j.nima.2019.162964.

18. N. P Shah, J. VanderZanden, and D. K. Wehe, "Design and construction of a 1-D, cylindrical, dual-particle, time-encoded imaging system," *Nucl. Instruments Methods Phys. Res. Sect. A Accel. Spectrometers, Detect. Assoc. Equip.*, vol. 954, 2020, doi: https://doi.org/10.1016/j.nima.2019.01.012.

19. J. Brennan *et al.*, "Demonstration of two-dimensional time-encoded imaging of fast neutrons," *Nucl. Instruments Methods Phys. Res. Sect. A Accel. Spectrometers, Detect. Assoc. Equip.*, vol. 802, pp. 76–81, 2015, doi: https://doi.org/10.1016/j.nima.2015.08.076.

20. R. H. Dicke, "Scatter-Hole Cameras for X-Rays and Gamma Rays," *Astrophys. J.*, vol. 153, pp. 101–106, 1968, doi: https://doi.org/10.1086/180230.
21. J. G. Ables, "Fourier Transform Photography: A New Method for X-Ray Astronomy," *Publ. Astron. Soc. Aust.*, vol. 1, no. 4, pp. 172–173, 1968, doi: https://doi.org/10.1017/s1323358000011292.
22. E. E. Fenimore and T. M. Cannon, "Coded aperture imaging with uniformly redundant arrays," *Appl. Opt.*, vol. 17, no. 3, p. 337, Feb. 1978, doi: https://doi.org/10.1364/ao.17.000337.
23. K. Amgarou and M. Herranz, "State-of-the-art and challenges of non-destructive techniques for in-situ radiological characterization of nuclear facilities to be dismantled," *Nuclear Engineering and Technology*, vol. 53, no. 11. 2021, doi: https://doi.org/10.1016/j.net.2021.05.031.
24. E. E. Fenimore and T. M. Cannon, "Coded aperture imaging with uniformly redundant arrays," *Appl. Opt.*, vol. 17, no. 3, pp. 337–347, Feb. 1978, doi: https://doi.org/10.1364/ao.17.000337.
25. S. Sun et al., "Development of a panorama coded-aperture gamma camera for radiation detection," *Radiat. Meas.*, vol. 77, pp. 34–40, Jun. 2015, doi: https://doi.org/10.1016/j.radmeas.2015.04.014.
26. M. L. Galloway et al., "Status of the High Efficiency Multimode Imager," in *IEEE Nuclear Science Symposium Conference Record*, 2011, pp. 1290–1293, doi: https://doi.org/10.1109/NSSMIC.2011.6154328.
27. D. Hellfeld, P. Barton, D. Gunter, A. Haefner, L. Mihailescu, and K. Vetter, "Real-Time Free-Moving Active Coded Mask 3D Gamma-Ray Imaging," *IEEE Trans. Nucl. Sci.*, vol. 66, no. 10, pp. 2252–2260, 2019, doi: https://doi.org/10.1109/TNS.2019.2939948.
28. D. Hellfeld, P. Barton, D. Gunter, L. Mihailescu, and K. Vetter, "A Spherical Active Coded Aperture for 4π Gamma-Ray Imaging," *IEEE Trans. Nucl. Sci.*, vol. 64, no. 11, pp. 2837–2842, 2017, doi: https://doi.org/10.1109/TNS.2017.2755982.
29. G. Amoyal, V. Schoepff, F. Carrel, M. Michel, N. Blanc de Lanaute, and J. C. Angélique, "Development of a hybrid gamma camera based on Timepix3 for nuclear industry applications," *Nucl. Instruments Methods Phys. Res. Sect. A Accel. Spectrometers, Detect. Assoc. Equip.*, vol. 987, 2021, doi: https://doi.org/10.1016/j.nima.2020.164838.
30. M. Jeong and G. Kim, "MCNP-polimi simulation for the compressed-sensing based reconstruction in a coded-aperture imaging CAI extended to partially-coded field-of-view," *Nucl. Eng. Technol.*, vol. 53, no. 1, pp. 199–207, Jan. 2021, doi: https://doi.org/10.1016/j.net.2020.02.011.
31. S. Siegel, R. W. Silverman, Y. Shao, and S. R. Cherry, "Simple charge division readouts for imaging scintillator arrays using a multi-channel PMT," *IEEE Trans. Nucl. Sci.*, vol. 43, no. 3 PART 2, 1996, doi: https://doi.org/10.1109/23.507162.
32. G. Daniel and O. Limousin, "Extended sources reconstructions by means of coded mask aperture systems and deep learning algorithm," *Nucl. Instruments Methods Phys. Res. Sect. A Accel. Spectrometers, Detect. Assoc. Equip.*, vol. 1012, 2021, doi: https://doi.org/10.1016/j.nima.2021.165600.
33. M. Jeong and M. D. Hammig, "Comparison of gamma ray localization using system matrixes obtained by either MCNP simulations or ray-driven calculations for a coded-aperture imaging system," *Nucl. Instruments Methods Phys. Res. Sect. A Accel. Spectrometers, Detect. Assoc. Equip.*, vol. 954, no. October, 2020, doi: https://doi.org/10.1016/j.nima.2018.10.031.
34. R. Zhang et al., "Reconstruction method for gamma-ray coded-aperture imaging based on convolutional neural network," *Nucl. Instruments Methods Phys. Res. Sect. A Accel. Spectrometers, Detect. Assoc. Equip.*, vol. 934, no. May, pp. 41–51, 2019, doi: https://doi.org/10.1016/j.nima.2019.04.055.
35. Q. Liu et al., "Neural network method for localization of radioactive sources within a partially coded field-of-view in coded-aperture imaging," *Appl. Radiat. Isot.*, vol. 170, Apr. 2021, doi: https://doi.org/10.1016/j.apradiso.2021.109637.
36. V. Y. Panin, F. Kehren, C. Michel, and M. Casey, "Fully 3-D PET reconstruction with system matrix derived from point source measurements," *IEEE Trans. Med. Imaging*, vol. 25, no. 7, pp. 907–921, 2006, doi: https://doi.org/10.1109/TMI.2006.876171.

Index

Printed in the United States
by Baker & Taylor Publisher Services